Dynamics of Elementary Atomic-Molecular Processes in Gas and Plasma

Proceedings of the Lebedev Physics Institute Academy of Sciences of Russia

Volume 165	**Solitons and Instantons, Operator Quantization,** Edited by V.L. Ginzburg
Volume 166	**The Nonlinear Optics of Semiconductor Lasers,** Edited by N.G. Basov
Volume 167	**Group Theory, Gravitation and Elementary Particle Physics,** Edited by A.A. Komar
Volume 168	**Issues in Intense-Field Quantum Electrodynamics,** Edited by V.L. Ginzburg
Volume 169	**The Physical Effects in the Gravitational Field Black Holes,** Edited by M.A. Markov
Volume 170	**The Theory of Target Compression by Longwave Laser Emission,** Edited by G.V. Sklizkov
Volume 171	**Research on Laser Theory,** Edited by A.N. Orayevskiy
Volume 172	**Phase Conjugation of Laser Emission,** Edited by N.G. Basov
Volume 173	**Quantum Mechanics and Statistical Methods,** Edited by M.M. Sushchinskiy
Volume 174	**Nonequilibrium Superconductivity,** Edited by V.L.Ginzburg
Volume 174 Supplemental Volume 1	**Thermodynamics and Electrodynamics of Superconductors,** Edited by V.L. Ginzburg
Volume 174 Supplemental Volume 2	**Electron Liquid Theory of Normal Metals,** Edited by V.P. Silin
Volume 175	**Luminescence Centers of Rare Earth Ions in Crystal Phosphors,** Edited by M.D. Galanin

Volume 175 **Luminescence and Anisotropy of Zinc Sulfide Crystals,**
Supplemental Volume Edited by M.D. Galanin

Volume 176 **Classical and Quantum Effects in Electrodynamics,** Edited by A.A. Komar

Volume 176 **Quantization, Gravitation and Group Methods in Physics,**
Supplemental Volume Edited by A.A. Komar

Volume 177 **Stoichiometry in Crystal Compounds and Its Influence on Their Physical Properties,** Edited by Yu. M. Popov

Volume 178 **The Delfin Laser Thermonuclear Installation: Operational Complex and Future Directions,** Edited by G.V. Sklizkov

Volume 179 **X-Ray Plasma Spectroscopy and the Properties of Multiply-Charged Ions,** Edited by I.I. Sobel'man

Volume 179 **Interaction of Ultrashort Pulses with Matter,**
Supplemental Volume 1 Edited by M.D. Galanin

Volume 179 **Electron-Excited Molecules in Nonequilibrium Plasma,**
Supplemental Volume 2 Edited by N.N. Sobolev

Volume 180 **Inelastic Light Scattering in Crystals,** Edited by M.M. Sushchinskiy

Volume 181 **Pulsed Metal and Metal Halogenide Vapor Lasers,** Edited by G.G. Petrash

Volume 182 **Luminescence of Wideband Semiconductors,** Edited by M.D. Galanin

Volume 183 **Invariants and the Evolution of Nonstationary Quantum Systems,** Edited by M.A. Markov

Volume 184 **Electron Processes in MIS-Structure Memories,** Edited by A.F. Plotnikov

Volume 185	**Injection Lasers in Optical Communication and Information Processing Systems**, Edited by Yu. M. Popov
Volume 186	**Photoproduction of Pions on Nucleons and Nuclei**, Edited by A.A. Komar
Volume 187	**Quantum Field Theory, Quantum Mechanics and Quantum Optics: Pt. 1. Symmetries and Algebraic Structures in Physics**, Edited by V.V. Dodonov and V.I. Man'ko
Volume 188	**Integral Systems, Solid State Physics and Theory of Phase Transitions. Pt. 2. Symmetries and Algebraic Structures in Physics**, Edited by V.V. Dodonov and V.I. Man'ko
Volume 189	**Clusters of Galaxies and Extragalactic Radio Sources**, Edited by A.D. Kuz'min
Volume 190	**Metal-Optics and Superconductivity**, Edited by A.I. Golovashkin
Volume 190 Supplemental Volume	**Superconductors with A15 Lattice and Bridge Contacts Based Upon Them**, Edited by M. M. Sushchinskiy
Volume 191	**Polarization of the Vacuum and a Quantum Relativistic Gas in an External Field**, A.Ye. Shabad
Volume 192	**Atomic and Ionic Spectra and Elementary Processes in Plasma**, Edited by I.I. Sobelman
Volume 193	**Research on Chemical Lasers**, Edited by A.N. Orayevskiy
Volume 194	**Biophysical Approach to Complex Biological Phenomena**, Edited by E. Volkov
Volume 195	**Physical Characteristics and Critical Temperature of High Temperature Superconductors**, Edited by M.M. Sushchinskiy
Volume 196	**Pulsars**, Edited by A.D. Kuzmin
Volume 197	**Reliability Problems of Semiconductor Lasers**, P.G. Eliseev
Volume 198	**Theory of Nonstationary Quantum Oscillators**, Edited by M.A. Markov

Volume 199	**Spatiotemporal Characteristics of Laser Emission,** M.V. Pyatakhin and A.F. Suchkov
Volume 200	**Laser Thermonuclear Targets and Superdurable Microballoons** Edited by A.I. Isakov
Volume 201	**Generation of Nonlinear Waves and Quasistationary Currents in Plasmas,** Edited by L.M. Kovrizhnykh
Volume 202	**Gas Dynamics,** Edited by Yu.I. Koptev
Volume 203	**Fiber Optics: Research and Development,** Edited by Ye.M. Dianov
Volume 204	**Sakharov Memorial Lectures in Physics,** Edited by L.V. Keldysh and V.Ya. Fainberg
Volume 205	**Squeezed and Correlated States of Quantum Systems** Edited by M.A. Markov
Volume 206	**Mathematical Physics, Applied Mathematics and Informatics,** Edited by Yu.I. Koptev
Volume 207	**Generalized Mayer Series in Classical Statistical Mechanics** Igor J. Ivanchik
Volume 208	**Physical Properties of High Temperature Superconductors,** Edited by Yu.I. Koptev
Volume 210	**Theory and Technology of High Temperature Superconductors,** Edited by Yu.I. Koptev
Volume 211	**Optics and Lasers** Edited by G.G. Petrash
Volume 212	**Dynamics of Elementary Atomic-Molecular Processes in Gas and Plasma,** Edited by V.A. Shcheglov
Volume 213	**Nonlinear Theory of Strong Electromagnetic Wave-plasma Interactions,** Edited by O.N. Krokhin
Volume 216	**Semiconductors and Insulators: Optical and Spectroscopic Research,** Edited by Yu.I. Koptev

Volume 217	**Nonlinear and Quantum Optical Phenomena in Nonequilibrium Media,** Edited by V.A. Shcheglov
Volume 218	**Radiative and Collisional Characteristics of Ions in Hot Plasmas,** Edited by I.I. Sobel'man
Volume 219	**The Muon Method in Science** V.P. Smilga and Yu.M. Belousov
Volume 220	**Quantum Electrodynamics with Unstable Vacuum** Edited by V.L. Ginzburg
Volume 221	**Laser Cathode-Ray Tubes** Edited by Yu. M. Popov
Volume 222	**Gaussian Beams and Optical Resonators** A. N. Oraevsky

Proceedings of the Lebedev Physics Institute,
Russian Academy of Sciences
Volume 212

Dynamics of Elementary Atomic-Molecular Processes in Gas and Plasma

V.A. Shcheglov (Ed.)

Nova Science Publishers, Inc.
New York.

Cover Design: Barbara Minerd

Art Director: Maria Ester Hawrys
Assistant Director: Elenor Kallberg
Graphics: Eddie Fung, Kerri Pfister,
Erika Cassutti and Barbara Minerd
Manuscript Coordinator: Gloria H. Piza
Book Production: Tammy Sauter, Gavin Aghamore
Circulation: Irene Kwartiroff and Annette Hellinger

Library of Congress Cataloging–in–Publication Data available upon request

ISBN 1-56072-153-7

6080 Jericho Turnpike, Suite 207
Commack, New York 11725
Tele. 516-499-3103 Fax 516-499-3146
E Mail Novasci1@aol.com

Printed in the United States of America

CONTENTS

BIMOLECULAR REACTION RATES IN THE THEORY OF BROWNIAN MOTION

S.A. Reshetnyak

INTRODUCTION

The methods of the theory of Brownian motion are widly used in the physical kinetics. Specifically, these methods were applied for calculations of the electron-ion recombination rates [1,2]. This paper is devoted to the determination of the bimolecular reaction rates of the type

$$A + B + N \rightleftarrows AB + N \quad (1)$$

on the basis of the theory of Brownian motion.

The basic work in this field belongs to Kramers [3], who considered the reaction (1) as a one-dimentional process of the random walk of some particle in the abstract potential field. Kramers found the kinetic equation of the Boltzmann equation type with the linear collision integral and estimated from this equation the rates of the direct reaction (1) in the limiting cases of high and low gas pressure.

In spite of the great significance of the work [3], a number of principal points remained unclarified and hampered the practical use of this model. First of all, the problems of the applications of the real interaction potentials of the reacting particles and the reduction of the two-particle interaction problem to the one-particle one remained unsolved, which is not trivial because of the correlation between the movement of the reacting system as a whole and the relative movement of its particles. Also there was unclarified the problem of the calculation of the inverse recombination process rate which is closly related to the determination of the equilibrium constant for the direct and inverse process rates and its agreement with the detailed balancing principle resulted from the quasi-classical statistical sums. Besides the one-dimensional relative movement of the interacting particles, it is nesessary to consider the rotation of the system around its center of mass, i.e. to consider the three-dimensional interaction. The calculation of the reaction rates is substantially complicated by the lack of the correct methods of construction of the nonequilibrium solutions of Kramers equation containing the interaction potential of an arbitrary type. One shoud note that the construction of such solutions must be based on the clear definition of the bound and free states of the reacting particles, also it is necessary to consider the influence of boundary conditions on the reaction rate.

At present there are a number of works which deal with one or other aspects of this problem as a whole. In the case of the high gas or liguid pressures the analysis is based on Smolukhovsky equation. The calculations of the first eigenvalue of Smolukhovsky equation were carried out, which characterizes the passing time of Brownian particle over the potential barrier, based on the variation principle [4] and the estimation of the first average passing time [5]. The difficulties associated with the necessity of constructing the nonequilibrium solution of Kramers equation resulted in the development of numerical methods of its analysis [6,7] and construction of the solution of the joint type in the limit of the low and medium gas pressure [8]. The application of this method for real interaction potential in the case of two identical particles is described in [9]. Here the task is simplified because of the independence on the relative movement of the reacting particles and the movement of the system center of mass.

This paper offers the succesive stochastic theory of the process (1) with the consideration of all questions on the basis of the three-dimentional kinetic equation for the two-particle distribution function. This theory includes the process of interaction between the electric charged particles. Also the limits for application of this method are thoroughly discussed.

When analysing the kinetic in the active laser media one inevitably comes across the great number of the processes that can be arbitrarily divided into two groups. The reactions that form the inverse population may be assigned to the first group. The rates of such processes must be defined with the high precision on the basis of the quantum-mechanical calculations for the reaction cross sections. Such calculation are rather cumbersome and, as a rule, specific. The processes which influence the conditions of producing the inverse population in an indirect way may be included in another group. Often it is nesessary to know only an order of the value of rates in these processes. In this sense the stochastic theory of bimolecular reactions (1) applied to the wide variety of the processes is rather reliable and effective tool for calculation of rates in the second group of reactions, and its development is an actual problem.

1. BASIC EQUATIONS AND ANALYSIS OF THE EQUILIBRIUM STATE

Consider a movement of two particles A and B with the interaction potential U(r) in gaseous or liquid medium represented as thermostat with constant temperature T. Fig. 1 shows a typical form of U(r). Particles in the thermostat, encountering the reacting ones, give or take from them the energy of order T. We shall consider the reactions that proceed at many collisions between reacting particles and the thermostat ones. The thermostat influence on the system results in the fluctuation-dissipative character of its movement. This character makes itself evident in the fact that the thermal noise randomly disturbs the movement of the reacting particles going on with a dissipation. In this case the movement of particles A and B is discribed by stochastic equations of the type:

$$\frac{d\mathbf{v}_i}{dt} = -m_i^{-1}\nabla U - \gamma_i \mathbf{v}_i + \mathbf{y}_i(t), \quad \frac{d\mathbf{r}_i}{dt} = \mathbf{v}_i, \qquad i = \mathrm{A}, \mathrm{B}, \tag{1.1}$$

where r_i and m_i are a position radius and a mass of i-th reacting particle; γ_i are dissipation coefficients which are assumed tobe given; y_i are fluctuation forces caused by the thermostat influence on the system.

Equations of the type (1.1) are called the Langevin equations. The random forces y_i are characterized by the correlation time τ_0 equal to the interaction time between reacting and thermostat particles [10]. Note that this method may be applied when the correlation time is small compared to the charachteristic time of the processes involved. The introduction of dissipation members proportional to the velocities of the reacting particles movement is justified only in the case when the observation time far exceeds the inverse frequencies of collisions between reacting and thermostat particles. At present section the random force is taken to be a stationary Gaussian random process: white noise with $\tau_o = 0$. Consideration of the finitness of the correlation time and discusion of conditions required for the realization of the method will be given below. So, let $y_i(t)$ be a white noise and the following conditions are satisfied:

$$\langle \mathbf{y}_{A,B}(t)\rangle = 0, \quad \langle \mathbf{y}_A(t)\,\mathbf{y}_B(t')\rangle = 0, \; \langle y_j(t)\,y_k(t')\rangle = 2D\delta_{jk}\delta(t-t'), \tag{1.2}$$

where the coefficient D characterizes the intensity of a random force and is determined by the thermostat temperature.

The reacting system state may be represented by the point (r_1, r_2, v_1, v_2) in the phase space. In this model the process of the reaction is considered similiarly to the Brownian motion of the point which presents the system state in the space. The system (1.1) supplemented by the conditions (1.2) is equivalent to the kinetic equation for the two-particle probability density F that the system is in a certian point of the phase space at a given moment. The transition from (1.1), (1.2) to the equation for F is performed similiarly to [11,12], so we immediately write the equation:

$$\begin{aligned} &\frac{\partial F}{\partial t} + \mathbf{v}_1\frac{\partial F}{\partial \mathbf{r}_1} + \mathbf{v}_2\frac{\partial F}{\partial \mathbf{r}_2} - m_1^{-1}\nabla_1 U\frac{\partial F}{\partial \mathbf{v}_1} - \\ &- m_2^{-1}\nabla_2 U\frac{\partial F}{\partial \mathbf{v}_2} = \gamma_1\frac{\partial}{\partial \mathbf{v}_1}\left(\mathbf{v}_1 F + m_1^{-1}T\frac{\partial F}{\partial \mathbf{v}_1}\right) + \\ &+ \gamma_2\frac{\partial}{\partial \mathbf{v}_2}\left(\mathbf{v}_2 F + m_2^{-1}T\frac{\partial F}{\partial \mathbf{v}_2}\right). \end{aligned} \tag{1.3}$$

Eq. (1.3) determines the two-particle distriution function of the reacting particles over states with the collision integral of the same type as in the Brownian motion theory. The equivalence of (1.3) and (1.1) - (1.2) shows up in the fact that the time-averaged value of any quantity in (1.1) agrees with the value of this quantity averaged over the ensemble or distribution function F.

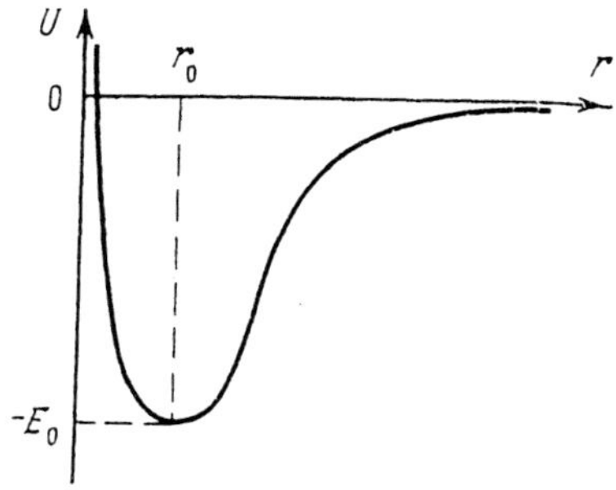

Fig.1. Form of the potential of interaction between reacting particles

Consider now free and bound states of the reacting particles A and B. For this purpose we discuss the quilibrium distribution function of Eq. (1.3):

$$F_0 = C \exp\left[-\frac{p_1^2}{2m_1 T} - \frac{p_2^2}{2m_2 T} - \frac{U(r)}{T} \right]. \tag{1.4}$$

This solution, like any non-stationary solution of Eq.(1.3), have to be normalized to some conserved value. There are two conserved values in the problem of dissociation or recombination of particles into a molecule: the total density of particles A and B both in bound and free states. Assume that N_{AB} is the density of bound states, and N_A and N_B are the densities of free particles A and B. Then the distribution function can be normalized either to $N_{AB} + N_A$ or to $N_{AB} + N_B$. The conserved values are defined by the initial conditions. Choose a constant C in (1.4) so that

{Eq. 1, p. 5}
$$\int F_0 dX_1 dX_2 = N_{AB} + N_A,$$

where $dX_1 = d^3p_1 d^3r_1$, $dX_2 = d^3p_2 d^3r_2$.

Let us define the boundary between bound and free states in such a way that A and B are free when the distance between them r > R and bound when r < R. Then in the equilibrium state:

$$N_{AB}^{(0)} = \int_{r<R} F_0\, dX_1\, dX_2 = C_1 \int_0^R \exp[-U(r)/T]\, d^3r, \tag{1.5}$$

$$N_A^{(0)} = \int_{r>R} F_0\, dX_1\, dX_2 = C_1 \int_R^L \exp[-U(r)/T]\, d^3r,$$

where after integration over momenta and location of one of the reacting particles the new normalization constant C_1 was introduced; $r = r_2 - r_1$; L is the total linear size of the system coinciding in order of magnitude with the average distance between free reacting particles as will be shown below.

Regarding the choice of the boundary value of R note the following. R must satisfy a single requirement that the reaction velocities should be independent on the choice of R.

Consider the most physically interesting case of low medium temperatures T compared to the bond energy E_0 between A and B particles. This case is also convenient in a mathematical sense because it allows to use the saddle-point method for the calculation of the integrals involved. Taking into account that the boundary between bound and free states for a gas and a liquid is always located in the region which size isless than the average distance between free particles, we get from (1.5)

$$N_{AB}^{(0)} = 4\pi C_1 \exp\left(\frac{E_0}{T}\right) \int_0^R \exp\left[-\frac{\mu\omega_0^2}{2T}(r - r_0)^2\right] r^2\, dr = 4\pi C_1 \exp\left(\frac{E_0}{T}\right) \frac{r_0^2}{\omega_0} \left(\frac{2\pi T}{\mu}\right)^{1/2},$$

$$N_A^{(0)} = 4\pi C_1 \int_R^L \exp\left[-\frac{U(r)}{T}\right] r^2\, dr \underset{R \ll L}{=} \frac{4}{3}\pi C_1 L^3 = C_1 V, \qquad (1.6)$$

where $\mu = m_1 m_2 (m_1 + m_2)^{-1}$ is the reduced mass of the reacting particles; r_0 and ω_0 are the equilibrium position and vibrational frequency; V is the system volume involved.

The following comment must be done concerning the calculation of the first integral in (1.5). Eq.(1.6) results from the representation of the potential as $U(x) = -E_0 + \mu w_0^2 (r-r_0)^2/2$ using the saddle-point method. This is valid if vibrations and rotations of the reacting system are independent and the amplitude r' of harmonic vibrations satisfies the condition

$$r' = r - r_0 > \omega_0^{-1} (T/\mu)^{1/2}. \qquad (1.7)$$

The interrelation between vibrations and rotations may not be considered at $T \ll \mu(\omega_0 r_0)^2$ [13].

Eqs.(1.6) result in the following ratio of densities in the equilibrium state:

$$\frac{N_{AB}^{(0)}}{N_A^{(0)}} = \frac{4\pi r_0^2}{V\omega_0}\left(\frac{2\pi T}{\mu}\right)^{1/2} \exp\left(\frac{E_0}{T}\right). \quad (1.8)$$

Using now formulas for quasi-classic statistical sums for the reacting particles, we get:

$$Z_A = (m_1 T/2\pi\hbar^2)^{3/2}, \qquad Z_B = (m_2 T/2\pi\hbar^2)^{3/2},$$

$$Z_{AB} = Z_T Z_R Z_V \exp\left(\frac{E_0}{T}\right) = \left[\frac{(m_1+m_2)T}{2\pi\hbar^2}\right]^{3/2} \frac{2\mu r_0^2 T}{\hbar^2} \frac{T}{\hbar\omega} \exp\left(\frac{E_0}{T}\right),$$

and find from it the ratio of densities in the equilibrium state

$$\frac{N_{AB}^{(0)}}{N_A^{(0)} N_B^{(0)}} = \frac{Z_{AB}}{Z_A Z_B} = \frac{4\pi r_0^2}{\omega_0}\left(\frac{2\pi T}{\mu}\right)^{1/2} \exp\left(\frac{E_0}{T}\right) = K_e. \quad (1.9)$$

Taking into account the agreement of (1.8) and (1.9), we find the condition for the phase volume $V = N_B^{-1}$. The limits for the phase volume can be explained as follows. The linear volume size ($L \sim N_B^{-1/3}$) are equal to the average distance between free particles B. That is why after the atom A has become free, i.e. the bound state has decayed, the atom gets into the field of action of another atom B on a distance$N_B^{-1/3}$, and the individuality of the pair interaction involved will be lost.

Except the interaction potential U(r) and the reacting particles masses, the other basic paramerters determining the reaction rateare the dissipation coefficients. If the interacting particles are neutral it is possible to specify their limitig values. Thus, for luquids or very dense gases these coefficients are determined by the Stokes formula:

$$\gamma_i = 6\pi a_i \eta m_i^{-1},$$

where η is the viscosity of medium, a_i are hydrodynamical radii of A and B particles.

For the rarefied gases the coefficients γ_i are determined from the expressions for diffusion coefficients in a usual space:

$$D_i = T/(m_i \gamma_i),$$

and with an accuray of the mass ratio are determined by the frequency of collisions between reacting and thermostat particles. Here $\gamma_i = \nu_i \mu_i m_i^{-1}$, $\mu_i = mm_i(m + m_i)^{-1}$, m is the thermostat particle mass.

In [9] the simple interpolation formula is presented for $\gamma_i(N)$, where N is the particles density in a thermostat. It is based on the given limiting values of γ_i and has the form

$$\gamma_i = 6\pi a_i \eta m_i^{-1} [1 - \exp(- N/N_i)], \quad (1.10)$$

where $N_i = 6\pi\eta a_i D_i N T^{-1}$. One can see from (1.10) that the friction coefficients are described by the Stokes formula at $N > N_i$ and by the diffusion coefficients of the reacting particles at $N < N_i$.

Expressing diffusion coefficientsD_i, obtained in the first Chapman-Ensckoge approximation, via the scattering cross sections of thermostat particles by the reacting particles, it is easy to obtain the formula

$$N_i = \frac{45\pi^2}{64\sqrt{2}} \left(\frac{m}{\mu_i}\right)^{1/2} \frac{a_i}{\bar{\sigma}\bar{\sigma}_i}, \quad (1.11)$$

where

$$\bar{\sigma} = \frac{1}{32} \int_0^\infty \sigma(x)\, x^7 \exp(- x^2/2)\, dx,$$

$$\bar{\sigma}_i = \frac{1}{8} \int_0^\infty \sigma_i(x)\, x^5 \exp(- x^2/2)\, dx,$$

Here $\sigma(x) = \int (1 - \cos^2\chi) d\sigma$ is the scattering cross section for the thermostat particles collisions that determines the medium viscosity; $\sigma_i(x) = \int (1 - \cos\chi) d\sigma$ is the diffusion scattering cross section at collision between the reacting and thermostat particles. The estimates made in [9] have shown that for the monatomic iodine in ethan medium $N_1 = N_2 = 1.4 \cdot 10^{21}$ cm^{-3}.

When interacting particles A and B are electrically charged the dissipation coefficients can be expressed via their mobility or collision freguences:

$$\gamma_i = q\,(m_i b_i)^{-1} = \mu_i m_i^{-1} \nu_i, \qquad b_i = q\,(\mu_i \nu_i)^{-1}, \tag{1.12}$$

where ions with identical charges q are considered; b_i are ion mobilities; ν_i are collision freguences of ions with environmental particles. If a thermostat medium is a mixture of different components then according to theBlank law, the ion mobilities are calculated by the formula

$$b^{-1} = \sum_j C_j b_j^{-1},$$

where b_j and C_j are the ion mobilities and the relative concentration of j-th component of the mixture.

The main goal of the paper is the determination of the reaction rates of the type (1), where N is the composition of the environment, i.e. of the thermostat. The interaction potential of A and B particles and their dissipation coefficients are assumed to be specified. By a reaction rate is meant a flow of the probability density trough the boundary R between bound and free states. In the equilibrium state the flow is zero and the main problem is the construction of the non-equilibrium distribution functions. The densities of the bound and free states of particles A and B at the initial moment are assumed not to correspond to the the thermostat temperature.

2. ONE-DIMENSIONAL KINETIC EQUATION WITH EFFECTIVE POTENTIAL

The construction of the non-equilibrium distribution of the Eq.(1.3) for the two-particle' probability density is a rather complex task. It may be simplified significantly if one separates the movement of the center of mass of the reacting system and will not consider it. However, it is nesessary to take into account the dissipative character of the reacting particles movement which results in a correlation between velocities of the relative movement of A and B particles and the center of mass. The simplest way of doing this is to describe the movement by the set of Langevin equations (1.1). Therefore,consider Eq.(1.1) averaged over some finite time interval t_0 which, generaly speaking, exeedes the random force correlation time τ_0. Introducing vectors of coordinates R and the center of mass velocity V, and also vector r of the relatve distance between A and and B particles:

$$\mathbf{r} = \mathbf{r}_2 - \mathbf{r}_1, \quad \mathbf{R} = (m_1\mathbf{r}_1 + m_2\mathbf{r}_2)M^{-1}, \quad M = m_1 + m_2, \quad \dot{\mathbf{r}} = \mathbf{v}, \quad \dot{\mathbf{R}} = \mathbf{V},$$

and dropping the averaging sign in (1.1), we have:

$$\frac{d\mathbf{v}}{dt} = -\mu^{-1}\nabla_r U - (m_1\gamma_2 + m_2\gamma_1)M^{-1}\mathbf{v} + (\gamma_1 - \gamma_2)\mathbf{V},$$
$$\frac{d\mathbf{V}}{dt} = -(m_1\gamma_1 + m_2\gamma_2)M^{-1}\mathbf{V} + \mu M^{-1}(\gamma_1 - \gamma_2)\mathbf{v}. \qquad (2.1)$$

It follows from (2.1) that in the case of the identical reacting particles ($\gamma_1 = \gamma_2$) their relative movement and the movement of the system center of mass are independent. If γ_1 and γ_2 are different, the relative movement influences the movement of the system center of mass. To account forthis influence, we consider the observation times $t \gg \gamma^{-1}$. Then, expressing from the second stationary Eq. (2.1) velocity V via v and substituting into the first equation, we find :

$$\frac{d\mathbf{v}}{dt} = -\mu^{-1}\nabla_r U - \gamma\mathbf{v}, \qquad (2.2)$$

where $\gamma = M\gamma_1\gamma_2/(m_1\gamma_1+m_2\gamma_2)$ is the reduced friction coefficient. Eq.(2.2) discribes the movement of one particle of mass μ with the friction coefficient γ in the field U(r). Without considering the connection between the movements involved the coefficient $\gamma = (m_1\gamma_2 + m_2\gamma_1)M^{-1}$ [14]. Another independent deduction of the formula for the reduced friction coefficient γ that immidiately follows from Eq.(1.3) for the two-particle distribution function will be given below in consideration of the limiting case of the large γ.»

Now we add to (2.2) the random force $\xi(t)$ and consider Brownian movement of one particle on the basis of the following Langevin equations:

$$\frac{d\mathbf{v}}{dt} = -\mu^{-1}\nabla_r U - \gamma\mathbf{v} + \xi(t),$$
$$\frac{d\mathbf{r}}{dt} = \mathbf{v} \qquad (2.3)$$

with the moments $\langle\xi(t)\rangle = 0$, $\langle\xi_i(t)\xi_\varkappa(t')\rangle = 2D\delta_{ik}\delta(t - t')$.

This setof equations is equivalent to the Kramers kinetic equation for the probability density f(t,r,v)

$$\frac{\partial f}{\partial t} + \mathbf{v}\frac{\partial f}{\partial \mathbf{r}} - \mu^{-1}\nabla U\frac{\partial f}{\partial \mathbf{v}} = \gamma\frac{\partial}{\partial \mathbf{v}}\left(\mathbf{v}f + T\mu^{-1}\frac{\partial f}{\partial \mathbf{v}}\right). \qquad (2.4)$$

The further simplification of Eq.(2.4) deals with the assumption that the rotation of the reacting system does not depend on its vibrations and is in the equilibrium with the thermostat. So introduce into (2.4) spherical coordinates of Brownian particle

$$\mathbf{r} = (r\sin\theta\cos\varphi,\ r\sin\theta\sin\varphi,\ r\cos\theta), \qquad \mathbf{v} = (v,\ v_\varphi,\ v_\theta)$$

and taking into account the fact that the spherical components of the velocity vector depend on angles, present Eq.(2.4) in the form

$$\frac{\partial f}{\partial t} + v\frac{\partial f}{\partial r} + \frac{v_\varphi}{r\sin\theta}\frac{\partial f}{\partial \varphi} + \frac{v_\theta}{r}\frac{\partial f}{\partial \theta} + \frac{v_\varphi^2 + v_\theta^2}{r}\frac{\partial f}{\partial v} + \left(\frac{\cos\theta v_\varphi^2}{r\sin\theta} - \frac{v_\theta v_\varphi}{r}\right)\frac{\partial f}{\partial v_\theta} -$$

$$-\left(\frac{v_\varphi v}{r} + \frac{\cos\theta\, v_\varphi v_\theta}{r\sin\theta}\right)\frac{\partial f}{\partial v_\varphi} - \mu^{-1}U'(r)\frac{\partial f}{\partial v} = \sum_i \gamma\frac{\partial}{\partial v_i}\left(v_i f + T\mu^{-1}\frac{\partial f}{\partial v_i}\right), \qquad (2.5)$$

where i = r, θ, φ.

Consider the function ρ which accounts forthe relative movement of the reacting particles along vector r of the distance between them, namely

$$\rho = r^2\int f d\Omega dv_\varphi dv_\theta, \qquad (2.6)$$

where $d\Omega = \sin\theta\, d\theta\, d\varphi$ is an element of the solid angle.

Then integrating Eq.(2.5) over angular variables and velocities v_φ and v_θ and multiplying the right and left part of (2.5) by r^2 we get:

$$\frac{\partial \rho}{\partial t} + v\frac{\partial \rho}{\partial r} + \frac{\partial}{\partial v}\left[\frac{\langle v_\varphi^2 + v_\theta^2\rangle}{r}\rho\right] - \mu^{-1}U'(r)\frac{\partial \rho}{\partial v} = \gamma\frac{\partial}{\partial v}\left(v\rho + T\mu^{-1}\frac{\partial \rho}{\partial v}\right),$$

where

$$\langle v_\varphi^2 + v_\theta^2\rangle = \frac{\int (v_\varphi^2 + v_\theta^2) f\, d\Omega\, dv_\varphi\, dv_\theta}{\int f\, d\Omega\, dv_\varphi\, dv_\theta}.$$

This equation also may be represented as the one-dimentional equation of the Brownian particle movement in a field with the effective potential U_e:

$$\frac{\partial \rho}{\partial t} + v\frac{\partial \rho}{\partial r} - \mu^{-1}\frac{\partial}{\partial v}\left(U'_e\rho\right) = \gamma\frac{\partial}{\partial v}\left(v\rho + T\mu^{-1}\frac{\partial \rho}{\partial v}\right), \qquad (2.7)$$

$$U_e = U(r) - 2\int \langle K\rangle r^{-1}\,dr,$$

where $\langle K\rangle = (\mu/2)\langle v_\varphi^2 + v_\theta^2\rangle$ is the average energy of the system rotations (more precisely of the rotational degrees of freedom).

Now we determine the average rotational energy using the set of Langevin equations (2.3). To do this we write these equations in the spherical coordinate system:

$$\frac{dv}{dt} = -\mu^{-1}U'(r) + \frac{v_\varphi^2 + v_\theta^2}{r} - \gamma v + \xi_r(t),$$

$$\frac{dv_\theta}{dt} = -\frac{vv_\theta}{r} + \frac{v_\varphi^2\cos\theta}{r\sin\theta} - \gamma v_\theta + \xi_\theta(t),$$

$$\frac{dv_\varphi}{dt} = -\frac{vv_\varphi}{r} - \frac{v_\varphi v_\theta\cos\theta}{r\sin\theta} - \gamma v_\varphi + \xi_\varphi(t),$$

$$\frac{dr}{dt} = v,\quad \frac{d\theta}{dt} = \frac{v_\theta}{r},\quad \frac{\partial\varphi}{\partial t} = \frac{v_\varphi}{r\sin\theta}. \qquad (2.8)$$

Multiplying the second and third equations of the system (2.8) by v_φ and v_θ, respectively, and summing them we arrive at the equation for the rotational energy

$$(2.9) \qquad dK/dt = -2(v/r + \gamma)\,K + \eta(t),$$

where $K = (\mu/2)\,(v_\theta^2 + v_\varphi^2)$, $\eta(t) = \mu(v_\theta\xi_\theta + v_\varphi\xi_\varphi)$ is the variation of the rotational energy per unit time caused by work of the random force $\xi(t)$.

Consider the averaged equation (2.9) not taking into account the interrelation between vibrations and rotations:

$$d\langle K\rangle/dt = -2\gamma\langle K\rangle + \langle\eta\rangle.$$

Taking into account that the average rotational energy takes the value T with time, we find its solution

$$\langle K\rangle = T + C\exp(-2\gamma t),$$

where C is fixed by the initial rotational energy of the system.

Hence it follows that the rotational degrees of freedom are thermolized in a time of order γ^{-1} and the relative movement of the reacting particles occures in the potential:

(2.10)
$$U_e(r) = U(r) - 2T \ln (r/r_0),$$

where r_0 is the dimensional factor determined by the zero energy of potential of the particle interactions. In this paper r corresponds to the position of the initial potential U(r) minimum.

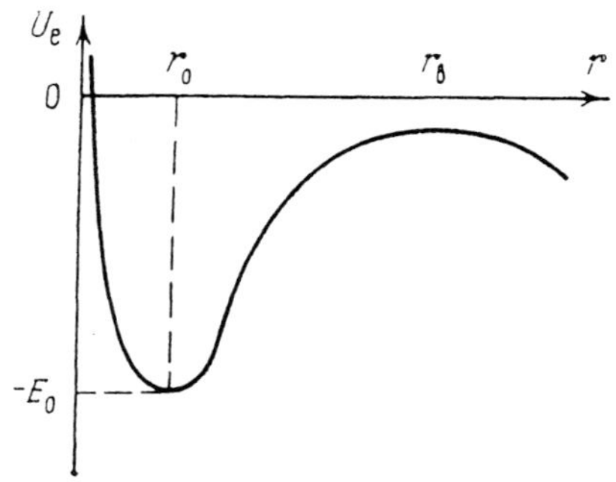

Fig. 2. Form of the effective interaction potential for reacting partices in the case of one-dimensional movement

Precisely the same expression for the effective potential $U_e(r)$ results from the value of $(v_\theta^2 + v_\varphi^2)$ averaged over Maxwell distribution of the rotational velocities of the reacting system. The appearence of the logarithm in (2.10) reflects the growth of d^3r phase volume element as r^2.

Hence, separating the center of mass movement and the relative movement of the reacting particles and supposing that the rotational degrees of freedom are in equilibrium with the translational ones, one can reduce the problem to the one-dimensional movement of a Brownin particle in the effective potential field (2.10). The consideration of rotations of the reacting particlessystem appears as a local maximum and lowered potential barier value of (2.10)(Fig. 2). As analysis shows, the position r_0 of the potential minimum and the vibrational frequency ω_0 near the equilibrium position for not very high temperatures are the same as in the case of the initial potential U(r)

(2.11) $$T \ll \mu(r_0\omega_0)^2.$$

This condition allows to assume the vibrational and rotational movements of the system to be independent. For neutral reacting particles the energy by which the potential barier height decreases is of order thermostat temperature.

3. CONSIDERATION OF FINITENESS OF THE RANDOM FORCE CORRELATION TIME. CONDITIONS FOR APPLICATION OF THE KRAMERS MODEL

In this section the main attention is paid to the construction of the kinetic equation for the probability density that corresponds to Langevin equations with the finite random force correlation time τ_0. Consider the set (2.3) in the one-dimensional case :

(3.1) $$\frac{dv}{dt} = -\mu^{-1}U_e'(r) - \gamma v + \xi(t), \qquad \frac{dr}{dt} = v,$$

where $\xi(t)$ is a random force with the correlation time τ_0. The average value of the force and the correlation function are defined by the formulas

$$\langle\xi(t)\rangle = t_0^{-1}\int_0^{t_0} \xi(t+\tau)\,d\tau,$$

(3.2) $$\langle\xi(t)\,\xi(t+\tau)\rangle = t_0^{-1}\int_0^{t_0} \xi(t+\tau')\,\xi(t+\tau+\tau')\,d\tau' = K(\tau),$$

where t_0 is the time interval far greater than τ_0. $\xi(t)$ is the stationary random process. For $\xi(t)$ of the white noise type $K(\tau) = 2D\delta(\tau)$. If the noise has a finite correlation time, then $K(\tau)$ may be specified in the form

(3.3) $$K(\tau) = (D/\tau_0)\exp(-|\tau|/\tau_0).$$

We present the kinetic equation for the distribution function as the continuity equation in the phase space:

$$\frac{\partial f}{\partial t}+\frac{\partial}{\partial r}(\dot{r}f)+\frac{\partial}{\partial v}(\dot{v}f)=0,$$

or
$$\frac{\partial f}{\partial t}+v\frac{\partial f}{\partial r}+\frac{\partial}{\partial v}\left[\left(-\mu^{-1}U'_e-\gamma v+\xi\right)f\right]=0. \quad (3.4)$$

Let us break up the distribution function into two parts:

$$f=\rho+\tilde{\rho},$$

where ρ = $\langle f\rangle$ is a part of the function f smoothed over the t_0 time scale, and $\tilde{\rho}$ is its fluctuating part associated with the random force $\xi(t)$ and satisfying to the equation $\langle\tilde{\rho}\rangle = 0$.

Using Eq.(3.4) and Eq.(3.4) averaged over the time scale t_0. we obtain

$$\frac{\partial\rho}{\partial t}+v\frac{\partial\rho}{\partial r}-\mu^{-1}U'_e\frac{\partial\rho}{\partial v}=\gamma\frac{\partial}{\partial v}(v\rho)-\frac{\partial}{\partial v}\langle\xi\tilde{\rho}\rangle,$$

(3.5)
$$\frac{\partial\tilde{\rho}}{\partial t}+v\frac{\partial\tilde{\rho}}{\partial r}-\mu^{-1}U'_e\frac{\partial\tilde{\rho}}{\partial v}-\gamma\frac{\partial}{\partial v}(v\tilde{\rho})=-\frac{\partial}{\partial v}(\xi\rho)-\frac{\partial}{\partial v}[\xi\tilde{\rho}-\langle\xi\tilde{\rho}\rangle].$$

It is required to find $\tilde{\rho}$ from the second equation of (35)and substitute it to the first one. If the thermostat temperature is low compared to the bond energy of reacting particles, the non-linear fluctuating terms in the second equation of (3.5) can be neglected and the equation can be presented in the form:

(3.6)
$$\frac{\partial\tilde{\rho}}{\partial t}+\widehat{\mathscr{D}}\tilde{\rho}=-\frac{\partial}{\partial v}(\xi\rho),$$

where

$$\widehat{D}\tilde{\rho}=\left(v\frac{\partial}{\partial r}-\mu^{-1}U'_e\frac{\partial}{\partial v}-\gamma\frac{\partial}{\partial v}v\right)\tilde{\rho}.$$

Fixing the moment t in (3.6), we shall seek the solution in the interval (t, t + τ), where $\tau \sim \tau_0 \ll t$. To do this we turn from (3.6) to the integral equation

(3.7)
$$\tilde{\rho}(t+\tau)=\tilde{\rho}(t)-\widehat{\mathscr{D}}\int_0^\tau\tilde{\rho}(t+\tau')\,d\tau'-\int_0^\tau\xi(t+\tau')\frac{\partial\rho}{\partial v}(t+\tau')\,d\tau'$$

and find its solution using the iteration method. In the case of white noise the second member in (3.7) may be dropped [12], hence at the first step of the iteration the solution for white noise should be taken as a check run:

(3.8) $$\widetilde{\rho}\,(t+\tau)=\widetilde{\rho}\,(t)-\frac{\partial\rho}{\partial v}\int\limits_0^{\tau}\xi\,(t+\tau')\,d\tau'.$$

Substituting (3.8) in the right part of (3.7) and repeating the mentioned operations, we arrive at the solution in the form of a power series in $\tilde{D}$ operator or, as it is easy to see, in a power series in the correlation time τ_0. Therewith the solution $\rho^{\sim}$ implicitly depends on the coordinate and velocity via the averaged distribution function ρ. Determine $<\xi\tilde{\rho}>$ at zeroth and first iteration steps. At zeroth step

(3.9) $$\langle\xi\widetilde{\rho}\rangle=-\frac{\partial\rho}{\partial v}\,t_0^{-1}\int\limits_0^{t_0}\xi\,(t+\tau)\,d\tau\int\limits_0^{\tau}\xi\,(t+\tau')\,d\tau'=-\frac{\partial\rho}{\partial v}\int\limits_0^{t_0}K\,(\tau)\,d\tau=-D\frac{\partial\rho}{\partial v},$$

where $D=\int_0^{t_0}K(\tau)d\tau$ is the intensity of the random force source. Formula (3.9) canbe easly obtained integrating (3.2) over the variable τ considering the inequality $t_0 \gg \tau_0$ that allows to treat t_0 as an infinite integration limit.

Substitution of (3.9) in (3.5) leads to theKramers equation (2.7)

$$\frac{\partial\rho}{\partial t}+v\frac{\partial\rho}{\partial r}-\mu^{-1}U_{\mathrm{e}}'\frac{\partial\rho}{\partial v}=\frac{\partial}{\partial v}\left(\gamma v\rho+D\frac{\partial\rho}{\partial v}\right).$$

Requiring the coicidence of the equilubrium distribution for this equation with the Maxwell-Boltzmann distribution

$$\rho_0\sim\exp\left(-\frac{\mu v^2}{2T}-\frac{U_{\mathrm{e}}}{T}\right),$$

we find the intensity of the random force source: $D=\gamma T\mu^{-1}$.

At first iteration step we have

(3.10) $$\widetilde{\rho}\,(t+\tau)=\widetilde{\rho}\,(t)+\int\limits_0^{\tau}\xi\,(t+\tau')\left[(\tau+\tau')\,\widehat{\mathscr{D}}\,\frac{\partial\rho}{\partial v}\,(t+\tau')-\frac{\partial\rho}{\partial v}\,(t+\tau')\right]d\tau'.$$

Let us expand the second component in the square brackets (3.10) as a power series in $(\tau'-\tau)$ and consider only the first twomembers of the expantion:

$$\frac{\partial\rho}{\partial v}(t+\tau') = \frac{\partial\rho}{\partial v}(t+\tau) + \frac{\partial^2\rho}{\partial t\,\partial v}(t+\tau)(\tau'-\tau) + \ldots \underset{t\gg\tau}{=} \frac{\partial\rho}{\partial v} + \frac{\partial^2\rho}{\partial t\,\partial v}(\tau'-\tau) + \ldots$$

Considering members of the first order of smallness in τ_0, we get

$$\widetilde{\rho}(t+\tau) = \widetilde{\rho}(t) - \frac{\partial\rho}{\partial v}\int_0^\tau \xi(t+\tau')\,d\tau' + \left(\frac{\partial}{\partial t} + \widehat{\mathscr{D}}\right)\frac{\partial\rho}{\partial v}\int_0^\tau \xi(t+\tau')(\tau-\tau')\,d\tau'.$$

From this we find the collision integral and the form of the kinetic equation for a finite correlation time:

$$\langle\xi\widetilde{\rho}\rangle = -D\frac{\partial\rho}{\partial v} + \left(\frac{\partial}{\partial t} + \widehat{\mathscr{D}}\right)\frac{\partial\rho}{\partial v}\int_0^{t_0} \tau K(\tau)\,d\tau, \tag{3.11}$$

$$\frac{\partial f}{\partial t} + v\frac{\partial f}{\partial r} + \frac{\partial}{\partial v}\left[(-\mu^{-1}U_e' - \gamma v)\,f\right] = D\frac{\partial^2 f}{\partial v^2} - \tau_0 D\frac{\partial}{\partial v}\left(\frac{\partial}{\partial t} + v\frac{\partial}{\partial r} - \mu^{-1}U_e'\frac{\partial}{\partial v} - \frac{\partial}{\partial v}\gamma v\right)\frac{df}{\partial v}, \tag{3.12}$$

where $\tau_0 D = \int_0^{t_0} \tau K(\tau)d\tau$. To demonstrate the validity of (3.11), one can multiply (3.2) by τ and integrate it.

It is easy to see that the Eq.(3.12) transforms into (2.7) providing that

$$\gamma\tau_0 \ll 1. \tag{3.13}$$

The random force correlation time is determined by the interaction time of reacting particles with thermostat ones. Therefore the condition (3.13) is true at least within the framework of the pair interaction approximation. Far more drastic condition appears if we take into consideration the fact that Eq.(2.7) was found by averaging over time interval more than τ_0. It places the certain constrains on the characteristic times of consideration of processes described by (2.7). It is immidiately obvious that the description of processes with times of γ^{-1} order is invalid. Hence, the reaction rates obtained within the framework of this model must satisfy the requirment:

$$\tau_R \gg \nu_i^{-1}, \tag{3.14}$$

where τ_R - the characteristic reaction time.

The essential difference between this approach and the classic theory of Brownian motion is that when calculating the reaction rates there is rather large time ofthe particle transition through the potential barrier. This allows to obtain the kinetic equation for the probability density smoothed over the time intervalthat is far exceeds the time between two subsequent collisions of the reacting particles with the thermostat ones. In the classic theory of Brownian motion the characteristic times are of order γ^{-1}. Therefore, to average over the great number of collisions the following requirements must be satisfied: $\tau_0 \ll \upsilon_i^{-1} \ll \gamma^{-1}$, but this is possible only in the case of heavy Brownian particles.

4. APERIODIC RELAXATION REGIME (THE CASE OF LARGE DISSIPATIONS)

Consider reactions proceeding in dense gases or liquids, assuming the friction coefficient to be large compared to the frequency of vibrations near the equilibrium position ($\gamma > \omega_0$). In this case the particle before getting at the minimum energy state performs arandom walk both along the distance vector r between the particle and center of force and over the sphere surface with r fixed. For the moments $t \gg \gamma^{-1}$ in the case considered, Eq.(2.4) allows further simplification associated with the transition from (2.4) to the Smolukhovsky equation. This transition is shown, for example, in [11,12]. Here to obtain formulas for the reduced dissipation coefficient and effective interaction potential by other method, the above-mentioned transition has not been performed from (8.4), but from Eq.(1.1) for the two-particle distribution function. The rates of dissociation and recombination processes are determined from the Smolukhovsky equation solution asymptotic in```time.

Let us separate out the coordinate and velocity of the system center of mass in Eq.(1.1) replacing variables in the same way as it had been done to obtain the system (2.1). As a result, instead of (1.1) we have:

$$\frac{\partial F}{\partial t} + \mathbf{v}\frac{\partial F}{\partial \mathbf{r}} + \mathbf{V}\frac{\partial F}{\partial \mathbf{R}} - \mu^{-1}\nabla_r U_e \frac{\partial F}{\partial \mathbf{v}} = G_1 \frac{\partial}{\partial \mathbf{v}}\left(\mathbf{v}F + T\mu^{-1}\frac{\partial F}{\partial \mathbf{v}}\right) +$$

$$+ G_2 \frac{\partial}{\partial \mathbf{V}}\left(\mathbf{V}F + TM^{-1}\frac{\partial F}{\partial \mathbf{V}}\right) + \frac{\mu}{M}(\gamma_2 - \gamma_1)\frac{\partial}{\partial \mathbf{V}}\left(\mathbf{v}F + T\mu^{-1}\frac{\partial F}{\partial \mathbf{v}}\right) +$$

(4.1) $$+ (\gamma_2 - \gamma_1)\frac{\partial}{\partial \mathbf{v}}\left(\mathbf{V}F + TM^{-1}\frac{\partial F}{\partial \mathbf{V}}\right).$$

where $G_1 = (\gamma_1 m_2 + \gamma_2 m_1)M^{-1}$, $G_2 = (\gamma_1 m_1 + \gamma_2 m_2)M^{-1}$.

For (4.1) the equilibrium distribution function has the form

(4.2) $$F_0 \sim \exp\left(-\frac{MV^2}{2T} - \frac{\mu v^2}{2T} - \frac{U_e}{T}\right).$$

The specificity of this case is that the friction coefficient is great ($\gamma > w_0$). The distribution function becomes Maxwellian rather rapidly in time of order γ^{-1} [11]. Hence, consider the following distribution function:

$$\rho(t, r) = \int F d^3v d^3V d^3R,$$

and write equations for the first moments of the probability density F

(4.3)
$$\frac{\partial \rho}{\partial t} + \frac{\partial}{\partial r_i}(P_i) = 0,$$

$$\frac{\partial P_j}{\partial t} + \frac{\partial}{\partial r_i}(P_{ij}) - \mu^{-1}\nabla_i U_e'(-\rho\delta_{ij}) = -G_1 P_j - (\gamma_2 - \gamma_1)Q_j,$$

$$\frac{\partial Q_j}{\partial t} + \frac{\partial}{\partial r_i}(M_{ij}) = -G_2 Q_j - \mu(\gamma_2 - \gamma_1)M^{-1}P_j,$$

where $$P_i = \int v_i F d^3v\, d^3V d^3R, \qquad Q_i = \int V_i F d^3v\, d^3V d^3R.$$

$$P_{ij} = \int v_i v_j F d^3v d^3V\, d^3R, \qquad M_{ij} = \int v_i V_j F d^3v\, d^3V d^3R.$$

The first equation (4.3) was obtained by the integrating (4.1) over the variables v, V, R, and second and third ones - from (4.1) previously multiplied by v_j and V_j, respectively.

For the moments of time $t \gg \gamma^{-1}$ it is possible to solve the stationary second and third equations of the system (4.3) and to express the second moments via the first ones by the formulas which are valid for the equilibrium distribution (4.2):

$$P_{ij} = T\mu^{-1}\rho\delta_{ij}, \qquad M_{ij} = 0.$$

As a result, the set (4.3) is reduced to the Smolukhovsky equation

(4.4) $$\frac{\partial\rho}{\partial t} + \frac{\partial P_i}{\partial r_i} = 0, \qquad P_i = -(\mu\gamma)^{-1}U'\rho + T\frac{\partial\rho}{\partial r_i},$$

where $\gamma = m\gamma_1\gamma_2(m_1\gamma_1 + m_2\gamma_2)^{-1}$ is the reduced friction coefficient described by the formula (2.2).

Expressing Eq.(4.4) in the spherical coordinate system and integrating over angular variables, we obtain:

$$\frac{\partial\rho}{\partial t} = (\mu\gamma)^{-1}\frac{1}{r^2}\frac{\partial}{\partial r}\left[r^2\left(U'\rho + T\frac{\partial\rho}{\partial r}\right)\right],$$

or

(4.5) $$\frac{\partial f}{\partial t} = (\mu\gamma)^{-1}\frac{\partial}{\partial r}\left(U'_e f + T\frac{\partial f}{\partial r}\right),$$

where

$$f = r^2\rho, \qquad U_e(r) = U(r) - 2T\ln(r/r_0).$$

Hence, if dissipations are great it would suffice to analyse the non-equilibrium distribution functions on the basis of Eq.(4.5) which involves the effective potential U_e. Formulas for the effective potential and the reduced coefficient γ coincide with (2.10) and (2.2).

Eq.(4.5) for the one-dimentional probability density corresponds to a certain Langevinequation that may be readily obtained from (3.1). In fact, not considering in (3.1) the relative movement velocity, we can reduce the set to the equation for the damped anharmonic oscillator with the fluctuating driving force

$$\ddot{r} + \gamma\dot{r} + \mu^{-1}U'_e(r) = \xi(t).$$

Taking into account the condition $\gamma > w_0$, we get the next Langevin equation that corresponds to (4.5) and describes the aperiodic relaxation regime:

$$\dot{r} = -(\mu\gamma)^{-1}U_e'(r) + \gamma^{-1}\xi(t).$$

To find the non-equilibrium solutions of (4.5) one have to specify the initial and boundary conditions. Let us associate the initial condition with the general density of bound states that does not correspond to their equilibrium density at the initial moment:

(4.6)
$$\int_0^R f(0, r)\, dr = N_{AB_0} \neq N_{AB}^{(0)},$$

where N_{AB_o} is the initial density of bound states of the reacting particles A and B. $N_{AB}^{(0)}$ is an equilibrium value.

The first boundary condition arises from consideration of the probability density flow j at the point r = 0. It is evident that the flow becomes zero because of the repulsion character of the potential at $r \to 0$:

(4.7)
$$j = -(\mu\gamma)^{-1}\left(U_e'f + T\frac{\partial f}{\partial r}\right)_{r=0} = 0.$$

The second boundary condition is conveniently written for the new unknown function φ specified by the equation:

(4.8)
$$f = \varphi \exp(-U_e/T).$$

In the equilibrium state the function φ takes the constant value $\varphi^{(0)}$ which agrees with the normalization of the equilibrium distribution $f^{(0)}$. Consider the equilibrium density $N_{AB}^{(0)}$ of bound states:

$$N_{AB}^{(0)} = \int_0^R \varphi^{(0)} \exp\left(-\frac{U_e}{T}\right) dr = \frac{\varphi^{(0)}}{r_0^2}\int_0^R \exp\left(-\frac{U_e}{T}\right) r^2\, dr = \varphi^{(0)}\left(\frac{2\pi T}{\mu\omega_0^2}\right)^{1/2} \exp\left(\frac{E_0}{T}\right).$$

Requiring the coincidence of this result with (1.9), we find

$$\varphi^{(0)} = 4\pi r_0^2 N_A^{(0)} N_B^{(0)},$$

where $N_A^{(0)}$ and $N_B^{(0)}$ are the equilibrium densities of free particles.

The meaning of the second boundary condition is that there is an equilibrium between the bound and free states near the boundary R and if r = R the function φ takes the equilibrium value corrsponding to the flowing densities N_A and N_B:

(4.9) $$\varphi\,|_{r=R} = 4\pi r_0^2 N_{\mathrm{A}} N_{\mathrm{B}}.$$

It is easy to verify by the substitution of (4.8) in (4.5) that in this case the mathemetical formulation of the problem involves the construction of non-equilibrium solutions of the following boundary-value problem:

$$\exp\left(-\frac{U_{\mathrm{e}}}{T}\right)\frac{\partial\varphi}{\partial t} = D\frac{\partial}{\partial r}\left[\exp\left(-\frac{U_{\mathrm{e}}}{T}\right)\frac{\partial\varphi}{\partial r}\right],\ 0<r<R,$$

$$\exp\left(-\frac{U_{\mathrm{e}}}{T}\right)\frac{\partial\varphi}{\partial r}\bigg|_{r=0} = 0,\ \ \varphi\,|_{r=R} = \varphi_R,$$

(4.10) $$\int_0^R \varphi(0, r)\exp\left(-\frac{U_{\mathrm{e}}}{T}\right)dr = N_{\mathrm{AB}_0},$$

where $D = T/\mu\gamma$.

In this paper we shall use as the nonequilibrium distribution functions asymptotic in the time (or quasi-stationary) solutions 4.10), which now have well developed algorithm of construction [15]. Here nonequilibrium solutions (4.10) are constructed using quasi-stationary distribution functions (QDF) and solving the eigenvalue (EV) and eigenfunction (EF) problem. Method QDF is based on the integro-differential equation equivalent to (4.10):

(4.11) $$\varphi = \varphi_0 + D^{-1}\int_0^{r}\exp\left(\frac{U_{\mathrm{e}}}{T}\right)dr'\int_0^{r'}\exp\left(-\frac{U_{\mathrm{e}}}{T}\right)\frac{d\varphi}{dt}\,dr''.$$

In the general case (4.11) is solved using the successive approximations method. If the thermostat temperature is low compared to the potential U_e depth, it is enough to make only the first approximation:

$$\varphi \simeq \varphi_0 + \beta(r)\dot{\varphi}_0, \tag{4.12}$$

where

$$\beta(r) = D^{-1}\int_0^{r}\exp\left(\frac{U_{\mathrm{e}}}{T}\right)dr'\int_0^{r'}\exp\left(-\frac{U_{\mathrm{e}}}{T}\right)dr''.$$

The time depence of parameter φ_0 is deduced from the second boundary condition for (4.10)

(4.13) $$\frac{d\varphi_0}{dt} = -\frac{\varphi_0 - \varphi_R}{\beta},$$

where $\beta = \beta(R)$.

For convenience of calculations we consider the process in which the dissipation plays the main role and $\varphi_0 \gg \varphi_R$. Let us determine the rate of variation of the bound states density:

$$\frac{dN_{AB}}{dt} = \int_0^R \exp\left(-\frac{U_e}{T}\right)\frac{\partial\varphi}{\partial t}\,dr = D\exp\left(-\frac{U_e}{T}\right)\frac{\partial\varphi}{\partial r}\bigg|_{r=R} = g(R)\,\dot{\varphi}_0,$$

where $$g(R) = \int_0^R \exp(-U_e/T)\,dr.$$

Inserting to these equations formulas for φ_R and (4.13), we immediately obtain the rate equation.

(4.14) $$\frac{dN_{AB}}{dt} = -\frac{g(R)}{\beta}\varphi_0 + \frac{4\pi r_0^2 g(R)}{\beta}N_A N_B,$$

which shows thatthe rate of recombination into the bound state is determined by the equation:

(4.15) $$k_r = \frac{4\pi r_0^2 g(R)}{\beta} = \frac{4\pi r_0^2 D g(R)}{\int_0^R \exp(U_e/T)\,g(r)\,dr}.$$

Expressing φ_0 via N_{AB} and substituting it into (4.14), we obtain

$$N_{AB} = \int_0^R \exp\left(-\frac{U_e}{T}\right)\left[\varphi_0 + \frac{\beta(r)}{\beta}(\varphi_R - \varphi_0)\right]dr \underset{\varphi_0 \gg \varphi_R}{=} \frac{\varphi_0}{\beta D}\int_0^R \exp\left(\frac{U_e}{T}\right)g^2(r)\,dr.$$

To find the rate of dissociation, consider the normalization:

(4.16) $$k_d = \frac{Dg(R)}{\int_0^R \exp\left(\frac{U_e}{T}\right)g^2(r)\,dr}.$$

The formulas obtained may be simplified if one takes into account the fact that the temperature T is low compared to the bond energy E_0. In fact, the integral g(R) is readily calculated:

$$g(R) = \left(\frac{2\pi T}{\mu\omega_0^2}\right)^{1/2} \exp\left(\frac{E_0}{T}\right).$$

The simple analysis shows that for the integrals in the denominators of Eq.(4.15) and (4.16) the integrating over (r_0, R) area gives the main contribution, hence

$$\int_0^R \exp\left(\frac{U_e}{T}\right) g^n(r)\, dr = g^n(R)\, J,$$

where

$$J = \int_{r_0}^R \exp(U_e/T)\, dr -$$

is the integral dependent on the choice of the interaction potential.

Considering the notes made above, the expressions for the rates of processes take the form:

$$k_r = \frac{4\pi r_0^2 T}{\mu\gamma J}, \quad k_d = \frac{\omega_0}{\gamma J}\left(\frac{T}{2\pi\mu}\right)^{1/2} \exp\left(-\frac{E_0}{T}\right). \tag{4.17}$$

These rates, as it should be, follow the detailed balancing principle:

$$\frac{k_d}{k_r} = \frac{N_A^{(0)} N_B^{(0)}}{N_{AB}^{(0)}} = \frac{\omega_0}{4\pi r_0^2}\left(\frac{\mu}{2\pi T}\right)^{1/2} \exp\left(-\frac{E_0}{T}\right) = K_e^{-1}. \tag{4.18}$$

Let us now specify the interaction potential of the reacting particles and calculate the integral J. Let it be the Lennard-Johnson potential:

$$U(r) = E_0\left[\left(\frac{r_0}{r}\right)^{12} - 2\left(\frac{r_0}{r}\right)^6\right]. \tag{4.19}$$

Then

$$J = r_0 \int_{r_0/R}^1 \exp\left[-\frac{2E_0}{T} x^6 \left(1 - \frac{x^6}{2}\right)\right] dx.$$

One can see that this integral will not depend on the choice of the boundary R between free and bound states, if the lower limit may be set equal to zero. To do this, the equation has to be valid:

$$\frac{2E_0}{T}\left(\frac{r_0}{R}\right)^6 \ll 1. \tag{4.20}$$

Hence, the boundary R should be chosen rather far away, but at the same time R has to be small compared to the average distance between free particles. Otherwise R gets beyond the scope of the phase volume involvedand Eq. (1.9) of the detailed balancing principle does not follow from the two-particle distribution function.

If the thermostat temperature is not too high, the upper integration limit in the formula for J may be substituted for the infinite one and the interaction integral may be presented in the form

$$J \simeq r_0 \int_0^\infty \exp\left[-\frac{2E_0}{T}x^6\right] dx = r_0 \left(\frac{T}{2E_0}\right)^{1/6} \Gamma(7/6), \tag{4.21}$$

where $\Gamma(x)$ is the gamma-function.

Show now how the formulas obtained above follow from the solution in the form of the series in EV and EF. Let us again consider only low dissociation degreesand specify Boltzmann initial distribution function with the bound states density which does not agree with its equilibrium value. Then the initial condition for function φ is

$$\varphi|_{t=0} = \varphi_0 = N_{AB_0}\omega_0 \left(\frac{\mu}{2\pi T}\right)^{1/2} \exp\left(-\frac{E_0}{T}\right). \tag{4.22}$$

To obtain zero boundary conditions we introduce new unknown function $\psi = \varphi - \varphi_R$ and consider the boundary-value problem

$$\begin{gathered}\exp\left(-\frac{U_e}{T}\right)\frac{\partial\psi}{\partial t} = D\frac{\partial}{\partial r}\left[\exp\left(-\frac{U_e}{T}\right)\frac{\partial\psi}{\partial r}\right],\ 0<r<R,\\ \exp\left(-\frac{U_e}{T}\right)\frac{\partial\psi}{\partial r}\Big|_{r=0} = 0, \qquad \psi|_{r=R} = 0,\\ \psi|_{t=0} = \varphi_0 - \varphi_R.\end{gathered} \tag{4.23}$$

Assuming $\psi \sim \exp(-\nu Dt)\psi$, we arrive to the problem of EV and EF

$$\frac{d}{dr}\left(h\frac{d\psi_n}{dr}\right) = -\nu_n h\psi_n,$$

(4.24) $$h\frac{d\psi_n}{dr}\bigg|_{r=0} = 0, \qquad \psi_n|_{r=R} = 0,$$

where ψ_n and ν_n are the eigenfunctions and eigenvalues, $h = \exp(-U_e/T)$.

The system of ψ_n functions satisfies the requirements of completeness and orthogonality with a weight h:

$$\int_0^R h\psi_n\psi_m\,dr = \delta_{nm}, \qquad h\sum_{n=1}^{\infty}\psi_n(r)\,\psi_n(r') = \delta(r - r').$$

Here it should be noted that EV $\nu_o = 0$ corresponds to EF $\psi_o \equiv 0$ because the equilibrium distribution for ψ function is an identical zero; $\psi_0 \equiv 0$ follows directly from the second boundary condition for (4.24).

The total solution may be represented as a series in EF and EV:

(4.25) $$\psi(t, r) = \sum_{n=1}^{\infty} C_n\psi_n(r)\exp(-\nu_n Dt),$$

where $$C_n = (\varphi_0 - \varphi_R)\int_0^R h\psi_n\,dr -$$

are coeffitients determined by the initial distribution function (4.23).

Since the thermostat temperature is far less than the potential barrier height the time of transition through barrier is great, and hence there is the minimum EV $\nu_1 \ll \nu_2, \nu_3, \ldots$ corresponding to this time. Therefore in the solution (4.25) at the metastable stage of relaxation ($\nu_1 Dt \ll 1$) it is suffice to consider only the first term of the series. In this work results from [15] were used to find the minimum EV ν_1 and EF ψ_1 corresponding to it. Present (4.24) as an integral equation:

(4.26) $$\psi = C - \nu\int_0^r h^{-1}(\xi)\,d\xi\int_0^{\xi} h(\eta)\,\psi(\eta)\,d\eta,$$

that may be solved by the iteration method; C is an arbitrary constant.

Consider the solution (4.26) appearing at the first approxi-mation:

$$\psi^{(1)} = C\left(1 - \nu^{(1)} \int_0^r h^{-1}(\xi)\, d\xi \int_0^\xi h(\eta)\, d\eta\right).$$

The boundary condition which has to be satisfied at r = R gives

$$\nu^{(1)} = \left[\int_0^R h^{-1}(\xi)\, d\xi \int_0^\xi h(\eta)\, d\eta\right]^{-1},$$

(4.27) $$\psi^{(1)} = C' \int_r^R h^{-1}(\xi)\, d\xi \int_0^\xi h(\eta)\, d\eta,$$

where a constant C` is determined by the condition of normalization:

$$\int_0^R h\, [\psi^{(1)}]^2\, dr = 1.$$

Now we find the rate of the bound states density variation at the metastable stage of the relaxation ($\nu Dt \ll 1$) using the solution (4.27)

(4.28) $$\frac{dN_{AB}}{dt} = \int_0^R \exp\left(-\frac{U_e}{T}\right) \frac{\partial\psi}{\partial t}\, dr = -\nu_1 D \left(\int_0^R h\psi_1\, dr\right)^2 (\varphi_0 - \varphi_R).$$

From (4.28) we find rates of the processes expressed in terms of ν_1 and ψ_1:

$$k_r = 4\pi r_0^2 D\nu_1 \left(\int_0^R h\psi_1\, dr\right)^2, \qquad k_d = \omega_0 D\nu_1 \left(\frac{\mu}{2\pi T}\right)^{1/2} \left(\int_0^R h\psi_1\, dr\right)^2 \exp(-E_0/T).$$

(4.29)

It is easy to show that formulas (4.29) agree with (4.17). To do this, one should calculate integrals in (4.29) by the saddle-point method which gives:

$$\nu_1 \simeq \left[\int_0^R \exp\left(\frac{U_e}{T}\right) g(r)\, dr\right]^{-1} = [g(R)\, J]^{-1},$$

$$\left(\int_0^R h\psi_1\, dr\right)^2 \simeq [\psi_1(r_0)]^2\, g^2(R) \simeq g(R).$$

Dwell on the temperature dependences of the reaction rates in the limit of great dissipations. In the first Chapman-Ensckog approximation the viscosity of medium $\eta \sim T^{1/2}$, therefore the rate of recombination into the bound state increases with temperature as $T^{1/3}$, while preexponential factor of the dissociation rate decreases as $T^{-1/6}$.

The Kramers result [3] may be obtained from Eq.(4.17) performing the following approximate calculation of integral J. The main contribution in J gives the integration region near the top of the potential barrier, therefore expressing the potential in the vicinity of r_B (the maximum position) in the form:

{Eq. 40, p. 20}

$$U_e(r) = U_b - \frac{1}{2}\mu\omega_b^2 (r - r_b)^2, \qquad U_e''(r_b) = -\mu\omega_b^2$$

and calculating J using by the saddle-point method, we get [3]

$$k_d \simeq \frac{\omega_0\omega_b}{2\pi\gamma}\exp\left(-\frac{Q}{T}\right), \tag{4.30}$$

where $Q = E_0 + U_B$ is the potential barier height.

Notice that Eq. (4.30) overestimates the dissociation rate In fact, let the interaction potential has the form (4.19). Then taking into account more accurate value of (4.21), we get from (4.17)

$$k_d = \frac{\omega_0}{\gamma r_0 \Gamma(7/6)}\left(\frac{2E_0}{T}\right)^{1/6}\left(\frac{T}{2\pi\mu}\right)^{1/2}\exp\left(-\frac{E_0}{T}\right). \tag{4.31}$$

The main parametres for the potential (4.19) are

$$U_b = -\frac{T}{3}\left(1 + \ln\frac{6E_0}{T}\right), \qquad r_b = r_0\left(\frac{6E_0}{T}\right)^{1/6}, \qquad \omega_b = \frac{1}{r_b}\left(\frac{12T}{\mu}\right)^{1/2}.$$

Inserting them into (4.30) one can readily see that the dissociation rate (4.30) is more than the rate (4.31) by a factor of 2 - 3.

5. VIBRATIONAL RELAXATION REGIME. (THE CASE OF SMALL DISSIPATIONS)

In this section the opposite limiting case of small damping coefficients ($\gamma < w_0$) is considered. In this case the system is isolated and performs vibrations near the equilibrium at the time between two succesive collisions of reacting particles with thermostat ones. On the phase plane (p, r) the system moves along some closed trajectory which parametres are completely determined by the energy

$$E = {}^1/_2\mu v^2 + U_e(r). \tag{5.1}$$

The period of revolution over the closed trajectory is given by the formula:

$$\theta(E) = (2\mu)^{1/2} \int_{r_1}^{r_2} [E - U_e(r)]^{-1/2}\, dr, \tag{5.2}$$

where $r_1(E)$ and $r_2(E)$ are the left and right positions of the return points (they may alsobe treated as vibrational amplitudes)

The area within the closed trajectory is equal with a precision of the numerical factor to:

$$I(E) = \frac{1}{2\pi}\int dp\, dr = \frac{1}{2\pi}\oint p\, dr = \frac{\sqrt{2\mu}}{\pi}\int_{r_1}^{r_2} [E - U_e(r)]^{1/2}\, dr. \tag{5.3}$$

From this follows the formula important for further calculations

$$\frac{dI}{dE} = \frac{1}{2\pi}\oint \frac{\partial p}{\partial E}\, dr = \frac{\theta(E)}{2\pi} = \frac{1}{\omega(E)}, \tag{5.4}$$

where $\omega(E)$ is the vibrational frequency.

At collisions between reacting and thermostat particles they perform transitions from a closed trajectory to another one. In this case the energy E or the action I(E) change moderatly and Brownian motion proceeds over the energy. We obtain the kinetic equation for the distribution function $\rho(I,t)$ from the Langevin equation for the anharmonic oscillator. To do this, we bring (3.1) to the form

$$dE/dt = -\gamma\mu v^2 + \mu v\xi(t).$$

Considering dE/dt as a large scale time derivative, i.e. an increment $\Delta t \gg \omega^{-1}$, we average this equation over the vibrational period:

(5.5)
$$dE/dt = -\gamma\mu\overline{v^2} + y(t),$$

where $y(t) = \overline{\mu v(t)\xi(t)}$ is the work done by the thermostat in a unit time.

Find the velocity squared averaged over the period of vibrations:

(5.6)
$$\mu\overline{v^2} = \frac{\mu}{\theta}\int_0^\theta v^2\,dt = \frac{\mu}{\theta}\oint v^2\frac{dr}{v} = \frac{1}{\theta}\oint p\,dr = \omega I.$$

The substitution of (5.6) into (5.5) results in the equation:

$$dE/dt = -\gamma\omega I + y(t).$$

Assuming the average value of the work y(t) to be the same both for harmonic and anharmonic oscillatorse, consider the harmonic oscillator for which wI = E. Then

$$dE/dt = -\gamma E + y(t).$$

Taking into account that in time the average oscillator energy must coincide with the thermostat temperature, for the stationary random process y(t) we have:

(5.7)
$$\langle y(t)\rangle = \gamma\langle E\rangle = \gamma T.$$

Hence the Langevin equation for the energy (5.5) takes the form:

(5.8)
$$dE/dt = -\gamma\omega I + \gamma T + \xi(t),$$

where $$\langle\xi(t)\rangle = 0,\quad \langle\xi(t)\xi(t')\rangle = D\delta(t-t').$$

Write the continuity equation for the probability density f(I,t):

(5.9)
$$\frac{\partial f}{\partial t} + \frac{\partial}{\partial I}(\dot{I}f) = \frac{\partial f}{\partial t} + \frac{\partial}{\partial I}\left[\omega^{-1}\frac{dE}{dt}f\right] = 0.$$

Inserting (5.8) into (5.9) and operating in the same way as in a section 3 we arrive at the following equations for the smoothed ρ and fluctuating $\tilde{\rho}$ parts of the distribution function f:

$$\frac{\partial\rho}{\partial t} = \frac{\partial}{\partial I}\left[\left(\gamma I - \frac{\gamma}{\omega} T\right)\rho - \omega^{-1}\langle\xi\tilde{\rho}\rangle\right],$$

(5.10)
$$\frac{d\tilde{\rho}}{dt} = -\frac{\partial}{\partial I}\left(\omega^{-1}\xi\rho\right).$$

Inserting the solution of the second equation of (5.10)

$$\tilde{\rho}(t+\tau) = \tilde{\rho}(t) - \frac{\partial}{\partial I}\left(\omega^{-1}\rho\int_0^{\tau}\xi(t+\tau')\,d\tau'\right)$$

into the first one, we get

$$\frac{\partial\rho}{\partial t} = \frac{\partial}{\partial I}\left[\gamma\left(I - T\omega^{-1}\right)\rho + \omega^{-1}\frac{\partial}{\partial I}\left(\omega^{-1}D\rho\right)\right].$$

Requiring now that the right part of the equation becomes an identical zero at $\rho_0 \sim \exp(-E/T)$ and taking into account the equality (5.4), we find the coefficient $D = \gamma T\omega I$ and the form of kinetic equation for the smoothed distribution function:

(5.11)
$$\frac{\partial\rho}{\partial t} = \gamma\frac{\partial}{\partial I}\left[I\left(\rho + \frac{T}{\omega}\frac{\partial\rho}{\partial I}\right)\right].$$

The same equation immediately follows [3] from the Kramers equation (2.7). Note previously that in the case of negligible damping coefficients ($\gamma \to 0$) Eq.(2.7) may be reduced to the linear homogeneous partial differential equation and its integration is based on the solution of the set of equations for characteristics

$$dt = \frac{dr}{v} = -\frac{\mu\,dv}{U_e'(r)}$$

or, as one can see, for equations of motion neglecting the thermostat influence on the system. The action I and vibrational frequency ω_0 may be determined in the same way using these equations. From formulas presented above it follows that the energy E (or I) completely defines the trajectory parameters. So in the case of small γ the distribution function ρ depends only on two variables E and t. Considering this fact, Eq.(2.7) is represented in the form

$$\frac{\partial\rho}{\partial t} = \gamma\left(\rho + T\frac{\partial\rho}{\partial E}\right) + \gamma\mu v^2\frac{\partial}{\partial E}\left(\rho + T\frac{\partial\rho}{\partial E}\right).$$

Taking into account the large-scale character of the time derivative and averaging over the vibration period, and using (5.4) and (5.6), we obtain (5.11).

Thus, for dissipation coefficients $\gamma < w_0$ the Kramers equation is reduced to the more simple equation (5.11) that is applied to not very high values of energy E. Actually, with increasing E the frequency $\omega(E)$ decreases due to anharmonicity of of vibrations and becomes zero on the top of the potential barrier. Hence, near the $U_e(r)$ maximum the relaxation has a purely aperiodical character and is described by the Smolukhovsky equation (4.10). Generaly for an arbitrary coefficient γ itis nesessary to solve Eq. (2.7) which is a rather complex mathemetical problem. At present the different appriximation methods of its analisys are used. Thus, the paper [8] offers the joint of the quasi-stationary solution of Eq. (2.7) found by Kramers in the vicinity of the local potential maximum withparabolic law of its variation and quasi-stationary solution of Eq. (5.11). This approach has been used to obtain the dissociation rate in the case of neutral interacting particles at small and medium dissipations γ. The results were confirmed by numerical calculations. Notice that for large γ this method allows to obtain only an approximate Kramers result (4.30) that, as it was shown above, is true within a factor of 2 or 3. Moreover, the role of boundary conditions and the possibility of description of the relaxation in the region where the potential is not described by the parabolic law are obscure.

In this paper Eq.(2.7) is analysed using the solution of the type of joint of two nonequilibrium distribution functions corresponding to the two limiting cases considered above. Consider Eq.(5.11) and reduce it to the form close to (4.10). Let us denote by r the right return point on the phase plane (the vibrational amplitude). Then the energy E is uniquely determined by the vibrational amplitude: $E = U_e(r)$. In this case, considering the equality (5.4), Eq. (5.11) can be put in a form

$$\frac{\partial\rho}{\partial t} = \frac{\gamma\omega}{U_e'}\frac{\partial}{\partial r}\left[I\left(\rho + \frac{T}{U_e'}\frac{\partial\rho}{\partial r}\right)\right].$$

Assuming $\rho = \exp(-U_e/T)\varphi$, we have:

$$\exp\left(-\frac{U_e}{T}\right)\frac{\partial\varphi}{\partial t} = \frac{\gamma\omega}{U'_e}\frac{\partial}{\partial r}\left[\frac{TI}{U'_e}\exp\left(-\frac{U_e}{T}\right)\frac{\partial\varphi}{\partial r}\right].$$

Introducing the new effective potential W and the diffusion coefficient D of the vibrational amplitude, represent this equation in the form similar to (4.10):

(5.12)
$$\exp\left(-\frac{W}{T}\right)\frac{\partial\varphi}{\partial t} = \frac{\partial}{\partial r}\left[D\exp\left(-\frac{W}{T}\right)\frac{\partial\varphi}{\partial r}\right],\ r > r_0,$$

where

(5.13)
$$D = \frac{\gamma T\omega I}{(U'_e)^2},\qquad W = U_e - T\ln\left(\frac{\overline{U}'_e}{\gamma\omega C}\right),$$

and C is some constant associated with the arbitrary choice of the energy of the potential W reading.

As r or vibrational amplitude grows the diffusion coefficient increases due to decreasing of the potential derivative squared and increasing of I(r). Let us determine the minimum value of D that is realized in the region of harmonic vibrations with the interaction potential $U_e(r) = -E_0 + 1/2\mu\omega_0^2(r - r_0)$, where r is the current coordinate of the movement along closed trajectory. In this case the trajectory is described by the ellipse equation:

$$\frac{p^2}{2\mu} + \frac{\mu\omega_0^2}{2}(r - r_0)^2 = \varepsilon$$

with the semiaxes $(2\varepsilon\mu)^{1/2}$ and $(2\varepsilon/\mu\omega_0^{\ 2})^{1/2}$, where ε is the energy measured from the potential minimum.

The ellipse area, divided by 2π, specify the integral I:

$$I(r) = \varepsilon(r)/\omega_0,\quad \varepsilon(r) = {}^1/_2\mu\omega_0^2(r - r_0)^2.$$

Thus we find for the harmonic vibrations region

(5.14)
$$D = \gamma T/2\mu\omega_0^2.$$

Alternatively, for the aperiodic region of relaxation $D = T/\mu\gamma$ and for $\gamma < \sqrt{2}\omega_0$ this diffusion coefficient exceeds (5.14). If the equality $\gamma = \sqrt{2}\omega_0$ is fulfilled, then both diffusion coefficients coincide and the relaxation regime becomes completely aperiodic. Since in this section the case $\gamma < \omega_0$ is considered there is the value of the return point $r = r_s$ such that the coefficient (5.13) is equal to $T/\mu\gamma$ due to the diffusion coefficient increasing with the increase of the vibrational amplitude. It is natural to take this value r_s as a boundary between the vibrational and aperiodic regions of relaxation

(5.15) $$\gamma = U'_{es}(\mu I_s \omega_s)^{-1/2}, \quad \gamma = \gamma(r_s),$$

where

$$U'_{es} = U'_e(r_s), \qquad I_s = I(r_s), \qquad \omega_s = \omega(r_s).$$

In addition to the joint of diffusion coefficients, we impose the condition of continuity on the potential functions U_e and W: It is achieved by the choice of the constant C in the formula (5.13):

(5.16) $$C = U'_{es}/\gamma\omega_s = (\mu I_s/\omega_s)^{1/2}, \quad C = C(r_s).$$

Thus, we will analyse the nonequilibrium solutions of the common kinetic equation (5.12) in the whole region involved ($0 < r < R$) with different diffusion coefficients and potentials for vibrational ($r \in (r_o, r_s)$) and aperiodic ($r \notin (r_o, r_s)$) relaxation regions. In this case the solutions of Eqs. (4.10) and (5.12) are automatically joined.

Consider the molecular density in the equilibrium state:

$$N_{AB}^{(0)} = \varphi^{(0)} \int_{r_0}^{R} \exp\left(-\frac{W}{T}\right) dr = g(R)\, \varphi^{(0)}.$$

Comparing this value with the one obtained from Eq.(1.9), we find

$$\varphi^{(0)} = g^{-1}(R)\, K_e N_A^{(0)} N_B^{(0)}.$$

At r = R the reacting particles become free and obey the equilibrium distribution with current values of free atoms concentrations. Hence the right boundary condition for (5.12) is:

(5.17) $$\varphi \,|_{r=R} = \varphi_R = g^{-1}(R)K_e N_A N_B.$$

The left boundary condition and the initial one remain as before:

(5.18) $$\exp\left(-\frac{W}{T}\right)\frac{\partial\varphi}{\partial r}\Big|_{r=r_0} = 0, \qquad \int_{r_0}^{R} \exp\left(-\frac{W}{T}\right)\varphi(0, r)\,dr = N_{AB_0},$$

where N_{AB_0} is the initial molecular concentration which remains practically constant if the dissociation degree is low.

Further one can almost faithfully repeat a deduction of the asymptotic in time solution of Eq.(4.10) and obtain the following reaction rates:

(5.19) $$k_r = B^{-1}K_e,$$

(5.20) $$k_d = B^{-1},$$

where

$$B = \int_{r_0}^{R} D^{-1}\exp\left(\frac{W}{T}\right) g(r)\,dr.$$

In the formula for B the main integration region is the vicinity of the potential maximum, hence one may substitude g(r) for g(R) and besides divide the integration region into vibrational and aperiodic ones. Then with regard to the weak dependence of I(r) compared to the exponential factor, we obtain:

(5.21) $$B \simeq (g_0 + g_A)(B_0 + B_A),$$
$$g_0 \simeq T\,(\gamma\omega_0 C)^{-1}[\exp(E_0/T) - \exp(-U_{es}/T)],$$

(5.22) $$g_A \simeq {}^1\!/_2\left(2\pi T/\mu\omega_0^2\right)^{1/2}\exp(E_0/T)\,[2 - \Phi(z_s) - \Phi(z_L)],$$
$$B_0 \simeq C I_s^{-1}[\exp(U_{es}/T) - \exp(-E_0/T)],$$
$$B_A \simeq \mu\gamma T^{-1} J_s,$$

where

$$J_s = \int_{r_s}^{R}\exp\left(\frac{U_e}{T}\right)dr, \qquad \Phi(z) = \frac{2}{\sqrt{\pi}}\int_0^z \exp(-t^2)\,dt, \qquad z_s = \left(\frac{\mu\omega_0^2}{2T}\right)^{1/2}(r_s - r_0),$$

$$z_L = \left(\mu\omega_0^2/2T\right)^{1/2}(r_0 - r_L),\ r_L -$$

r_L is the left return point corresponding to r_s.

At the given damping coefficient γ one should solve Eq.(5.15) for r_s and then find the reaction rates (5.19) and (5.20). This problem may be simplified if the rates are considered as the functions parametrically dependent on r_s. Actually, in view of the definition (5.15), the coefficient γ is uniquely fixed specifying r_s, hence the reaction rates must be determined depending on r_s.

For small damping coefficients γ the interface lies near the local maximum of potential and with increasing of γ it moves in the direction of the potential minimum. .

One can see that in the limiting case of small dissipations ($\gamma \to 0$, $r_s \to r_b$) the inequalities $g_O \ll g_A$ and $B_O \gg B_A$ are true and hence:

$$(5.23) \qquad B = g_0 B_0,\ k_d = \gamma\omega_0 I_b T^{-1} \exp\left[-(E_0 - U_{eb})/T\right],$$

where $I_b = I(r_b)$, $U_{eb} = U_e(r_b)$ are potential values at the local maximum point. For the potential (4.19) U_{eb} corresponds tothe thermostat temperature in order of magnitude. In the harmonic approximation $I_b \sim E_0 w_0^{-1}$ and one can easily deduce the Kramers estimate [3] from (5.23): {Eq. 61, p. 25}

In another limiting case ({Eq. 62, p. 25}) the inverse inequalities are true: $g_O \ll g_A$ and $B_O \ll B_A$. Here the relaxation regime is completely aperiodic with the diffusion coefficient $D = I/\mu\gamma$, so formulas (5.21) and (5.22) transform into (4.17):

$$(5.24) \qquad B \simeq g_A B_A, \qquad k_d \simeq \frac{\omega_0}{\gamma J_s}\left(\frac{T}{2\pi\mu}\right)^{1/2} \exp\left(-\frac{E_0}{T}\right).$$

Rates (5.24) contain the interaction integral J_s coinciding with the integral J in formulas (4.17). We shall show it on the example of the Lennard-Johnson potential (4.19). Choosing R in accordance with the inequality (4.20) and neglecting the repulsive branch of potential, we get

$$(5.25) \qquad J_s = r_0 \left(T/2E_0\right)^{1/6} \Gamma(7/6)\, E_6(z),$$

where

$$E_n(z) = \Gamma^{-1}\left(\frac{n+1}{n}\right)\int_0^z \exp(-t^n)\,dt -$$

is the generalized error integral,

$$z = \frac{r_0}{r_s}\left(\frac{2E_0}{T}\right)^{1/6}.$$

At $r_s \to r_0$ formula (5.25) goes into (4.21).

Thus, when the r_s parameter is varied from its minimum value r_0 to r_b, the continuous transition occures from the limiting case (5.24) to another limiting case (5.23). In the case (5.24) reaction rates are inversly related to γ, while for (5.23) they are proportional to γ, so there is the friction coefficient γ_0 for which the reaction rates have a maximum value. Further calculations will be performed to obtain the formula for γ_0. In this case approximate analytic expresions for ω and I are required.

We present Eq. (5.3) for the period of vibrations in the form

(5.26)
$$\theta(r) = \frac{\pi}{\omega_0} + \int_{r_0}^{r} \frac{\sqrt{2\mu}\,d\xi}{\sqrt{U_e(r) - U_e(\xi)}},$$

where the term π/ω_0 is the passing time between the left return point and r_0 in the approximation of the harmonic vibrations law for the left branch of potential. Let r^* be an inflection of the potential function. Assuming vibrations to be harmonic in the $r_0 < r < r^*$ region and taking into account that the region of integration near the return point r makes the main contribution to the integral, expand the potential $U_e(\xi)$ in a series in powers of $(\xi - r)$. Then with consideration for the first three terms of the expantion we find:

(5.27)
$$\theta(r) \simeq \frac{2\pi}{\omega_0} + \frac{4}{\Omega}\ln\left(\sqrt{x} + \sqrt{x+1}\right),$$

where {Eq. 65, p. 26}. Here $\ddot{U}_e(r) < 0$, so the designation $\ddot{U}_e = -\mu\Omega^2$ is used:

In the region of the local potential maximum there is the logarithmic divergence of the period of vibrations

(5.28)
$$\theta(r) \simeq \frac{2}{\omega_b}\ln\frac{2(r_b - r^*)}{r_b - r}, \qquad \omega_b = \Omega(r_b).$$

Now we estimate the integral I(r);

(5.29) $$I(r) \simeq \frac{\varepsilon(r)}{2\omega_0} + \frac{\sqrt{2\mu}}{\pi} \int_{r_0}^{r} \sqrt{U_e(r) - U_e(\xi)}\, d\xi,$$

where the first term is a part of the area enclosed by the trajectory to the left of the point r_0; $\mathcal{E}(r) = E_0 + U_e(r)$.

Once again, breaking up the integration area into two parts and using the formula for potential in the vicinity of its maximum: $U_e(r) = U_{eb} - \frac{1}{2}\mu\omega_b^2(r_b - r)^2$, we get the estimation:

(5.30) $$I(r) \sim \frac{\varepsilon(r) + \varepsilon(r^*)}{2\omega_0} + \frac{\mu\omega_b}{2\pi}\left[a\sqrt{a^2 - x^2} - x^2 \ln\frac{a + \sqrt{a^2 - x^2}}{x}\right],$$

where $x = r_b - r,\ a = r_b - r^*,\ r^* < r < r_b.$

The direct calculations of integrals (5.2) and (5.3) show that formulas (5.28) and (5.30) are accurate to within several tens of percent. In addition, the estimation (5.30) is inadequate in the case of the Coulomb interaction potential.

Rates of the processes considered have a maximum at some value of the density of the thermostat particles (or γ_0). The investigation of Eqs.(5.21) and (5.22) for the extremum is cumbersome due to implicit dependence of the boundary r_s on the coefficient γ. It is obvious that the extremum is reached at such r_s values that B_0 becomes equal to B_A: $\frac{T}{U'_{es}} \exp\left(\frac{U_{es}}{T}\right) \sim J_s$.

Taking into account that the boundary r_s is situated near the local maximum of the potential function, we get the condition for extremum:

$$r_s \sim r_b - \frac{T}{\mu\omega_b^2 r_b} = r_b\left(1 - \frac{T}{\mu\omega_b^2 r_b^2}\right).$$

Considering formulas (5.15) and (5.28), the following dissipation coefficient corresponds to the given value of the change-over point r_s:

(5.31) $$\gamma_0 \sim \left[\frac{T^2}{\pi\mu\omega_b I_b r_b^2} \ln \frac{2(r_b - r^*)\mu\omega_b^2 r_b}{T}\right]^{1/2}.$$

It follows from Eq.(5.31) that as temperature increases γ_0 grows approximately as $T^{1/2}$, i.e. the maximum of the reaction rate moves into the region of high gas pressure.

With the aim to determine the error of the above-described approximate solution of the Kramers equation, the comparision was made with numerical simulation [8] of the dissociation rate of the bound state of two neutral particles interacting by the law $U_e(r) = V_0[\exp(-r/a) - \exp(-r/b)]^2$, $V_0 = E_0\alpha^{2\alpha/\alpha-1}(\alpha-1)^{-2}$, $\alpha = a/b = 20$, $E_0/T = 4$.

The dissociation rate was determined from formulas (5.2), (5.3), and (5.20). The results of calculations are shown in Fig.3. One can see that the solution (2.7) of the type of joint of two limiting cases in the region of the dissociation rate velocity maximum differs from the exact one by a factor of 2. To obtain a higher accuracy in the whole region of the coefficient γ variation, it is nesessary to construct the noneqiulibrium solution of Eq. (2.7). Notice that the analysis suggested (2.7) has in the region of the dissociation rate maximum lower precision compared to[8], but it contains the right result for the reaction rate in the case of large dissipations $\gamma > \omega_0$ and, besides, it does not require to solve complex transcendental equation.

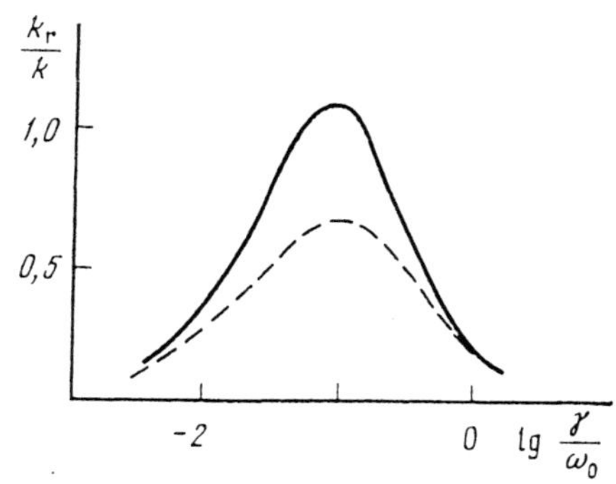

Fig. 3. A comparision of the analytical solution (solid line) with the numerical analysis [8] (dashed line); $K = (\omega_0/2\pi)\exp(-E_0/T)$

At present the existence of the reaction rate maximum for not too high gas pressures (on the order of atmospheric pressure) has been found in the case of the ion-ion recombination. It may be explained first of all on the basis of Eq.(5.31) by the long-range character of the Coulomb interaction between the reacting particles and their polarization interaction with the thermostat particles. The ion-ion recombination is considered in the next section.

3. THREE-PARTICLE ION-ION RECOMBINATION

The three-particle ion-ion recombination is one of the most significant processes in discharges through electronegative gases and also for the analysis of kinetics of active media for eximer lasers. Direct measurements of the rate of this process involve the great number of side reactions occuring in discharges. That is why there is practicaly no information in the literature on the direct measurements of the three-particle ion-ion recombination rates and there is a number of works in which the rate of this process is estimated only at the order of magnitude. Early works devoted to the calculation of the ion-ion recombination rate were made by Thomson [16] and Langevin (see [17]) for low and high gas densities, respectively. Natanson [18] and Flannery [19] generalized these theories for the case of arbitrary gas densities (pressures). Though these models have a qualitative charecter, they adequately explain the available experimental data. It will be shown in this section that the consistent theory of the three-article ion-ion recombination which involves the limiting cases of Thomson and Langevin is based on the Brownian motion theory in the reacting systems or on the Kramers model.

Consider the plasma satisfying condition for the ideal gas:

$$r_T \ll N_i^{-1/3} \ll r_D, \tag{6.1}$$

where $r_T = q^2/T$ is the Thomson radius or the Landau length at which the ion interaction energy becomes equal to the gas temperature, r_D is the Debye shield radius, $N_i^{-1/3}$ is the average distance between ions whose density is N_i.

Assume that the transfer of ions from the free state to the bound one occurs at distances satisfying the inequality $r_T < R < < N_i^{-1/3}$. Then, according Eq.(6.1), the Debye shield of the Coulomb interaction potential of ions may be neglected. The Coulomb term of interaction between ions A^+ and B^- is approximated by the function of type

$$U(r) = A/r^n - q^2/r, \; n > 1, \tag{6.2}$$

where parameters A and n characterize the repulsion part of the term. These parameters are associated with the potential well depth E_0 of the Coulomb term and position of its minimum r_0:

In this case (6.2) takes the form

$$E_0 = \frac{n-1}{n}\frac{q^2}{r_0}, \qquad A = \frac{q^2}{n} r_0^{n-1}. \tag{6.3}$$

The effective interaction potential $U_e(r)$ in one-dimensional problem is defined by formulas (5.2) and (6.3) and has the following basic parameters:

$$U(r) = \frac{E_0}{n-1}\left[\left(\frac{r_0}{r}\right)^n - n\frac{r_0}{r}\right]. \tag{6.4}$$

Formulas (6.4) were obtained for temperatures $T \ll E_0$. As is

$$r_b = r_T/2, \qquad x_b = \frac{n}{n-1}\cdot\frac{E_0}{2T} = \frac{r_T}{2r_0} \gg 1, \qquad \omega_b = \left(\frac{2T}{\mu r_b^2}\right)^{1/2},$$

$$U_{eb} = -2T(1 + \ln x_b), \qquad \omega_0 = \left(\frac{nE_0}{\mu r_0^2}\right)^{1/2}.$$

evident from Eq.(6.4), the long-range character of Coulomb potential shows up in that the local maximum $U_e(r)$ is located at great distances which are practically equal to the Thomson radius and, hence, I_b is much more than the action integral for the Lennard-Johns potential. In addition, the value U_{eb} of the potential barrier decrease is far greater than the medium temperature. All this as a whole results in lower dissipation coefficients (5.31) for which the recombination rate has a maximum.

In the case of the Coulomb term (6.3) the estimation (5.27) for the period of vibrations of reacting ions near the equilibrium position is true. However, estimations of integrals I_b and J_s must be remade. Neglecting the repulsion part of the potential and centryfugal energy, we have:

$$I_b \simeq \frac{\sqrt{2\mu}}{\pi}\int_{r_0}^{r_b}\sqrt{U_{eb}-U_e(\xi)}\,d\xi = \frac{2r_b}{\pi}\sqrt{\mu T}\int_0^{\ln x_b}\sqrt{\exp(t)-1-t\exp(-t)}\,dt \simeq$$

$$\simeq \frac{r_b}{\pi}\sqrt{2\mu T}\int_0^{\infty} t\left(1+\frac{t}{6}\right)\exp(-t)\,dt = \frac{4\sqrt{2}}{3\pi}r_b(\mu T)^{1/2}, \quad (6.5)$$

and similarly for the integral J_s:

$$J_s = \int_{r_s}^{R}\exp\left(\frac{U_e}{T}\right)dr \simeq \int_{r_s}^{\infty}\exp\left(-\frac{r_T}{r}-2\ln\frac{r}{r_0}\right)dr \simeq \frac{r_0^2}{r_T}\left[1-\exp\left(-\frac{r_T}{r_s}\right)\right] \simeq$$

$$\simeq r_0^2/r_T. \quad (6.6)$$

One can see from Eq.(6.6) that J_s practically does not depend on r_s because the second term may be neglected even for the maximum possible value of $r_s = r_T/2$.

The reduced dissipation coefficient γ in the one-particle equation (2.2) has the form:

$$(6.7) \quad \gamma = \gamma_1\gamma_2(m_1+m_2)(m_1\gamma_1+m_2\gamma_2)^{-1} = q[\mu(b_1+b_2)]^{-1}.$$

If the ions mobilities are expressed in terms of their collision frequences withthermostat particles, then Eq.(6.7) may be presented in the form:

$$(6.8) \quad \gamma = v_T\lambda^{-1},$$

where $v_T = (T/\mu)^{1/2}$ is the average thermal velocity of reacting ions, $\lambda = \lambda_1(\mu/\mu_1)^{1/2}+\lambda_2(\mu/\mu_2)^{1/2}$, λ_1 and λ_2 are free path lengths of ions the thermostat medium.

If the thermostat particles are of the type of reacting ions, then the scattering of ions by the thermostat particles results from the resonance recharging process. Otherwise their scattering cross section is determined by the polarization interacton and

$$\lambda_1 = \lambda_2 = (\sigma N)^{-1} = \frac{(T/\alpha q_e)^{1/2}}{2{,}21\pi N},$$

where σ is the polarization interaction cross section, α is the polarizability of thermostat particles, q_e is the electron charge.

Consider now the limiting values of the recombination velocity (5.19) for large and small dissipation coefficients γ. As we have seen before, in the case of large dissipations ($r_s \to r_0$) one can set $B \simeq g_A B_A$ in the denominator of Eq.(5.19). Then, taking into account formulas (6.6) and (6.7), we get

$$k_r = 4\pi T r_T (\mu\gamma)^{-1} = 4\pi q(b_1 + b_2). \tag{6.9}$$

The result (6.9) coincides precisely with the Langevin formula [17] for the ion-ion recombination velocity in the limit of high gas densities. In principle, it is determined by the flow of ions of a single type through the surface of the sphere of radius r_T with an ion of another type in its center, resulting from their electric drift in the field of the central ion.

In other limiting case of the small dissipations ($r_s \to r_b$), considering formulas (6.4), (6.5), and (6.8), we get from (5.19)

$$k_r = {}^4/_3 \sqrt{\pi}\, e^2 v_T \lambda^{-1} r_T^3. \tag{6.10}$$

The formula (6.10) agrees with the recombination velocity obtained in the Thomson model [16] in the limit of low gas pressure accurate to the numerical factor:

$$k_r \sim N\bar{v}\sigma b^3,$$

where b is the critical radius at which the ion interaction energy is of order of their thermal energy ($b \sim r_T$); σ is the cross section of collisions between charged and thermostat particles of the same order of magnitude asthe elastic scattering cross section, $\bar{v}$ is the relative velocity of collisions between ions and thermostat particles.

In the Thomson model the recombination occures as a result of the reacting ions approaching each other over distances $\sim r_T$ and transfer of the energy $\sim$ T at the elasic collision between an ion and a buffer particle. In this case the process is considered to be accomplished. But actually the bound system of reacting ions appears with the energy close to the dissosiation boundary which can at the next collision either transit again to the free ions state or decrease its energy once again by T. AT $T \leq E_0$ the recombination is the random walk process over energy of the bound ion system. If within a given theory the interaction of one of the ions with the thermostat particle is simulated by the elastic collision with the transfer of energy of order T, then this process is inelastic for the bound ion system. Since the radius of the polarization interaction between an ion and thermostat particle is much less than the bound state dimension ($\sigma^{1/2} \ll r_T$) and for the energies close to the dissociation boundary the period of vibrations is much more than their interaction time

$$\theta \sim \omega_b^{-1} \gg \sqrt{\bar{\sigma}}/\bar{v},$$

the dissipation of the energy is completely determined by pair collisions. So, if the coefficient γ is expressed via ion mobilities or frequences of the elastic collisions with buffer particles (accurate to the ratio of their masses considering the part of the energy transfered), then the estimation of the bound state energy dissipation is true.

This paper presents calculations of the ion-ion recombination rates for specific processes:

$$Kr^+ + F^- + Ar \rightarrow (KrF)^* + Ar, \tag{6.11}$$

$$Xe^+ + Cl^- + Ne \rightarrow (XeCl)^* + Ne, \tag{6.12}$$

for which is known [20] thatat T = 300 K maxima of reaction rates are more than 10^{-6} cm^3/s and are realized at pressures of the atmospheric order ($P_0 \simeq 0{,}1$ MPa). In [20] parameters of Coulomb terms are also presented: $r_0 = 2.3$ Å, $E_0 = 4.86$ eV for (6.11) and and $r_0 = 2{,}9$ Å, $E_0 = 3{\cdot}83$ eV for (6.12). The polarizabilities of thermostat particles are taken from the handbook [21]: $\alpha_{Ar} = 1.6$ Å^3, $\alpha_{Ne} = 0{\cdot}4$ Å^3. The exponent for the repulsive branch of the potential is uniquely specified by r_0 and E_0: n = 4.5 for (6.11) and n = 4.4 for (6.12). The results of calculations are presented in Fig.4. The maximum recombination rate for (6.11) is equal to $6 \cdot 10^{-6}$ cm^3/s for the density of thermostat particles $N_0 = 2 \cdot 10^{19}$ cm^{-3} or the pressure $P_0 = 0.1$ MPa. The maximum recombination rate for (6.12) is equal to $8 \cdot 10^{-6}$ cm^3/s at $N_0 = 5 \cdot 10^{19}$ cm^{-3} and P_0 =

= 0.2 MPa. Positions of the rate maxima (6.11) and (6.12) obtained from Eq.(5.31) correspond to $\lg(\gamma_0/\omega_0) = -3.7$ and $\lg(\gamma_0/\omega_0) = -3.6$.

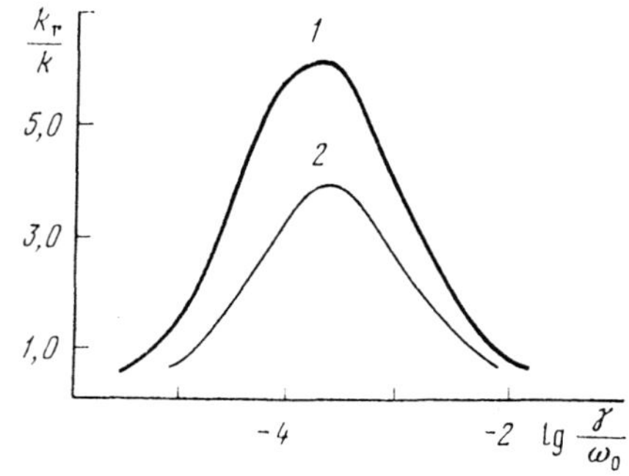

Fig. 4. The rate of three-particle recombination of K_R-processes (6.11) and (6.12) (curves 1 and 2, respectively); $K = r_0^2(8\pi T/\mu)^{1/2}\cdot 10^4$.

DISCUSSION

The formulas obtained above allow to estimate rates of processes for the wide class of reactions.The main data used for the estimation are: the interaction potential parameters, cross sections of the elastic collisions between reacting and thermostat particles, and also the density and tempetature of the thermostat.

In the framework of this approach one need not introduce the some capture radius which is initialy determined from pure physical considerations. The boundary between bound and free states used in the model obeys to a single natural requirement: the results of calculations should not depend on the choice of this boundary. This requirement is similar to the specifying the number m_0 of energy levels in problems of the on-level kinetics. In this case the calculation results do not depend on the choice of m_0 lying above "the bottleneck" [2]. In addition, it was shown that the acting mass law agreed with the quasi-classical statsums as well as detailed balancing principle for the rates of direct and reverse processes follow from the equilibrium distribution function. Thus, this approach involves all main features of on-level kinetics for reacting systems.

It is known that the classical approach is valid if the vibrational quantum energy is small compared to the medium temperature. Taking into account that in the case of recombination "the bottleneck" is located in the vicinity of the top of potential barrier, this requirement is not rigorous and has the form

$$\hbar\omega_{n,\ n+1}(r_b) < T,$$

where $h\omega_{n,n+1}(r_b)$ is a vibrational quantum in the region of the interaction energy $r \sim r_b$. Note that in this case the choice of the method of studying recombination process is not crucial because the rate of the reverse dissociation process is evaluated from the equilibrium constant.

The statistic approach may be used if in the course of reactions the great number of collisions occures between reacting and thermostat particles (the inequality (3.14)). Considering any specific process, one should test if this condition is fulfilled. Consider Eq.(3.14) as applied, for example, to the ion-ion recombination process in the limit of large and small gas densitiess. For the polarization interaction between ions and thermostat particles in the limit of high gas pressures we have

$$N \gg \frac{4\pi r_T}{\sigma^2}\left(\frac{N_i}{N}\right), \qquad v_T\lambda^{-1} \gg \omega_0.$$

In the limit of low gas pressures

$$N_i \ll \frac{3}{4\sqrt{\pi}\, e^2}\left(\sqrt{\frac{\mu}{\mu_1}} + \sqrt{\frac{\mu}{\mu_2}}\right)^2 r_T^{-3}, \quad v_T\lambda^{-1} \ll \omega_0.$$

One can see, that the application of the statistical approach places restrictons on the plasma density and temperature and also its ionizaton degree.

In this paper the study was madefor relatively low gas temperatures compared to the bond energy of interacting particles. First, this case is most often implemented in practice and, second, it is convenient for calculations because it allows to use the saddle-point method for calculations of integrals using the potential expansion in the vicinity of its minimum, neglecting the coupling between vibrations and rotation of the system. Since rates of processes were obtained for the arbitrary interaction potentials, the generalization for arbitrary temperatures is trivial and lies in the abandonment of the saddle-point method.

The effective potential (2.10) of the one-dimensional problem odtained within the approximation of the Maxwell distribution over the rotation velocities takes into account classically exactly the centrifugal rotation energy [22]. To verify it, consider the quantum-mechanical one-dimensional potential for two particles interacting according the law U(r):

$$U(r) + \frac{\hbar^2}{2\mu r^2} K(K+1),$$

where K is the quantum number of the total angular momentum.

Considering weak vibrational excitations near the equilibrium r_0, we express this potential as a series in $\zeta = r - r_0$ powers with regard to the first terms of the expansion:

$$U(\xi) = U_0 + \frac{1}{2}\mu\omega_0^2\xi^2 + \frac{\hbar^2}{2\mu r_0^2} K(K+1) - \frac{\hbar^2}{\mu r_0^3} K(K+1)\xi.$$

Assigning constant third term to the U_0 and using the average value

$$\overline{K(K+1)} = \frac{\sum_{K=0}^{\infty} K(K+1)(2K+1)\exp[-(B_e/T)K(K+1)]}{\sum_{K=0}^{\infty}(2K+1)\exp[-(B_e/T)K(K+1)]} = \frac{T}{B_e} = \frac{2\mu r_0^2 T}{\hbar^2},$$

we have

$$U(r) = U_0 + \frac{1}{2}\mu\omega_0^2\xi^2 - \frac{2T}{r_0}\xi.$$

The same formula results from expandingthe potential (2.10) in ζ powers.

In conclusion we notice that the improvement of this approach for determination of chemical reaction rates is primarily associated with the correct construction of non-equilibrium solutions of the Kramers equation (2.7) which may be presented in the form

$$\partial\rho/\partial t = \hat{L}\rho,$$

where the operator $\hat{L}$ is non-Hermitian and have at least one complex eigenvalue which imaginary part is responsible for vibrations of reacting particles nearthe eqiulibrium, while the real one describes the dissipation process. The main problem in the construction of the asymptotic in time solution lies in the determination of the inverse operator $\hat{L}^{-1}$.

ABSTRACT

The complete theory is developed of determining bimolecular reacton rates on the basis of the classical stochastic Christionsen-Kramers treatment which represent a chemical reaction by the random walk process of the type of Brownian motion of a state-point in a phase space. Particularly,the formula was obtained for the ion three-body recombination rate which contains the limiting results of Thomson and Langevin. The conditions for the theory application are discussed in detail, one of which is the necessity of the large number of collisions between reacting particles during the reaction.

REFERENCES

[1] A.V. Gurevich and L.P. Pitaevskii, Recombination coefficient in the dense low-temperature plasma, Zh. Eksp. Teor. Fiz. 46, (1964) 1281.

[2] L.M. Biberman, V.S. Vorob`ev, and I.T. Yakubov, Kinetics of non-equilibrium low-temperature plasma, Nauka, Moscow, (1982) 375 p.

[3] H.A. Kramers, Brownian motion in a field of force and the diffusion model of chemical reactions, Physica (The Hague), 7, (1940) 284.

[4] R.S. Larson and M.D. Kostin, Friction and velocity in Kramers theory of chemical kinetics, J. Chem. Phys. 72, (1980) 1392.

[5] V. Seshadri, B.J. West, and K. Lindenberg, Analytic theory of Ibid, p. 1145.

[6] J.T. Hynes, R. Kapral, and G.M. Torrie, Stochastic trajectory simulation of iodine recombination in liquids, Ibid, p. 177.

[7] C.L. Brooks, M.W. Balk, and S.A. Adelman, Dynamics of liquid state chemical reactions: Vibrational energy relaxation of molecular iodine in liquid solution, Ibid, 79, (1983) 784.

[8] B. Carmeli and A. Nitzan, Non-markovian theory of activated rate processes, Phys. Rev. A - Gen. Phys. 29, (1984) 1481.

[9] M.G. Sceats, J.M. Dawes, and D.P. Millar, Aton recombination in dense gases and fluids - a stochastic model, Chem. Phys. Lett, 114, (1984) 63.

[10] R. Balesku, The Equilibrium and Nonequilibrium Statistical Mechanics, Mir, Moscow, (1978) v. 2.

[11] A.I. Akhiezer and S.V. Peletminskii, Methods of Statistical Physics, Nauka, Moscow, (1977).

[12] Yu. L. Klimontovich, Statistical Physics, Nauka, Moscow, (1982).

[13] Yu.V. Rumer and M.Sh. Rivkin, Thermodynamics, Statistical Physics and Kinetics, Nauka, Moscow, (1977).

[14] S.A. Reshetnyak and L.A. Shelepin, The QDF method in diffusion theory of chemical reactions, Proc. Lebedev Phys. Inst., 144, (1984).

[15] S.A. Reshetnyak, S.M. Kharchev, and L.A. Shelepin, Asymptotic methods in the theory of linear kinetic equations, Teor. Mat. Fiz., 49, (1981) 131.

[16] J.J. Thomson, Recombination of gaseous ions, the chemical combination of gases and monomolecular reactions, Philos. Mag., 47, (1924) 337.

[17] B.M. Smirnov, Physics of Weakly-Ionized Gas, Nauka, Moscow, (1972).

[18] G.L. Natanson, On the theory of volume ion recombination, Zh. Tekh. Fiz., 29, (1959) 1373.

[19] M.R. Flannery and T.P. Yang, Ionic recombination of rare-gas atomic ions X^+ with F^- in a dense gas X. Appl. Phys. Lett., 32, (1978) 327.

[20] Eximer Lasers, (Ed C. Rodes), Mir, Moscow, (1981).

[21] A.A. Radzig and B.M. Smirnov, Handbook on Atomic and Molecular Physics, Atomizdat, Moscow, (1980).

[22] L.D. Landau and E.M. Lifshiz, Statistical Physics, Nauka, Moscow, (1964).

USE OF THE MONTE-CARLO METHOD (METHOD OF CLASSICAL TRAJECTORIES) FOR CALCULATION OF RATE CONSTANTS AND ENERGY DISTRIBUTION OF PRODUCTS IN TWO-CHANNEL CHEMICAL REACTIONS

N.A. Konoplev, A.A. Stepanov, and V.A. Shcheglov

INTRODUCTION

Although the lasing wavelengths of available chemical lasers cover almost the whole near and middle infrared, the lack of continuous tunability of their output frequency in the wide range makes one to search for new chemical reactions producing the nonequilibrium distribution of the products that can be used to obtain oscillation at frequencies different from those of already known lasers. In this regard the two-channel reactions of the type

(1) $$A + BC(v_0, J_0) \rightarrow \begin{cases} AB(v, J) + C \\ AC(v, J) + B, \end{cases}$$

proceding with the simultaneous production of vibrationally excited molecules of two types AB and AC in the active medium appear to be rather interesting. However, for the effective use in chemical lasers such reactions must possess a number of features: (1) the thermal effect of the reaction (1) have to be sufficiently high for both AB and AC channels; 92) the appreciable fraction of the released energy have to be localizeded in the vibrations of products; (3) the reaction rate constants have to be high in both channels. To this type of reactions one may assign, for example, two-channel reactions of atomic hydrogen with chlorine monobromid

(2) $$H + BrCl(v_0, J_0) - \begin{cases} HBr(v, J) + Cl \\ HCl(v, J) + Br, \end{cases}$$

and with fluorine monochloride

(3) $$H + ClF.(v_0, J_0) \rightarrow \begin{cases} HCl(v, J) + F \\ HF(v, J) + Cl \end{cases}$$

and some other.

*Throughout this chapter @ indicates that the Subsequent numbers or letters are subscripts

To evaluate the prospects of using a particular reaction to obtain the oscillation at corresponding frequencies, it is very imprtant to know the character of the energy distribution of molecular products over vibrational and rotational degrees of freedom and also values of the rateconstants for individual reaction channels. Though at present the most complete information on these problems is given mainly by the experimental methods of investigation,sometimes answers to a number of important questions may be obtained on the basis of quantum-mechanical and classical trajectory calculations. This paper presents the results of the investigation of such reactions on the basis of the Monte-Carlo method (which has been widely used for calculations of the elementa chemical proccesses since the sixties [1-3]), which were obtained FIAN in 1978 - 84 [4-9].

1. DESCRIPTION OF THE CLASSICAL TRAJECTORIES METHOD

In the method of classical trajectories the calculation of the elementary chemical reactions is based on the equations of classical mechanics with a specified potential of interaction between individual atomss. The latter, as a rule, are assumed to be structureless. Usually, the translation of the calculation results into the "quantum-mechanical language" is based on the analysis of the final state of products and calculation of the kinetic energy of the relative movement of particles and also total rotational and vibrational energies of the molecules produced.

The name of the method Monte-Carlo arises from the fact that the classical equations of motion(in the Hamiltonian form) are numerically integrable (as a rule, in the center of mass system) along the trajectory with an arbitrary initial orientation of particles in the space and the results obtained are then averaged over all orientations. To calculate multiple integrals arising at such averaging, it is convenient to use the Monte-Carlo method because in this case the orientations of particles may be specified arbitrary with a given distribution law. Consider now the calculation method in more detail.

1.1. Equations of motion

When calculating the elementary proccesses of the type (1), it is convenient without loss of generality to introduce the Cartesian system (x, y, z) in the following way (see Fig. 1) Let z axis be parallel to the initial orientation of the velocity vector V@R of the relative motion of an atom A and molecule BC. BC. Draw the yOz plane through the V@R vector direction and the AO′line connecting atom A with a center of mass of molecule BC. In this case both the impacting atom A and the center of mass of molecule BC will be placed in the initial state at the plane x = 0 (in the yOz plane).

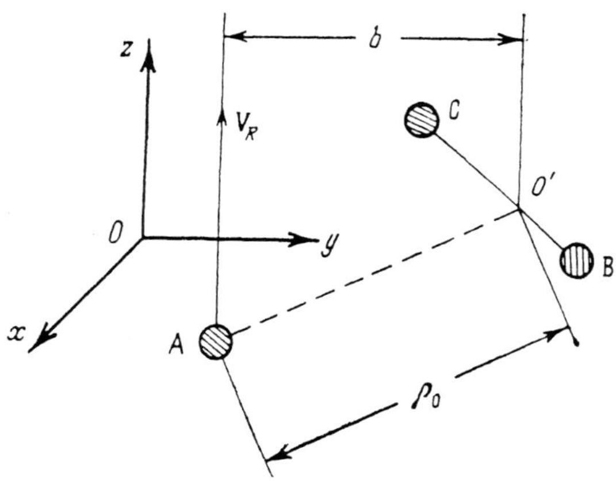

Fig.1. The choice of the coordinate system for the description of collision procces between an atom A and molecule BC

In the case of the three-dimentional motion the system of three particles is described by eighteen independent variables, however, six of them correspond to the motion of system as a whole. Denote for convenience coordinates of the atom A @(x@A, y@A, z@A) by x@i (where i = 1, 2, 3 corresponds to the coordinates x, y, and z, respectively, coordinates of the atom B by x@i+3, and of the atom C by x@i+6. The Cartesian components of momenta of these atoms are designated as p@xi, p@xi+3, p@xi+6 (i = 1, 2, 3).

In Cartesian coordinates the Hamiltonian of the three-particle system has the form

$$H\left(x_i, p_{x_i}\right) = \frac{1}{2}\sum_{i=1}^{3}\left(\frac{1}{m_A}p_{x_i}^2 + \frac{1}{m_B}p_{xi+3}^2 + \frac{1}{m_C}p_{x_{i+6}}^2\right) + U\left(x_1, x_2, \ldots, x_9\right), \quad (4)$$

where U is the potential of interactionbetween particles, hence, the classical equations of motion are written in the form

$$\frac{dx_i}{dt} = \frac{\partial H}{\partial p_{x_i}}, \qquad \frac{dp_{x_i}}{dt} = -\frac{\partial H}{\partial x_i}, \quad (5)$$

here i = 1, 2, ..., 9.

To separate the motion of the center of mass of three-particle system and to describe the relative motion of atoms, it is convenient to introduce generalized coordinates q@i, q@i+3, and q@i+6 (i = 1, 2, 3) which are related with Cartesian ones by relations

$$\begin{aligned} q_i &= x_{i+6} - x_{i+3}, \\ q_{i+3} &= x_i - \frac{1}{m_{BC}}\left(m_B x_{i+3} + m_C x_{i+6}\right) \\ q_{i+6} &= \frac{1}{M}\left(m_A x_i + m_B x_{i+3} + m_C x_{i+6}\right) \end{aligned} \quad (6)$$

Where the following notations are used: m@BC and @M, where m@k is a mass of k-th atom. It is obvious that coordinates q@i in (6) characterize the relative motion of atoms B and C, q@i+3 describe the relative motion between atom A and the center of mass of molecul BC, and q@i+6 characterize the motion of the center of mass of the system as a whole. The inverse relation between the Cartesian coordinates and the generalized ones is of the form

$$\begin{aligned} x_i &= \frac{m_{BC}}{M}q_{i+3} + q_{i+6}, \\ x_{i+3} &= -\frac{m_C}{m_{BC}}q_i - \frac{m_A}{M}q_{i+3} + q_{i+6}, \end{aligned} \qquad x_{i+6} = \frac{m_B}{m_{BC}}q_i - \frac{m_A}{M}q_{i+3} + q_{i+6}. \quad (7)$$

It isknown that the generalized momenta p@qi, conserving at the canonical transformation of coordinates of the type (6) the equations of motion in the form(5), are related with Cartesian ones p@xi by the relation

$$p_{x_i} = \sum_{k=1}^{9} p_{q_k}\frac{\partial q_k}{\partial x_i}. \quad (8)$$

Therefore, taking into account (6), we have

$$(9)\qquad p_{x_i} = p_{q_{i+3}} + \frac{m_A}{M} p_{q_{i+6}},$$

$$p_{x_{i+3}} = - p_{q_i} - \frac{m_B}{m_{BC}} p_{q_{i+3}} + \frac{m_B}{M} p_{q_{i+6}},$$

$$p_{x_{i+6}} = p_{q_i} - \frac{m_C}{m_{BC}} p_{q_{i+3}} + \frac{m_C}{M} p_{q_{i+6}}$$

and inversely

$$(10)\qquad p_{q_i} = \frac{1}{m_{BC}} (m_B p_{x_{i+6}} - m_C p_{x_{i+3}}),$$

$$p_{q_{i+3}} = \frac{m_{BC}}{M} p_{x_i} - \frac{m_A}{M} (p_{x_{i+3}} + p_{x_{i+6}}),$$

$$p_{q_{i+6}} = p_{x_i} + p_{x_{i+3}} + p_{x_{i+6}}.$$

In new variables the expression (4) for Hamiltonian may be written in the form:

$$(11)\qquad H(q_i, p_{q_i}) = \frac{1}{2} \sum_{i=1}^{3} \left(\frac{1}{\mu_{BC}} p_{q_i}^2 + \frac{1}{\mu_{A,BC}} p_{q_{i+3}}^2 + \frac{1}{M} p_{q_{i+6}}^2 \right) + U(q_1, q_2, \ldots, q_6),$$

where BC and A,BC

The equations of motion in new variables have the form similiar to (5):

$$(12)\qquad \frac{dq_i}{dt} = \frac{\partial H}{\partial p_{q_i}}, \qquad \frac{dp_{q_i}}{dt} = - \frac{\partial H}{\partial q_i} \quad (i = 1, 2, \ldots, 9).$$

Notice that the center of mass motion of three-particle system in (12) is separated explicitly (at i = 7, 8, 9). However, one need not use these equations to describe relative motions of individual atoms and, hence, they will not be used later on.

The interaction potential U depending on the generalized coordinates q@1, q@2 ..., q@6 is determined, as a rule, on the basis of quantum-mechanical calculations. In the case of univalent bonds between atoms, the semiemperical LEPS approximation (the London - Eiring - Polany - Sato formula (see [10]) is widely used, according to which the potential energy surface is specified in the form

$$U(R_{12}, R_{13}, R_{23}) = Q_{12} + Q_{13} + Q_{23} - \frac{1}{\sqrt{2}} [(I_{12} - I_{23})^2 + (I_{12} - I_{13})^2 + (I_{23} - I_{13})^2]^{1/2}. \quad (13)$$

Here the subscripts 1, 2 and 3 correspond to atoms A, B and C, respectively; R@ij are interatomic distances and Q@ij(R@ij) and I@ij(R@ij) are Coulomb and exchange integrals

$$Q_{ij}(R_{ij}) = \frac{1}{2}\left[E_{ij}^{(1)}(R_{ij}) + \frac{1 - K_{ij}}{1 + K_{ij}} E_{ij}^{(2)}(R_{ij})\right],$$

$$I_{ij}(R_{ij}) = \frac{1}{2}\left[E_{ij}^{(1)}(R_{ij}) - \frac{1 - K_{ij}}{1 + K_{ij}} E_{ij}^{(2)}(R_{ij})\right].$$

In the latter relations K@ij areemperical fitting parameters and E@ij(1) and E@ij(2) are the Morse and anti-Morse functions, respectively, and

$$E_{ij}^{(1)}(R_{ij}) = D_{ij}\{\exp[-2\alpha_{ij}(R_{ij} - R_{ij}^0)] - 2\exp[-\alpha_{ij}(R_{ij} - R_{ij}^0)]\},$$

$$E_{ij}^{(2)}(R_{ij}) = A_{ij}\{\exp[-2\alpha_{ij}(R_{ij} - R_{ij}^0)] + 2\exp[-\alpha_{ij}(R_{ij} - R_{ij}^0)]\},$$

where D@ij is the dissociation energy of the i - j connection (counted from the potential enegry curve minimum), while A@ij are usually assumed to be A@ij = 1/2 D@ij. In the above formulas R@ij are equilibrium lengths of bonds between corresponding atoms (in a molecule) and ij are parameters characterizing the range of action of the interaction forces.

As a rule, values of αij are not given in the handbooks and are calculated from the spectroscopic constants:

$$\alpha_{ij} = \frac{(\omega_e)_{ij}}{2R_{ij}^0[(B_e)_{ij} D_{ij}]^{1/2}},$$

where (ωe)ij is the vibrational frequency and (B@e)ij is the rotational constant.

Notice that if values of (B@e)ij are absent in the literature, they may be determined from the formula

$$(B_e)_{ij} = \frac{h}{8\pi^2 c (I_e)_{ij}}.$$

Here h is the Plank constant, (I@e)ij = @(R@ij)2 isthe inertia moment of a molecule,μ is its reduced mass and C is the velocity of light.

The interatomic distances expressed via the generalized coordinates may be presented in the form

$$(14)\quad R_{12}^2 = \sum_{i=1}^{3}\left(\frac{m_C}{m_{BC}} q_i + q_{i+3}\right)^2,$$

$$R_{13}^2 = \sum_{i=1}^{3}\left(\frac{m_B}{m_{BC}} q_i - q_{i+3}\right)^2, \quad R_{23}^2 = \sum_{i=1}^{3} q_i^2.$$

Hence, based on (11) - (14) the equations of motion for the system of interacting particles may be written in the following way:

$$(15)\quad \frac{dq_i}{dt} = \frac{1}{\mu_{BC}} p_{q_i},$$

$$\frac{dq_{i+3}}{dt} = \frac{1}{\mu_{A,BC}} p_{q_{i+3}},$$

$$\frac{dq_{i+6}}{dt} = \frac{1}{M} p_{q_{i+6}}$$

and

$$(16)\quad \frac{dp_{q_i}}{dt} = -\left[\frac{1}{R_{12}}\frac{m_C}{m_{BC}}\left(\frac{m_C}{m_{BC}} q_i + q_{i+3}\right)\frac{\partial U}{\partial R_{12}} + \frac{1}{R_{13}}\frac{m_B}{m_{BC}}\left(\frac{m_B}{m_{BC}} q_i - q_{i+3}\right)\frac{\partial U}{\partial R_{13}} + \right.$$

$$\left. + \frac{q_i}{R_{23}}\frac{\partial U}{\partial R_{23}}\right],$$

$$\frac{dp_{q_{i+3}}}{dt} = -\frac{1}{R_{12}}\left(\frac{m_C}{m_{BC}} q_i + q_{i+3}\right)\frac{\partial U}{\partial R_{12}} + \frac{1}{R_{13}}\left(\frac{m_B}{m_{BC}} q_i - q_{i+3}\right)\frac{\partial U}{\partial R_{13}},$$

$$\frac{dp_{q_{i+6}}}{dt} = 0, \quad i = 1, 2, 3.$$

1.2. Initial conditions

The initial state of the system "atom + two-atomic molecule" is completely determined by the specification of nine initial parameters:

(1) $\rho 0$ is the initial distance between an atom A and the center of mass of a molecule BC,

(2) b is the aiming collision parameter (see Fig. 1),

(3) φ and θ are azimuth and polar angles determining the space orientation of a molecule BC,

(4) E@R (or V@R) is the energy (velocity) of the relative movement of A and BC,

(5) E@rot (or J@0) is the rotational energy (the rotational quantum number) of a molecule BC,

(6) E@vib (or @0) is the vibrationalenergy (the oscillation quantum number) of a molecule BC,

(7) η is the angle determining the orientation of the angular momentum vector L@BC of a molecule BC in the space,

(8) ψ is the angle characterizing the initial vibrational phase of a molecule BC.

Using the parameters mentioned above one can specify all initial conditions required for the equations of motion (15) and (16). So, for coordinates q@i+3 and the generalized momenta p@qi+3 (i = 1, 2, 3) characterizing the relative motion of A and BC these conditions is of the form

(17) $$q_4(0) = 0, \quad q_5(0) = -b, \quad q_6(0) = -\sqrt{\rho_0^2 - b^2}$$

and

(18) $$p_{q_4}(0) = p_{q_5}(0) = 0, \quad p_{q_6}(0) = \mu_{A,BC} V_R.$$

Here ρ0 is the initial distance between atom A and the center of mass of BC, V@R is the initial velocity of the relative motion of A and BC related with the relative motion energy E@R by the equation

$$V_R = (2E_R/\mu_{A,BC})^{1/2}.$$

Now consider initial conditions for coordinates q@i and momenta p@i (i = 1, 2, 3) determining the initial state of molecule BC.

Since molecule BC may be oriented in space arbitrary, first we choose the method of a simple and unique determination of the BC axis position in the space. To do it we construct in the center of mass of BC (at the point 0´ in Fig. 1) the local Cartesian system of coordinates x, y, z. oriented parallel to the basic one x, y, z. It is obvious that the angle between z´axis and the molecule BC axis counted off the z´axis direction may be chosen as a polar angle φ. We shall count the azimuthal angle off the x´O´y´ plane clockwise, if to see from the end of z´axis (Fig. 2). Notice that the position of BC axis in space is determined uniquely if angle θ varies within the interval $(0, \pi)$ and angle φ - within the interval $0, 2\pi)$.

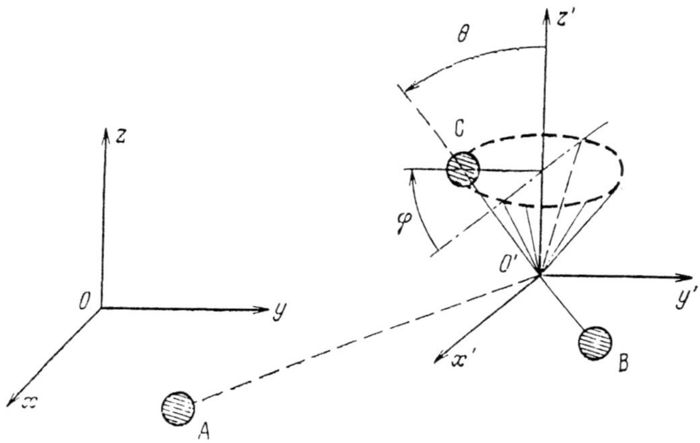

Fig. 2. Definition of azimuthal φ and polar θ angles in the space orientation of a molecule BC

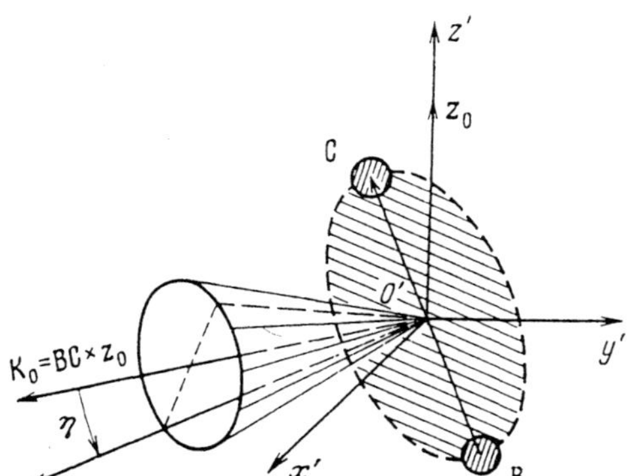

Fig. 3. Definition of an angle η characterizing the direction of the angular momentum vector of molecule BC

Now the initial conditions for q_i may be presented in the form

$$
\begin{aligned}
q_1(0) &= R_{23}(0)\cos\varphi\sin\theta,\\
q_2(0) &= R_{23}(0)\sin\varphi\sin\theta,\\
q_3(0) &= R_{23}(0)\cos\theta,
\end{aligned}
\tag{19}
$$

where $R_{23}(0)$ - the initial distance between atoms B and C in the initial molecule, which for relatively small vibrational amplitude (for small values of the oscillation quantum number 0) can be determined from the relation

$$R_{23}(0) = R_{23}^0 + {}^1/_2\,(R_{max} - R_{min})\sin\psi, \quad 0 \leqslant \psi \leqslant 2\pi. \tag{20}$$

where R_{23}^0 is the equilibrium length of bond between atoms in the molecule BC, ψ is the initial phase of vibrations, R_{max} and R_{min} are maximum and minimum values of the bond length in BC for a given value of the vibration-rotation energy.

Since R_{max} and R_{min} characterize the position of the turning points on the potential energy curve $U(R_{23})$ they may be determined as roots of the transcendental equation

$$D_{23}\{1 - \exp[-\alpha_{23}(R_{23} - R_{23}^0)]\}^2 = E_{vJ} - E_{rot}, \tag{21}$$

where $E_{vJ} = E_{vib} + E_{rot}$ is the total vibration-rotation energy of the molecule BC.

Note that if the values of the oscillation quantum number 0 and rotational quantum number J_0 of molecule BC are given as the initial data then the vibration-rotation energy is defined by the known spectroscopic formula

$$E_{vJ} = \omega_e(v_0 + {}^1/_2) - \omega_e x_e (v_0 + {}^1/_2)^2 + [B_e - \alpha_e(v_0 + {}^1/_2)]\,J_0(J_0+1),$$

where ω_e, $\omega_e x_e$, B_e, and e are spectroscopic constants, and

$$E_{rot} = \frac{h^2 J_0(J_0+1)}{8\pi^2 \mu_{BC}(R_{23}^0)^2}.$$

Consider the initial conditions for generalized momenta p@qi. First, we introduce the definition of the angle η characterizing the orientation of the angular momentum vector of molecule BC in the space. For this purpose denote the vector of the angular velocity of rotation of molecule BC by $\boldsymbol{\omega}$. It is obvious that the vector of angular momentum L@BC will be directed along $\boldsymbol{\omega}$ and perpendicular to the plane of rotation passing through the molecule axis. We shall demonstrate that the direction of $\boldsymbol{\omega}$ (and, hence, of L@BC) is uniquely determined by specifying the angle between the angular velocity vector $\boldsymbol{\omega}$ and some auxiliary vector K@O = @BC x z@O, where z@O is a unit orth along z´axis.

The direction of vector K@O is known because we know the orientation of BC axis in the space. Lets separate around the vector K@O a solid angle with an opening 2@η at a cone vertex located at thepoint O´ (Fig. 3). Vectors K@O and $\boldsymbol{\omega}$ are perpendicular to the molecule BC axis and emanate from the point O´, i.e. they lay in a plane perpendicular to the molecule axis. Therefore, by drawing through the vector K@O direction the plane perpendicular to the molecule axis, we determine the vector $\boldsymbol{\omega}$ direction (and, hence, the vector L@BC direction as well) as the line of intersection ot this plane with the surface of the solid angle circumscribed about the vector K@O. Hence, if the direction of the angle reading is specified, the vector $\boldsymbol{\omega}$ direction is determined uniquely

THe total momentum p@q of the relative movement of atoms B and C ia a sum of two components: the rotational component of the momentum p@qrot and vibrational component p@qvib. Therefore, we have

$$p_{q_i}(0) = p_{q_i}^{\rm rot}(0) + p_{q_i}^{\rm vib}(0), \qquad i = 1, 2, 3. \tag{22}$$

It can be shown that rotational components p@qirot(0) in (22) are defined as

$$\begin{aligned} p_{q_1}^{\rm rot}(0) &= p_{\rm rot}(\cos\varphi\cos\theta\sin\eta + \sin\varphi\cos\eta), \\ p_{q_2}^{\rm rot}(0) &= p_{\rm rot}(\sin\varphi\cos\theta\sin\eta - \cos\varphi\cos\eta), \\ p_{q_3}^{\rm rot}(0) &= -\, p_{\rm rot}\sin\theta\sin\eta, \end{aligned} \tag{23}$$

where

$$p_{\rm rot}^2 = L_{\rm BC}^2/\left(R_{23}^0\right)^2,\ L_{\rm BC}^2 = (h^2/4\pi^2)\,J_0\,(J_0+1)$$

and h is Plank constant.

It is obvious that the vibrational component of the momentum can be written in the form

$$p_{q_1}^{\rm vib}(0) = p_{\rm vib}\delta_\psi \cos\varphi \sin\theta,$$

(24)

$$p_{q_2}^{\rm vib}(0) = p_{\rm vib}\delta_\psi \sin\varphi \sin\theta,$$

$$p_{q_3}^{\rm vib}(0) = p_{\rm vib}\delta_\psi \cos\theta.$$

Here

$$\delta_\psi = \mathrm{sign}\,(\cos\psi) = \begin{cases} +1, & \cos\psi \geqslant 0, \\ -1, & \cos\psi < 0, \end{cases}$$

and

$$p_{\rm vib} = \{2\mu_{\rm BC}[E_{\rm vib} - U_{\rm BC}(R_{23}(0))]\}^{1/2},$$

in this case

$$E_{\rm vib} = \omega_e(v_0 + {}^1/_2) - \omega_e x_e (v_0 + {}^1/_2)^2$$

and

$$U_{\rm BC}(R_{23}(0)) = D_{23}\{1 - \exp\left[-\alpha_{23}\left(R_{23}(0) - R_{23}^0\right)\right]\}^2.$$

1.3. Analysis of the final state of products

When integrating equations of motion along the trajectory we shall determine the end of integration by the following way. Lets denote the characteristic size of the reaction vessel by $\rho 0$ and the distance between the leaving particle and the center of mass of the molecule formed - by ρ. It is obvious that the condition $\rho > \rho 0$ means that after the collision of the initial particles the reaction products fly apart over a distance at which the interaction between them may be neglected.

In the case of the nonreactive collision (the final system is "atom A + BC") the distance between atom A and molecule BC is defined by

$$\rho^2_{A,BC} = \sum_{i=1}^{3} q^2_{i+3}.$$

If the collision is followed by the formation of molecule AB we have

$$\rho^2_{C,AB} = \frac{1}{m^2_{AB}} \sum_{i=1}^{3} \left(\frac{m_B M}{m_{BC}} q_i - m_A q_{i+3} \right)^2,$$

where m@AB = m@A + m@B.

If the collision results in the formation of molecule AC then

$$\rho^2_{B,AC} = \frac{1}{m^2_{AC}} \sum_{i=1}^{3} \left(\frac{m_C M}{m_{BC}} q_i + m_B q_{i+3} \right)^2,$$

and m@AC = m@A + m@C.

Consider all these three cases in more detail.

1.3.1. The case of the nonreactive collision

If the collision has occured without the realignment of bonds inthe system of particles then the velocity of relative motion of atom A and molecule BC is determined by the expression:

$$(V'_R)^2 = \left(\frac{M}{m_A m_{BC}} \right)^2 \sum_{i=1}^{3} p^2_{q_{i+3}},$$

while the kinetic energy of the relative translational motion is equal to

$$E'_R = {}^1\!/_2 \mu_{A,BC} (V'_R)^2.$$

The vibration-rotation energy of molecule BC is calculated by the formula

$$E^{BC}_{vJ} = \frac{m_{BC}}{m_B m_C} \sum_{i=1}^{3} p^2_{q_i} + U_{BC},$$

where U_{BC} is the potential vibrational energy of molecule BC determined by the relation

$$U_{\mathrm{BC}} = D_{23}\{1 - \exp[-\alpha_{23}(R_{23} - R_{23}^{0})]\}^2.$$

The molecule angular momentum squared is

$$L_{\mathrm{BC}}^2 = (q_1 p_{q_2} - q_2 p_{q_1})^2 + (q_2 p_{q_3} - q_3 p_{q_2})^2 + (q_3 p_{q_1} - q_1 p_{q_3})^2,$$

while the rotational energy is defined by the relation

$$E_{\mathrm{rot}}^{\mathrm{BC}} = \frac{L_{\mathrm{BC}}^2}{2I_{\mathrm{BC}}} = \frac{L_{\mathrm{BC}}^2}{2\mu_{\mathrm{BC}} \sum_{i=1}^{3} q_i^2}.$$

Here I_{BC} is the molecule moment of inertia relative the rotation axis.

The vibrational energy of molecule BC is determined by the formula

$$E_{\mathrm{vib}}^{\mathrm{BC}} = \frac{\left(\sum_{i=1}^{3} q_i p_{q_i}\right)^2}{2\mu_{\mathrm{BC}}\left(\sum_{i=1}^{3} q_i^2\right)} + U_{\mathrm{BC}}.$$

It is easy to verify that

$$E_{vJ}^{\mathrm{BC}} = E_{\mathrm{vib}}^{\mathrm{BC}} + E_{\mathrm{rot}}^{\mathrm{BC}}.$$

1.3.2. The case of the molecule AB formation

In this case the velocity of the relative motion of C and AB is

$$(V_R')^2 = \left(\frac{M}{m_{\mathrm{AB}}}\right)^2 \sum_{i=1}^{3} \left(\frac{1}{m_{\mathrm{BC}}} p_{q_{i+3}} - \frac{1}{m_{\mathrm{C}}} p_{q_i}\right)^2,$$

hence, the kinetic energy of the relative translational motion is equal to

$$E_R' = {}^1/_2 \mu_{\mathrm{C,AB}} (V_R')^2,$$ where $\mu_{C,AB} = m_C m_{AB}/M$.

The vibration-rotation energy of molecule AB is determined by the relation

$$E_{vJ}^{AB} = \sum_{i=1}^{3}\left\{\sqrt{\frac{m_A}{2m_B m_{AB}}}\,p_{q_i} + \frac{M}{m_{BC}}\sqrt{\frac{m_B}{2m_A m_{AB}}}\,p_{q_{i+3}}\right\}^2 + U_{AB}.$$

Here U@AB is the potential vibrational energy of molecule AB.

Furthermore, the angular momentum of molecule AB squared is

$$L_{AB}^2 = L_x^2 + L_y^2 + L_z^2,$$

and

$$L_x = \frac{m_A m_C}{m_{AB} m_{BC}}\left(q_2 p_{q_3} - q_3 p_{q_2}\right) + \frac{m_B m_C M}{m_{AB} m_{BC}^2}\left(q_2 p_{q_6} - q_3 p_{q_5}\right) +$$
$$+ \frac{m_A}{m_{AB}}\left(q_5 p_{q_3} - q_6 p_{q_2}\right) + \frac{m_B M}{m_{AB} m_{BC}}\left(q_5 p_{q_6} - q_6 p_{q_5}\right),$$
$$L_y = \frac{m_A m_C}{m_{AB} m_{BC}}\left(q_3 p_{q_1} - q_1 p_{q_3}\right) + \frac{m_B m_C M}{m_{AB} m_{BC}^2}\left(q_3 p_{q_4} - q_1 p_{q_6}\right) + \frac{m_A}{m_{AB}}\left(q_6 p_{q_1} - q_4 p_{q_3}\right) +$$
$$+ \frac{m_B M}{m_{AB} m_{BC}}\left(q_6 p_{q_4} - q_4 p_{q_6}\right)$$

and

$$L_z = \frac{m_A m_C}{m_{AB} m_{BC}}\left(q_1 p_{q_2} - q_2 p_{q_1}\right) + \frac{m_B m_C M}{m_{AB} m_{BC}^2}\left(q_1 p_{q_5} - q_2 p_{q_4}\right) + \frac{m_A}{m_{AB}}\left(q_4 p_{q_2} - q_5 p_{q_1}\right) +$$
$$+ \frac{m_B M}{m_{AB} m_{BC}}\left(q_4 p_{q_5} - q_5 p_{q_4}\right).$$

The rotational energy of molecule AB is determined by the formula

$$E_{rot}^{AB} = \frac{L_{AB}^2}{2I_{AB}} = \frac{L_{AB}^2}{2\mu_{AB}\sum_{i=1}^{3}\left(q_{i+3} + \frac{m_C}{m_{BC}}\,q_i\right)^2},$$

where I@AB is th moment of inertia of molecule AB relative the rotation axis, μAB = m@Am@B/m@AB.

The vibrational energy of molecule AB is equal to

$$E_{\text{vib}}^{\text{AB}} = \frac{1}{2}\,\mu_{\text{AB}}\left[\sum_{i=1}^{3}\left(q_{i+3} + \frac{m_{\text{C}}}{m_{\text{BC}}}\,q_i\right)^2\right]^{-1}\left\{\sum_{i=1}^{3}\left(\frac{m_{\text{C}}}{m_{\text{BC}}}\,q_i + q_{i+3}\right)\times\right.$$

$$\left.\times\left(\frac{1}{m_{\text{B}}}\,p_{q_i} + \frac{M}{m_{\text{A}}m_{\text{BC}}}\,p_{q_{i+3}}\right)\right\}^2 + U_{\text{AB}},$$

here E@ABvj = E@ABvib + E@ABrot.

To determine the oscillation v and rotationl J quantum numbers of the newly formed molecule AB one can use quantum-mechanical expressions for L@2 and E@vJ

$$L^2 = \frac{h^2}{4\pi^2}\,J\,(J+1)$$

and

$$E_{vJ} = \omega_e\,(v + {}^1/_2) - \omega_e x_e\,(v + {}^1/_2)^2 + [B_e - \alpha_e\,(v + {}^1/_2)]\,J\,(J+1).$$

If particular values of L@2AB and E@vJ are known, the corresponding values of v and J may be calculated from the latter expressions (rounded off to the nearest integer).

1.3.3. The case of the molecule AC formation

In this case the velocity of the relative motion of B and AC is equal to

$$(V'_R)^2 = \frac{M^2}{m_{\text{AC}}^2}\sum_{i=1}^{3}\left(\frac{1}{m_{\text{BC}}}\,p_{q_{i+3}} + \frac{1}{m_{\text{B}}}\,p_{q_i}\right)^2,$$

so that

$$E'_R = {}^1/_2\mu_{\text{B,AC}}\,(V'_R)^2,$$

where μB,AC = m@Bm@AC/M.

The vibration-rotation energy of molecule AC is determined by the expression

$$E_{vJ}^{AC} = \sum_{i=1}^{3} \left\{ \sqrt{\frac{m_A}{2m_C m_{AC}}}\, p_{q_i} - \frac{M}{m_{BC}} \sqrt{\frac{m_C}{2m_A m_{AC}}}\, p_{q_{i+3}} \right\}^2 + U_{AC},$$

where U@AC is the potential vibrational energy of molecule AC.

The angular momentum squared is determined from the formula

$$L_{AC}^2 = L_x^2 + L_y^2 + L_z^2,$$

here

$$L_x = \frac{m_A m_B}{m_{AC} m_{BC}} \left(q_2 p_{q_3} - q_3 p_{q_2}\right) - \frac{m_B m_C M}{m_{AC} m_{BC}^2} \left(q_2 p_{q_6} - q_3 p_{q_5}\right) - \frac{m_A}{m_{AC}} \left(q_5 p_{q_3} - q_6 p_{q_2}\right) +$$
$$+ \frac{m_C M}{m_{AC} m_{BC}} \left(q_5 p_{q_6} - q_6 p_{q_5}\right),$$

$$L_y = \frac{m_A m_B}{m_{AC} m_{BC}} \left(q_3 p_{q_1} - q_1 p_{q_3}\right) - \frac{m_B m_C M}{m_{AC} m_{BC}^2} \left(q_3 p_{q_4} - q_1 p_{q_6}\right) - \frac{m_A}{m_{AC}} \left(q_6 p_{q_1} - q_4 p_{q_3}\right) +$$
$$+ \frac{m_C M}{m_{AC} m_{BC}} \left(q_6 p_{q_4} - q_4 p_{q_6}\right),$$

and

$$L_z = \frac{m_A m_B}{m_{AC} m_{BC}} \left(q_1 p_{q_2} - q_2 p_{q_1}\right) - \frac{m_B m_C M}{m_{AC} m_{BC}^2} \left(q_1 p_{q_5} - q_2 p_{q_4}\right) - \frac{m_A}{m_{AC}} \left(q_4 p_{q_2} - q_5 p_{q_1}\right) +$$
$$+ \frac{m_C M}{m_{AC} m_{BC}} \left(q_4 p_{q_5} - q_5 p_{q_4}\right).$$

The rotational energy of molecule AC is

$$E_{rot}^{AC} = \frac{L_{AC}^2}{2I_{AC}} = \frac{L_{AC}^2}{2\mu_{AC} \sum_{i=1}^{3} \left(q_{i+3} - \frac{m_B}{m_{BC}} q_i\right)^2},$$

where μAC = m@Am@C/m@AC is the reduced molecule mass.

The vibrational energy of molecule AC is calculated in the following way

$$E_{vib}^{AC} = \frac{1}{2} \mu_{AC} \left[\sum_{i=1}^{3} \left(q_{i+3} - \frac{m_B}{m_{BC}} q_i\right)^2 \right]^{-1} \left\{ \sum_{i=1}^{3} \left(\frac{m_B}{m_{BC}} q_i - q_{i+3}\right) \times \right.$$
$$\left. \times \left(\frac{1}{m_C} p_{q_i} - \frac{M}{m_A m_{BC}} p_{q_{i+3}}\right) \right\}^2 + U_{AC},$$

here

$$E_{vJ}^{\mathrm{AC}} = E_{\mathrm{vib}}^{\mathrm{AC}} + E_{\mathrm{rot}}^{\mathrm{AC}}.$$

1.4. Determination of rate constants

Since the method of calculation of the rate constants of the reaction (1) is the same for both channels, we shall restrict our consideration to one channel with the formation of a molecule AB(v, J).

Lets deduce the general formula for the rate constant k(v@0, J@0; v, J) of the reaction

$$\mathrm{A} + \mathrm{BC}\ (v_0,\ J_0) \rightarrow \mathrm{AB}(v,\ J) + \mathrm{C}.$$

First of all, the total reaction rate (which we denote as R(v@0, J@0, v, J)) is determined, obviously, by the effective number of collisions between A and BC(v@0, J@0) going on in the unit volume of the reaction vessel in a unit time and resulting in the formation of molecules AB(v,J). The rate constant of this process is related with the total reaction rate by the expression

$$R = k\bar{n}_{\mathrm{A}}\bar{n}_{\mathrm{BC}},$$

where n@A and n@BC are densities of corresponding particles in the unit volume.

If we denote the distribution function of molecules BC over vibration-rotation states v@0, J@0 by F@BC(v@0, J@0), then the quantity n@BC(v@0, J@0) = $\bar{\text{n}}$@BCF@BC(v@0, J@0) determines the density molecules BC in the specified quantum state.

Let us denote the distribution functions of atoms A and molecules BC over translational motion velocities by f@A(V@A) and f@BC(V@BC), respectively. Then the expressions @dnA and @dnBC give us the numbers of atoms A and molecules BC(v@0, J@0) having the traslational motion velocities in the intervals V@A,

V@A + dV@A and V@BC, V@BC + dV@BC, therewith @dVi. The velocity of the relative motion of atom A and BC molecule is denoted by V@R = V@A - V@BC.

First, let us determine thenumber of collisions z_1 experienced per a second by BC(v@0, J@0) molecule having the translational velocity in the interval V@BC, V@BC + dV@BC and colliding with A atoms having velocities in the interval V@A, V@A + dV@A and flying relative BC at the aiming parameters in the interval b, b + db. Obviously, this number is equal to the number of A atoms located in a given moment in the volume confined between two cylinders with radii b and b + db and the height equal to the velocity of relative movement of A and BC: V@R = @V@A - V@BC (see Fig. 4):

$$z_1 = dn_{\mathrm{A}} 2\pi b db V_R.$$

As a result, the total number of collisions (in the unit volume) of all A atoms (with aiming parameters in the interval b, b + db and velocities in the interval V@A, V@A + dV@A) with all BC molecules (having velocities in the interval V@BC, V@BC + dV@BC) proves to be equal to

$$z = z_1 dn_{\mathrm{BC}}(v_0, J_0) = 2\pi b db V_{\mathrm{R}} dn_{\mathrm{A}} dn_{\mathrm{BC}}(v_0, J_0).$$

Lets assume that P@r characterizes the possibility of molecule AB formation at the specified quantum state (v, J) per a collision between A and BC(v@0, J@0). Then the effective number of collisions z resultig in the reaction with the formation of AB(v, J), is z@r.

Integrating this expression (considering the formula for z) over all possible values of velocities and aiming parameters, we find the total rate of the reaction with formation of AB molecule:

$$R(v_0, J_0; v, J) = 2\pi \bar{n}_{\mathrm{A}} \bar{n}_{\mathrm{BC}} F_{\mathrm{BC}}(v_0, J_0) \int_0^\infty d\mathbf{V}_{\mathrm{A}} \int_0^\infty d\mathbf{V}_{\mathrm{BC}} \int_0^\infty b\, db\, [V_R f_{\mathrm{A}}(\mathbf{V}_{\mathrm{A}}) \times$$
$$\times f_{\mathrm{BC}}(\mathbf{V}_{\mathrm{BC}}) P_r(v_0, J_0; v, J; V_R, b)].$$

Hence, according to the definition of the rate constant for a given process we obtain

$$k(v_0, J_0; v, J) = 2\pi F_{\mathrm{BC}}(v_0, J_0)\int_0^\infty d\mathbf{V}_{\mathrm{A}}\int_0^\infty d\mathbf{V}_{\mathrm{BC}} \times$$

$$\times \int_0^\infty b\,db\,\{V_R f_{\mathrm{A}}(\mathbf{V}_{\mathrm{A}})\,f_{\mathrm{BC}}(\mathbf{V}_{\mathrm{BC}})\,P_r(v_0, J_0; v, J; V_R, b)\}.$$

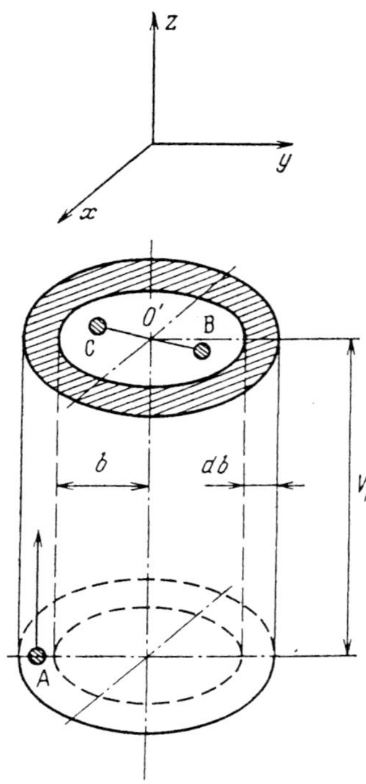

Fig.4. On determination of the number of collisions between A and BC in the unit volume

Introducing the effective cross section of the reaction by the expression

$$\sigma(v_0, J_0; v, J; V_R) = 2\pi\int_0^\infty P_r(v_0, J_0; v, J; V_R, b)\,b\,db, \tag{25}$$

We rewrite the expression for k in the form

$$k(v_0, J_0; v, J) = F_{BC}(v_0, J_0) \int_0^{\infty} d\mathbf{V}_A \int_0^{\infty} d\mathbf{V}_{BC} \times \tag{26}$$

$$\times [V_R f_A(\mathbf{V}_A) f_{BC}(\mathbf{V}_{BC}) \sigma(v_0, J_0; v, J; V_R)].$$

The total rate constant @k(v, J) independent on the initial values ofnumbers v@0 and J@0 is determined by summation of (26) over all possible values of v@0 and J@0, hence,

$$@k(v, J) = \sum_{v_0, J_0} k(v_0, J_0; v. J).$$

Note that for the thermal equilibrium conditions f@A(V@A) and f@BC(V@BC) are the Maxwell distribution functions over translational motion velocities, i.e.

$$f_i(\mathbf{V}_i) d\mathbf{V}_i = \left(\frac{m_i}{2\pi kT}\right)^{3/2} \exp\left(-\frac{m_i V_i^2}{2kT}\right) d\mathbf{V}_i,$$

where i = A, BC; m@BC = m@BC + m@C, T is the gas temperature.

Hence, for the thermal equilibrium conditions (which are commonly realzed in practice) the relation (26) becomes

$$k(v_0, J_0; v, J) = F_{BC}(v_0, J_0) \cdot \left(\frac{m_A m_{BC}}{4\pi^2 k^2 T^2}\right)^{3/2} \int_0^{\infty} d\mathbf{V}_A \exp\left(-\frac{m_A V_A^2}{2kT}\right) \int_0^{\infty} d\mathbf{V}_{BC} V_R \times$$

$$\times \sigma(v_0, J_0; v, J; V_R) \exp\left(-\frac{m_{BC} V_{BC}^2}{2kT}\right).$$

Let us substitute variables V@A, V@BC for V@R, V@*, where V@R = V@A - V@BC is the velocity of the relative movement of A and BC and V@* = (m@AV@A + m@BCV@BC)/M is velocity of the center of mass of three-particle system (M = m@A + m@BC), then

$$\mathbf{V}_A = \mathbf{V}_* + \frac{m_{BC}}{M} \mathbf{V}_R$$

and

$$\mathbf{V}_{BC} = V_* - \frac{m_A}{M} \mathbf{V}_R.$$

After simple transformations we obtain

$$k(v_0, J_0; v, J) = F_{BC}(v_0, J_0)\cdot\left(\frac{m_A m_{BC}}{4\pi^2 k^2 T^2}\right)^{3/2} \int_0^\infty d\mathbf{V}_* \exp\left(-\frac{MV_*^2}{2kT}\right) \int_0^\infty d\mathbf{V}_R \times$$
$$\times V_R \sigma(v_0, J_0; v, J; V_R) \exp\left(-\frac{\mu_{A,BC} V_R^2}{2kT}\right),$$

where $\mu_{A,BC} = m_A m_{BC}/M$ is the reduced mass of the system of colliding particles.

Introducing spherical coordinates, we get

$$d\mathbf{V}_* = V_*^2 \sin\theta \, d\varphi \, d\theta \, dV_*$$

and

$$d\mathbf{V}_R = V_R^2 \sin\theta \, d\varphi \, d\theta \, dV_R.$$

Besides, since

$$\int_0^{2\pi} d\varphi \int_0^{\pi} \sin\theta \, d\theta = 4\pi$$

and

$$\int_0^\infty \exp\left(-\frac{MV_*^2}{2kT}\right) V_*^2 \, dV_* = \frac{\sqrt{\pi}}{4\,(M/2kT)^{3/2}},$$

then

$$k(v_0, J_0; v, J) = \sqrt{\frac{\pi}{2}}\, F_{BC}(v_0, J_0)\cdot\left(\frac{\mu_{A,BC}}{kT}\right)^{3/2} \int_0^\infty \exp\left(-\frac{\mu_{A,BC} V_R^2}{2kT}\right) \times$$
$$\times \sigma(v_0, J_0; v, J; V_R)\, V_R^3 \, dV_R.$$

Going, at last, from velocity V_R to the kinetic energy E_R of the relative motion of A and BC, we obtain the final expression for the reaction rate constant

$$k(v_0, J_0; v, J) = \left(\frac{2}{kT}\right)^{3/2} \frac{1}{\sqrt{\pi\mu_{A,BC}}} F_{BC}(v_0, J_0) \int_0^\infty \exp\left(-\frac{E_R}{kT}\right) \times \tag{27}$$
$$\times \sigma(v_0, J_0; v, J; E_R)\, E_R \, dE_R.$$

Note that in practice the range of integration over the energy is substituted by some finite region(E_{min},E_{max}) whose bounds are determined, as a rule, on the basis of the trajectory calculations (E_{min} characterizes usually the activation energy of the corresponding reaction because at $E < E_{min}$ reactive collisions are absent).

According to (27), the calculation of thereaction rate is reduced in fact to the problem ofdetermination of the reaction cross section $\sigma(v_0, J_0; v, J; E_R)$, which is determined from the trajectory calculations.

According to (25), to calculate $\sigma(v_0, J_0; v, J; E_R)$ one have to know the probability $P_r(v_0, J_0; v, J; E_R, b)$ of the reaction for various values of E_R and b. In principle, the reaction probability may be determined on the basis of the analysis of the great number of trajectory calculations at fixed values of E_R and b. But it is obvious that for every individual trajectory the probability will also depends on the initial conditions specified for a certain trajectory, i.e. on the angles φ, θ, η and ψ, determining the mutual orientation of particles in the space. In this case the reaction probability $P_r(E_R, b, \varphi, \theta, \eta, \psi)$ for every individual trajectory is equal to unity or zero (depending on whether or not the reaction with the AB(v, J) formation has occured at the collision of particles). Because the reaction probability in (25) is averaged over all possible orientatons of particles, we have to assume that

$$(28) \quad P_r(v_0, J_0; v, J; E_R, b) = \frac{1}{2\pi}\int_0^{2\pi} d\varphi \frac{1}{2}\int_0^{\pi} \sin\theta\, d\theta \frac{1}{2\pi}\int_0^{2\pi} d\eta \frac{1}{2\pi}\int_0^{2\pi} d\psi\, P_r(v_0, J_0; v, J; E_R, b, \varphi, \theta, \eta, \psi).$$

The calculation of the reaction probability by formula (28) is convenient to perform by the Monte-Carlo method (see, for example, [11]). The essence of the method consists in the following.

Let us substitute the variables:

$$\xi_\varphi = \varphi/2\pi, \quad \xi_\theta = \cos\theta, \quad \xi_\eta = \eta/2\pi \text{ и } \xi_\psi = \psi/2\pi.$$

Considering this, (28) can be transformed to the form

$$P_r(v_0, J_0; v, J; E_R, b) = \frac{1}{2}\int_0^1 d\xi_\varphi \int_{-1}^1 d\xi_\theta \int_0^1 d\xi_\eta \int_0^1 d\xi_\psi P_r(v_0, J_0; v, J; E_R, b, \xi_\varphi, \xi_\theta, \xi_\eta, \xi_\psi).$$

Considering ξ_φ, ξ_η, ξ_ψ and ξ_θ to be random numbers uniformly distributed on the intervals (0, 1) and (-1, +1), the value of the multiple integral in the right part of the latter expression may be defined by the relation

$$P_r(v_0, J_0; v, J; E_R, b) \simeq \frac{1}{N}\sum_{i=1}^{N} P_{r_i}(v_0, J_0; v, J; E_R, b, \xi_{\varphi_i}, \xi_{\theta_i}, \xi_{\eta_i}, \xi_{\psi_i}) = N_r/N, \quad (29)$$

where N is the total number of trajectories calculated at fixed values of E@R and b, and N@r is the number of the reactive collisions resulting in the AB(v, J) formation.

Note that it is convenient to consider the angles ξ_φ, ξ_η, ξ_ψ and ξ_θ themselves to be random rather than φ, η, ψ and θ. In this case, when specifying the initial conditions (at the integration of equations of motion) one should assume for angles φ, η and ψ the uniform distributions in the (0, 2π) interval and for cos θ - the uniform distribution in the (-1, +1) interval.

The similar method is used to determine the effective cross section of the reaction by the formula (25). In practice the upper limit of integration over the aiming parameter is substituted for some finite value b@max at which the reaction probability becomes zero. As a rule, the particular value of b@max depends on the form of the potential energy surface and is determined from the trajectory calculations.

Considering in (25) the aiming parameter b to be a random number and assuming that b@2 has the uniform distribution in the (0, b@2max) interval, we get for the effective cross section $\sigma(v_0, J_0; v, J; E_R)$ similarly to (29):

$$\sigma(v_0, J_0; v, J; E_R) = \frac{\pi b^2_{\max}}{N_b} \sum_{j=1}^{N_b} P_{rj}(v_0, J_0; v, J; E_R, b_j),$$

where N_b is the number of random points in the $(0, b^2_{max})$ interval.

In principle, the Monte-Carlo method may also be used to determie the reaction rate constant from (27). However, in the latter case it is more appropriate to use one of the standard methods of the numerical integration. We have calculated the integral in (27) using the Simpson formula.

Let us consider again the formula (27) which involves the expression for the distribution function $F_{BC}(v_0, J_0)$ over the vibration-rotation states of BC molecule. In the case of thermal equilibrium conditions this function is

$$F_{BC}(v_0, J_0) = \frac{1}{Q} q_{J_0}(2J_0 + 1)\exp\left(-\frac{E_{v_0J_0}}{kT}\right),$$

where

$$Q = \sum_{v_0, J_0} q_{J_0}(2J_0 + 1)\exp\left(-\frac{E_{v_0J_0}}{kT}\right)$$

is the statistical sum of the vibration-rotation states, q_{J0} is the factor caused by the nuclear spin of molecule because of the requirement that the wave function has to be symmetric (specifically, for the heteronuclear molecules q_{J0} is always equal to 1), E_{v0J0} is the vibration-rotation energy of a given state.

If the energy of the rotational quantum of molecule is low (compared to the average thermal motion energy), the effective cross section of the reaction, as a rule, does not depend (or depends only slightly) on the particular value of the rotational quantum number. Hence, if one neglects the vibration-rotation interaction (which is rather weak in this case), the distribution function $F_{BC}(v_0, J_0)$ may be presented as a product of two independent normalized functions

(30) $$F_{BC}(v_0,\ J_0) = F_{vib}(v_0)F_{rot}(J_0),$$

where F@vib(v@0) and F@rot(L@0) characterize vibrational and rotational distributions of BC molecules, respectively. To determine the total rate constant k(v, J) of AB(v, J) molecules formation, one have to sum (27) over all possible initial values of v@0 and J@0. Assuming that the effective cross section of the reaction does not depend on J@0, we get for the total rate constant according to (30) the expression

$$k(v,\ J) = \sum_{v_0, J_0} \cdot k(v_0,\ J_0;\ v,\ J) \simeq \left(\frac{2}{kT}\right)^{3/2} \frac{1}{\sqrt{\pi\mu_{A,BC}}} \sum_{v_0} F_{vib}(v_0) \int_0^\infty \exp\left(-\frac{E_R}{kT}\right) \times$$
$$\times\ \sigma(v_0;\ v,\ J;\ E_R)\, E_R\, dE_R.$$

In particular, if the oscillation quantum energy of molecule BC is rather great (compared to the average thermal motion energy) then at the thermal eqiilibrium condiions vibrational levels with v@0 > 0 are virtually not occupied and one can assume that F@vib(0) = 1 and F@vib(v@0 > 0) = 0. In the latter case the summation over v@0 is found to be unnecessary, and the total rate constant is determined by the most simple relation

(31) $$k(v,\ J) \simeq \left(\frac{2}{kT}\right)^{3/2} \frac{1}{\sqrt{\pi\mu_{A,BC}}} \int_0^\infty \exp\left(-\frac{E_R}{kT}\right) \sigma(v,\ J;\ E_R)\, E_R\, dE_R.$$

2. STUDY OF TWO-CHANNEL REACTIONS INVOLVING HYDROGEN (DEUTERIUM) ATOMS AND CHLORINE MONOBROMIDE

In this section we shall dwell on results of the trajectory calculations for the case of two-channel reaction (2) and also the reaction which occures at the substitution of a hydrogen atom for its more heavy isotope, deuterium atom:

(32) $$D + BrCl\,(v_0,\ J_0) \rightarrow \begin{cases} DCl\,(v,\ J) + Br, \\ DBr\,(v,\ J) + Cl. \end{cases}$$

The values of spectroscopic constants and energies of some bonds for the reactions involved are presented in Table 1.

Notice that the low value of the rotational constant of BrCl molecule results in the weak influence of the rotational quantum number J@0 on the character of distribution offinal products both in the case of reaction (2) and reaction (32). The influence of the oscillation quantum number v@0 is more essential according to calculations and so, in general, one have to consider the real distribution of BrCl molecules over vibrational states (at a given gas temperature T). Nevertheless, because the averaging of rate constants over all possible values of v@0 has required too many trajectory calculations, we did not carry out such averaging.

molecular constant, \ bond,	$^{1}H-^{79}Br$	$^{79}Br-^{35}Cl$	$^{1}H-^{35}Cl$	$^{2}D-^{79}Br$	$^{2}D-^{35}Cl$
R^{0}_{ij}, Å	1.415	2.136	1.2746	1.4115	1.2744
α_{ij}, Å^{-1}	1.812	1.973	1.870	1.83	1.85
D_{ij}, Cal/mol,	90.30	52.07	106.36	90.39	106.4
ω_e, cм$^{-1}$	2649.68	444,276	2990.95	1885.33	2144.77
$\omega_e x_e$, cm@-1	45.52	1,843	52.818	22.73	26.92
B_e, cм$^{-1}$	8.469	0,1526	10.591	4,34	5.55

Table 1. Spectroscopic and molecular constants for reactions (2) and (32)

Note that the integration of the equations of motion was performed by the Kutta-Merson method of 5-th precision order with automatic choosing of the integration step. The precision of integration was controlled by the conservation law for the system total energy and, as a rule, was about 10@-3 to 10@-4.

When calculating the reaction rateconstants the integration in (31) was performed by the Simpson formula (13 to 15 points over E@R were used), the integrals in (25) were calculated by 15 random values of the aiming parameter b, while the multiple integrals in (28) were determined by 120 random values of sets of angles and

The minimum value of the relative translational energy E@R of reagents (at which the individual reactive collisions in one of the channels for both reactions are still observed) at the interaction potential used by us was about 2 Cal/mole. In general, this value slightly exceeds the value of the activation energy E@a = 1@0.5 kcal/mol of the reaction (2) estimated in [12] on the basis of measurements of the distribution of products over vibrational and rotational states. However, taking into account that there are no other data on E@a values for reactions (2) and (32), such correspondence may be considered as rather satisfactory. Ror this reason the interaction potential was not fitted in the calculations (the fitting parameters were assumed to be equal to K@ij = 0).

The maximum value of the aiming parameter (estimated on the basis of trajectory calculations) was b@max = 3,4@ for the reaction (2) and b@max = 3,4@ for thereaction (32). The interval of the relative translational motion energies studied was 2.3 to 9 kcal/mole in the case of reaction (2) and 2 to 10 kcal/mole for the reaction (32).

Fig.5 shows the typical picture of the reaction (2) dynamics which was obtained from the analysis of calculations of some individual trajectories. Note that Fig.5 presents the case of the nonreactive collision when the A atom and BC molecule bounce apart. Figs. 5,b and 5,c shows the cases corresponding to the reactive collisions with the formation of HBr molecule (Fig.5,b) and HCl molecule (Fig. 5,c). In two latter cases one can easily trace the formation of the intermediatethree-particle complex and the subsequent evolution from it of a new molecular structure at the stage when particles are flying apart. In addition, one can see that with the formation of a new molecule the vibrational frequency changes.

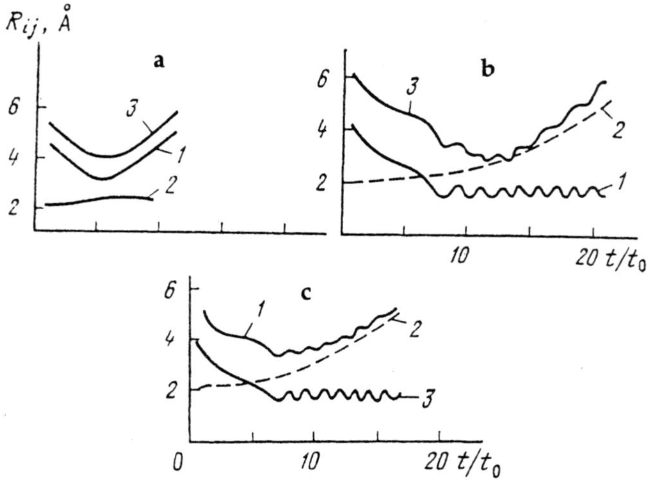

Fig. 5. Time dependence of the internuclear distances R_{ij} along the trajectory of movement of colliding particles A and BC ($t_0 = 10^{-14}$ sec; 1 - R_{12}, 2 - R_{23}, 3 - R_{13})

a: nonreactive collision of H and BrCl; b: reactive collision with the formation of HBr molecule; c: reactive collision with the formation of HCl molecule

Now let us dwell on some characteristic calculation results obtained for each above-mentioned reaction.

2.1. The reaction involving hydrogen atom

Fig. 6 shows temperature dependencies of the total rate constants of the reaction (2) in each channel. Note that the calculations give a somewhat higher value of the rate constant of HCl molecule formation, however, one can see that in fact already at temperatures T > 500 K the rate constants in both channels become practically equal. The approximation of the calculation results by the least square method results in the following type of temperature dependence of rate constants (in the interval T = 100 - 1600 K):

$$k_{HBr} = 4{,}45 \cdot 10^{13} \cdot T^{0{,}17} \cdot \exp(-2670/RT) \text{ см}^3/(\text{ mol. c})$$

and

$$k_{HCl} = 6{,}16 \cdot 10^{13} \cdot T^{0{,}15} \cdot \exp(-2490/RT) \text{ см}^3/(\text{ mol. c}).$$

The results obtained here differ somewhat from the results of the paper [12] where it has been shown that the ratio of therate constants k@HCl/k@HBr = 0.4 (from relations given above it follows that k@HCl/k@HBr < 1,5). One can note, however, that according to the trajectory calculations the formation of HBr molecules becomes preferential as the oscillation quantum number v@0 increases. Therefore, it is possible that the consideration of the real distribution of BrCl molecules over vibrational states at fixed temperature would lead to a better agreement with the experiment.

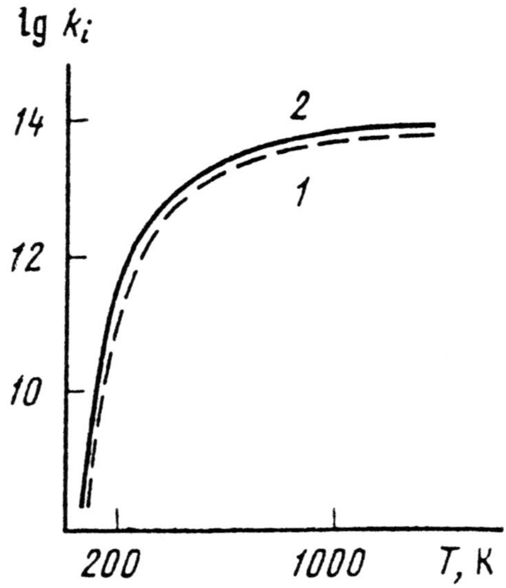

Fig. 6. Temperature dependencies of the rate constants of the reaction (2) for channels with formation of HBr (1) and HCl (2)

Fig.7, a, b shows the molecular products distribution over vibrational levels in each of two channels of the reaction (2). In this figure the experimental results from [12] are also shown by dotted line. Here one can point out a rather satisfactory agreement between calculation results and the experiment, especially, for the channel with HCl molecules formation.

Really, according to [12] the total released energy share localized in the HBr molecules vibrations amounts to 58 % and in the HCl molecules vibrations to 55 %. The calculation at T = 300 K for these cases gives 45 % and 59, respectively. From calculations it follows that the average number of excited quanta @ per a HBr molecule is equal to 2.4 and per a HCl molecule ε = 4.16.

The analysis of HBr and HCl molecules distribution over rotational sublevels showed essentially greater disagreement between calculations and the experiment. Thus, according to [12] the fraction of energy exciting rotational states of HBr and HCl is equal to 12 and 9%, respectively. However, the calculations for these two cases gives only 3 and 2 %, respectively. And though in a pure qualitative respect the character of the molecule distribution over rotational states in both cases is approximately the same (the shift of the spectrum of excited rotational states into the region of higher J values with decreasing of the number v of the vibrational level), nevertheless, the experimentally measured fraction of energy conversing to the rotation of molecular products is four times greater than the energy fraction obtained from the trajectory calculations (see Figs. 8, 9).

The essential disagreement with the experiment in molecules distribution over the rotational states is probably associated with a lack of knowledge of the interaction potential details, especially at rather great distances (the high rotational states are mainly populated due to reactive collisions which occure at great values of the aiming parameter).

Also it is interesting to note that experimentally measured distribution of HCl molecules over the rotational states on vibrational levels with v = 1 -3 has a well-defined bimodal character. Such peculiarity is explained by the existence of two different paths of the reaction with formation of HCl when the hydrogen atom approaches BrCl molecule from different sides. The calculations showed that at v = 2 such bimodal distribution is, in principle, possible (Fig. 9). However, the similar phenomenon has not been observed at othher values of v.

2.2. The reaction involving deuterium atom

Fig. 10 shows the influence of the isotopic effect on the character of distribution ofmolecular products over vibrational levels. For comparision we also show in this figure by the dotted line the distributions for the light isotope, hydrogen atom.

One can see that in the case of a heavy isotope the maximum of the distribution function in both reaction channels shifts into the region of higher values of the oscillation quantum number. Table 2 presents relative fractions of energy releasing in the form of vibrational and rotational energy of the molecular products and also average numbers of vibrational quanta εi per a molecule formed.

The approximation of calculation results for the temperature dependences of the reaction (32) rate constants in the temperature interval T = 100 - 1600 K by the least squares method gives the following form of the temperature dependences of rate constants:

$$k_{\mathrm{DBr}} = 1{,}3 \cdot 10^{12} \cdot T^{0{,}57} \cdot \exp\left(-2363/RT\right) \quad \text{см}^3/(\text{mol. c})$$

and

$$k_{\mathrm{DCl}} = 4{,}3 \cdot 10^{12} \cdot T^{0{,}46} \cdot \exp\left(-2075/RT\right) \quad \text{см}^3/(\text{mol. c}).$$

A correlation with the reaction (2) shows that for more heavy isotope the reaction rate constants in both channels decrease about 2 - 3 times (at T = 300 K). As the temperature increases this difference decreases.

The rotational distributions of DBr and DCl molecules have practically the Boltzmann character (with the temperature corresponding to the energy stored in the form of vibrational energy of molecular products). The bimodal distribution over rotational sublevels was not observed.

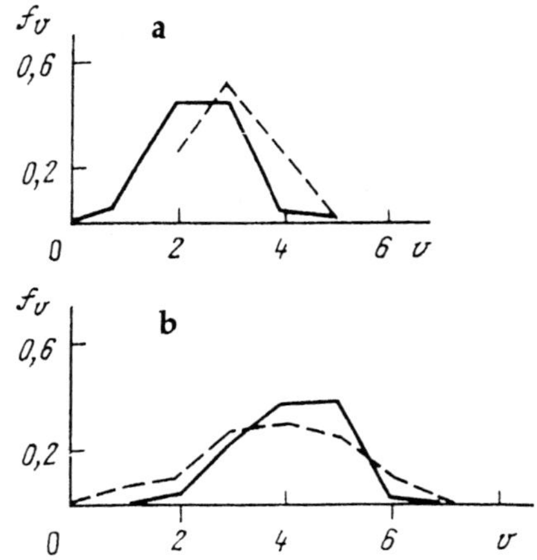

Fig. 7. Distribution over the vibrational levels of HBr (a) and HCl (b) molecules at T = 300 K (solid lines: calculations, dashed lines: the experiment [12])

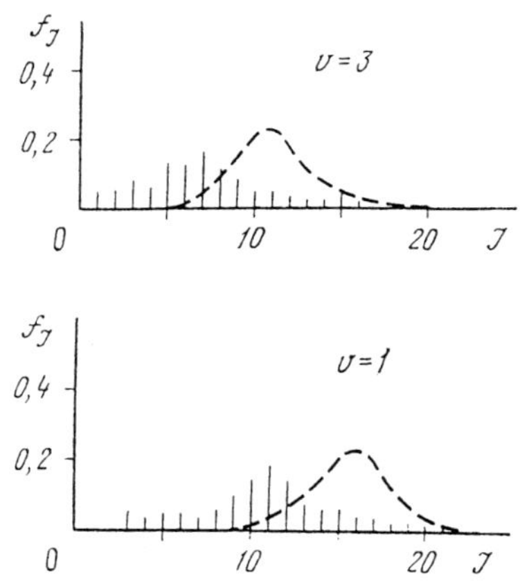

Fig. 8. Distribution of HBr molecules over rotational states in the reaction (2) at @ = 3 and 1 (experimental results [12]are presented by a dashed line [12])

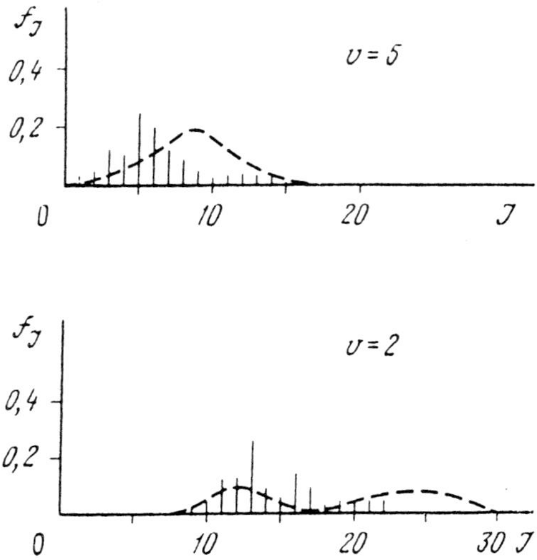

Fig.9. Distribution of HCl molecules over rotational states in the reaction (2) at @= 5 and 2 (experimental results [12] are presented by a dashed line)

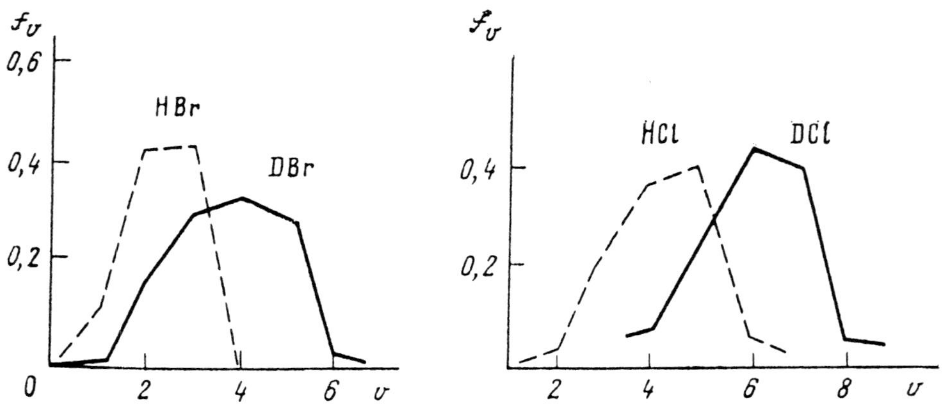

Fig.10. Distribution of the molecular products among vibrational levels in reactions (2) (dashed lines) and (32) (solid lines) at T = 300 K

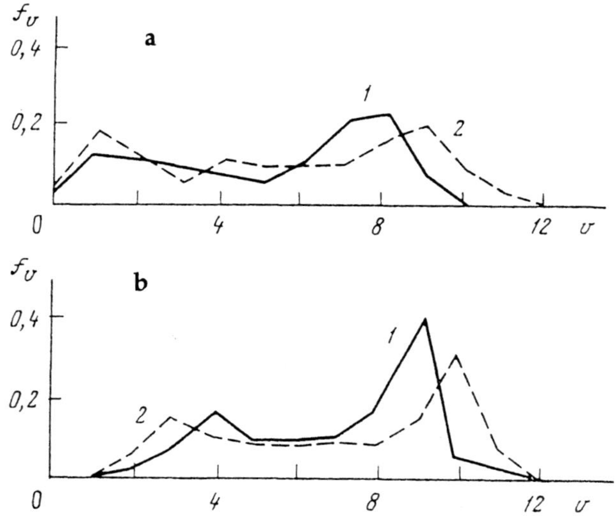

Fig.11. Bimodal character of the distribution of DBr (a) and DCL (b) molecules over vibrational levels at the vibrational excitation of BrCl molecules (T = 300 K, J@0 = 50, v@0= 1 (curves 1), 2 (curves 2))

The strong influence of the vibrational excitation of BrCl molecules on the character of the molecular products distribution over the vibrational states engages our attention. The calculations showed that at the low vibrational excitation (v_0 = 1 - 2) maxima of the distribution functions both for DBr and DCl are displaced into the regionbf higher vibrational numbers. This is explained by the fact that with v_0 increasing the intermediate three-particle complex begins to decompose faster and therewith a large share of the energy is left in the vibrations of the molecular products. With further increasing of BrCl vibrational excitation the chatacter of the distribution changes. Specifically, the bimodal distribution appears of the molecular products over vibrational levels (Fig. 11, a, b). The qualitative explanation of this phenomenon is similar to the case of the bimodal distribution över rotational sublevels: with v_0 increasing two reaction paths appear because a deuterium atom may approach a BrCl molecule from both sides. Not ice that in the case of a hydrogene atom the significantly higher vibrational excitation of BrCl is needed to observe the same effect. The calculations also show that at $v_0 > 10$ the bimodal character of distribution in thereaction (32) disappears transforming to the practically uniform distribution of the molecularproducts over vibrational levels.

parameter of the energy distribution \ molecular product	HBr	DBr	HCl	DCl
η_{vib}, %	45	47	59	62
η_{rot}, %	3	5	2	3
ε_i	2,4	3,8	4,16	6,06

Table 2. Energy distribution among the products in reactions (2) and (32)

3. TWO-CHANNEL REACTIONS INVOLVING HYDROGEN (DEUTERIUM) ATOM AND FLUORINE MONOCHLORIDE

Let us consider now thetwo-channel reaction (3) and the reaction involving instead of hydrogen atom its heavy isotope, deuterium:

$$\text{D} + \text{ClF}(v_0, J_0) \rightarrow \begin{cases} \text{DCl}(v, J) + \text{F}, \\ \text{DF}(v, J) + \text{Cl}. \end{cases} \qquad (33)$$

Table 3 gives the values of spectroscopic and molecular constants for this case.

In both cases the calculations were performed for 60 sets of randomly chosen orientational angles of the initial three-particle system and the phase of vibrations of ClF molecule, for 9 randomly chosen values of the aiming parameter (b = 0 - 2,5 Å and for 9 values of the energy of the reagents relative motion (in the case of the reaction (3)E@R = 1,5 - 6 kcal/mole, in the case of the reaction (33) E@R = 2,5 - 6,5 kcal/mole). The rotational and vibrational numbers of ClF reagent molecule were assumed to be constant at calculations (J@0 = 15 and v@0 = 0, respectively). The total number of trajectories studied for each reaction mentioned was 60 x 9 x 9 = 4860.

Table 4 gives the calculated temperature dependences of the reaction (3) rate constants. It also presents experimental values of these constants from [13, 14]. It follows from the table that the agreement between calculations and the experiment may be considered satisfactory.

molecular \ bond	$^1H-^{19}F$	$^1H-^{35}Cl$	$^{35}Cl-^{19}F$	$^2D-^{19}F$	$^2D-^{35}Cl$
R^0_{ij}, Å	0,9168	1,2746	1,6281	0,9171	1,2744
α_{ij}, Å^{-1}	2,232	1,870	2,320	2,22	1,85
D_{ij},kcal/mole	139,8	106,0	60,1	140,5	106,4
ω_e, см$^{-1}$	4141,03	2990,95	787,5	3001,0	2144,77
$\omega_e x_e$, cm@-1	90,439	52,818	7,0	47,97	26,92
B_e, см$^{-1}$	20,949	10,591	0,517	11,1	5,55

Table 3. Spectroscopic and molecular constants for reactions (3) and (33)

method and conditions of the determination \ rate constant	calculation			experiment
	T, K			
	300	500	900	300—400
k_{HCl}, cm@3/(mole sec)	$2\cdot10^{12}$	10^{13}	$2,4\cdot10^{13}$	$2.2\cdot10^{12}$
k_{HCl}/k_{HF}	7,8	5,9	5,3	6,1

Table 4. Calculated and experimental values of the reaction (3) rate constants at various temperatures.

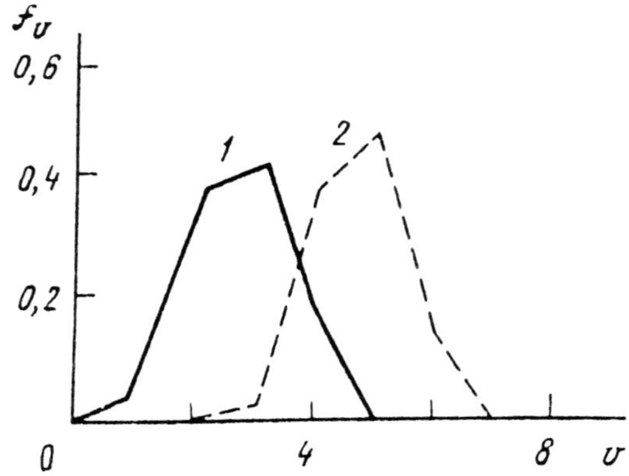

Fig. 12. Relative distributions of the molecular products among vibrational levels in the reaction (3) at T = 400 K (1 - HCl, 2 - HF)

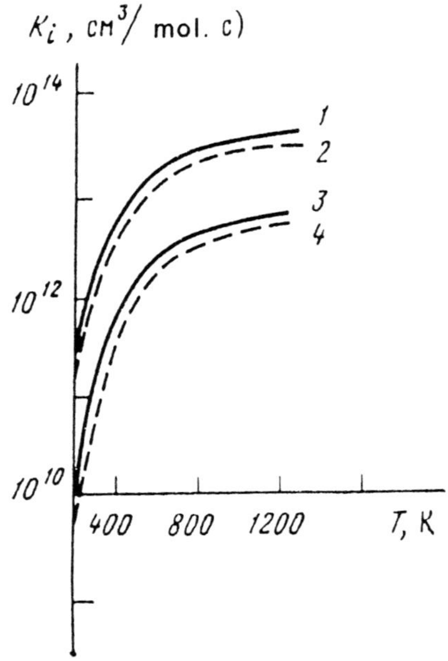

Fig. 13. Temperature dependences of rate constants of the reaction (3) (solid lines) and (33) (dashed lines)
line) 1 -k@HCl, 2 - k@DCl, 3 -k@HF, 4 - k@DF

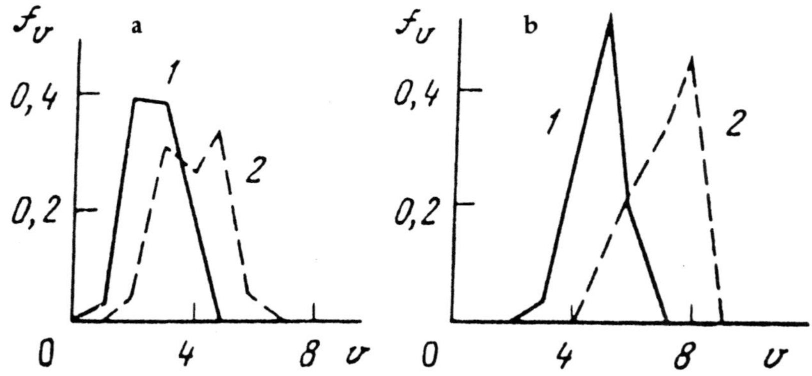

Fig. 14. Distributions of the molecular products in the reaction (3) (solid line) and (33) (dashed line) among vibrational levels at T = 400 K

a: 1 - HCl, 2 - DCl; b: 1 - HF, 2 - DF

The approximation of the calculation results gives the following temperature dependences for the reaction (3) rate constants in the tempetature interval 200 - 1000 K:

$$k_{HCl} = 2.8 \cdot 10^{16} \cdot T^{-0,79} \cdot \exp(-1510/T) \text{ см}^3/(\text{mol. c})$$

and

$$k_{HF} = 4.4 \cdot 10^{16} \cdot T^{-1,05} \cdot \exp(-1820/T) \text{ см}^3/(\text{mol. c}).$$

The calculations showed that in the case considered the influence of the rotational energy of ClF reagent is rather weak as well (special calculations were performed for J@0 = 9, 15, and 20). For this reason there was no need to average the results over J@0. The influence of the ClF vibrational energy was more noticeable (the calculations were performed for @0 = 0, 1, and 2). On the other hand, the share of the excited ClF molecules is rather small at the temperatures below 1000 K, and, hence, the excited states of ClF give a small contribution to the calculation results.

temperature / the ratio of constants	T, K		
	200	600	1200
k_{HCl}/k_{DCl}	1,31	1,30	1,29
k_{HF}/k_{DF}	1,41	1,34	1,31

Table 5. The ratio of the rate constants for reactions (3) and (33) at various temperatures

Fig. 12 shows the distribution of reaction (3) products among vibrational states. In the average in the course of the reaction 2,75 excitation quanta are accounted for a HCl oscillator, while for a HF oscillator ε = 4.8. About 47 to 48% of the reaction energy transform into the vibrational energy of HCl molecules, and 2,5 to 4 % transforminto the rotational energy (in the temperature range 300 - 500 K). For the case of HF molecules formation these figures are 58 to 60% and 1 to 2% respectively. Note that these results are nicely correlated with the experimental data in [15].

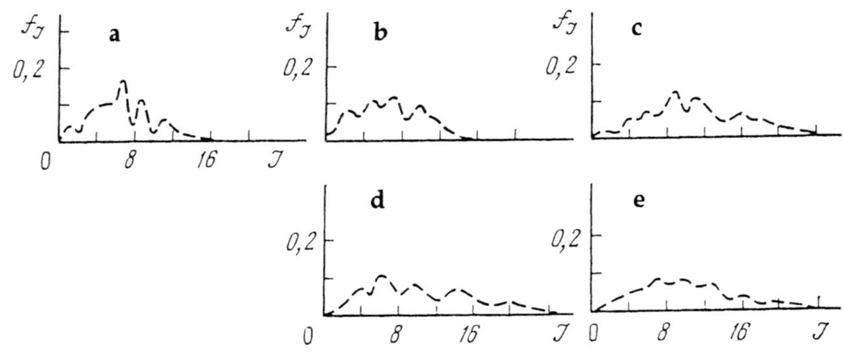

Fig. 15. Distribution of the molecular products of reactions (3) (HCl, a - b) and (33) (DCl, c - e) among rotational sublevels at T = 400 K (v = 2 (a), 3 (b, c), 4 (d), and 5 (e))

The substitution of the atomic hydrogen for the deuterium slightly reduces the values of reaction (33) rate constants in both channels in comparison with reaction (3) (See Table 5 and Fig. 13 where results are presented comparing the calculations for both cases). This is explained by the change in the impacting atom mass. However, the qualitative character of temperature dependences of rate constants does not change.

The most significant difference appears in the distribution of the molecular products over vibrational and rotational levels. To make it obvious Fig. 14 presents the comparison of calculated distributions of molecular products over vibrational levels in reactions (3) and (33) at thegas temperature T = 400 K. One can see that in a given case the influence of the isotopic effect shows up mainly in the fact that the maximum of the distribution function in both channels of the reaction is displaced into the region of higher values of the vibrational quantum number. Table 6 also presents the information on the relative energy fractions (of the value of the total energy released in the reaction course) that are accounted for the vibrational and the rotational degrees of freedom of the molecules formed (T = 400 K). This table also presents average numbers of vibrational quanta accounted for a molecule of each type.

molecular product / distribution parameter	HCl	DCl	HF	DF
η_{vib}, %	47,6	50,0	63,5	67,3
η_{rot}, %	3,5	4,5	0,85	1,85
ε_i	2,75	4,05	4,9	7,28

Table 6. Energy distribution among the products of reactions (3) and (33).

One should note that in principle the share of the vibrational energy η_{vib} decreases, while the share of the rotational energy increases as the temperature increases. Nevertheless, the temperature dependence of η_{vib} is rather weak.

Fig. 15 shows the distribution of the molecular products over rotational sublevels in the reactions mentioned above by the example of HCL and DCl molecules. Since theshare of the energy transforming into the rotational energy in the case of DCl is greater, while the rotational quantum value, on the contrary, is less than for HCl, the spectrum of rotational states populated at the DCl molecules formation turns out to be more wide.

4 .ANALYSIS OF REACTIONS $F + Cl_2 \rightarrow ClF + F$ and $Cl + F_2 \rightarrow ClF + F$

The interest in the study of reactions

$$F + Cl_2(v_0, J_0) \rightarrow ClF(v, J) + Cl \quad (34)$$

and

$$Cl + F_2(v_0, J_0) \rightarrow ClF(v, J) + F \quad (35)$$

first of all arises from the fact that both reactions are involved in the kinetic cycle in a chemical HF - HCl laser [16]. In addition, the reaction (35) is followed by a detectable heat release (the thermal effect of this reaction is $\Delta H = -22$ kcal/mole) and for this reason it may be of its own interest. Finally, there are no specific data for the above-mentioned reactions in the literature (except estimated values of constants [17]) .

molecular constant \ bond,	$^{35}Cl-^{35}Cl$	$^{35}Cl-^{19}F$	$^{29}F-^{19}F$
R^0_{ij}, Å	1,988	1,6281	1,417
α_{ij}, Å^{-1}	2,033	2,32	2,902
D_{ij}, $\frac{kcal}{mol}$	57,9	60,1	38,02
ω_e, см$^{-1}$	564,9	787,5	924
$\omega_e x_e$, см$^{-1}$	4	7	16
B_e, см$^{-1}$	0,2407	0,517	0,882

Table 7. Spectroscopic and molecular constants for reactions (34) and (35).

Table 7 gives spectroscopic and molecular constants for reactions (34) and (35). Note that vibrational and rotational numbers of the molecular reagents (Cl@2 and F@2, respectevly) were assumed to be fixed (v@0 = 0 and J@0 = 25 and 20). Other characteristic parameters have the values: b@max = 2,7 and 2,5 @A, E@min = 5 and 2,5 kcal/mole, and E@max = 10 and 8 kcal/mole (for reactions (34) and (35), respectively).

$$k_{34} = 2{,}71 \cdot 10^{13} \cdot T^{-0{,}0284} \cdot \exp(-5621/RT) \text{ см}^3/(\text{mol. c})$$

and

$$k_{35} = 2{,}485 \cdot 10^{13} \cdot T^{0{,}143} \cdot \exp(-2366/RT) \text{ см}^3/(\text{mol. c}).$$

Fig. 16 shows the calculated temperature dependences of the reaction rate constants. The approximation of these dependencies by the least squares method (in the temperature range 200 to 1400 K) gives

Fig. 17 presents the distribution of ClF molecular products among the vibrational levels in thereaction (35). The calculations show that this distribution proves to be essentially nonequilibrium, and in the course of the reaction the vibrational levels with v = 7 - 11 become mainly populated, while low levels are practically not populated.

In the reaction (34) the vibrational levels with $c\ v > 0$ are practically not occupied due to very low heat release (the thermal effect ΔH = -2 kcal/mole), and the reaction rate is found to be essentially lower while the threshold is twice higher than for the reaction (35)

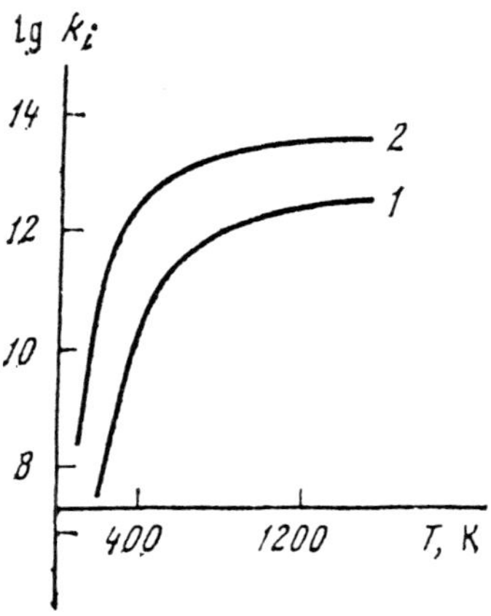

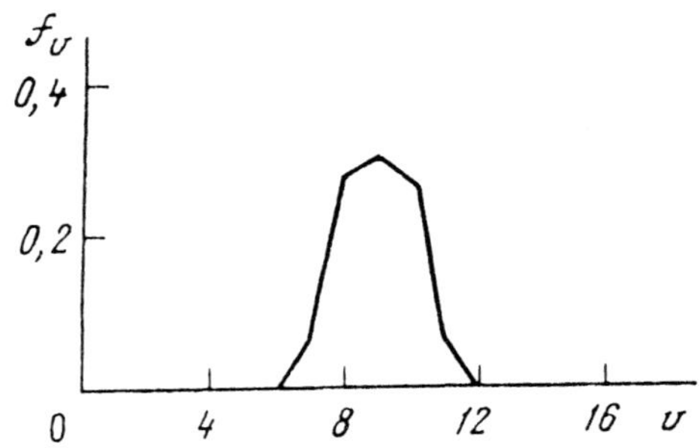

Fig. 16. Temperature dependencies of the rate constants of reactions (34) and (35): 1 - k_{34}, 2 - k_{35}

Fig. 17. Distribution of ClF molecular products among vibrational levels at T = 300 K in the reaction (35).

It is also follows from calculations that in the case of reaction (35) most of the energy (about 90-95%) is released as the vibrational and rotational energy of ClF molecules. For example, at T = 300 K the share of the vibrational energy is η_{vib} = 82,7% (the average number of quanta per a molecule is ε = 8,9) and that of the rotational energy is η_{rot} = 11,9%.

One may note that in this respect the reaction (35) compares well, for example, with the reaction $F + H_2 \rightarrow HF + H$ which is used in chemical continuous HF lasers.

CONCLUSION

Thus, the results presented in this paper show that the calculations of the two-channel elementary chemical processes based on the method of classical trajectories (the Monte-Carlo method) allow to obtain practically complete information on the character of a particular reaction course, to find the type of temperature dependencies of rate constants and the character of the product distribution among translational, rotational, and vibrational degrees of freedom.

Nevertheless, one should note that though the procedure of such calculations is rather simple (in comparison with the quantum-mechanical calculations), the acquisition of the information mentioned above requires tostudy the great number of trajectories (from several thousands to several tens of thousands) and a considerable volume of machine time.

At last, as many authors have shown, the calculations are very sensitive to the type of the potential energy surface, and, hence, the satisfactory agreement with the experiment is achieved, as a rule, by the special fitting of the interaction potential parameters. This, obviously, presents a difficult problem, especially if the experimental data for the reaction involved are absent.

ABSTRACT

The method of classical trajectories (the Monte-Carlo method) is applied for calculation of the rate constants and energy distribution among products in two-channel chemical reactions. The results of calculations are compared with available experimental data.

REFERENCES

[1] M. Karplus, R.N. Porter, and R.D. Sharma, Exchahge reactions with activation energy. 1. Simple barrier potential for (H, H@2), J. Chem. Phys. 43, (1965) 3259.

[2] D. Banker, The method of classical trajectories, In Computational Methods in the Physics of Atmospheric and Molecular Collisions, (Eds. B. Older et al.), Mir, Moscow, (1974) p. 277.

[3] R.N. Porter and L.M. Raff, Classical trajectory methods in molecule collisions, In Dynamic Molecular Collisions, B, N.Y.:L., (1976) 1.

[4] A.A. Stepanov and V.A. Shcheglov, Rate constants of the formation and energy distribution amongproducts in the two-channel reaction H + ClF → HCl + F; HF + Cl, Krat. Soobshch. po Fizike, 11, (1979) 8.

[5] A.A. Stepanov and V.A. Shcheglov, Influence of the isotopic effect on the energy distribution among products of the reaction of hydrogen atoms with chlorine monofluoride molecules, Ibid, 3, (1980) 14.

[6] N.A. Konoplev, A.A. Stepanov, and V.A. Shcheglov, Rate constants of the formation and energy distribution among product in reactions F + Cl_2 → ClF + Cl, Cl + F_2 → ClF + F, Ibid, 5, (1981) 19.

[7] N.A. Konoplev, A.A. Stepanov, and V.A. Shcheglov, Calculation of rate constants and energy distribution among products in the two-channel reaction H + BrCl → HBr(HCl) + Cl(Br) by the method of classical trajectories, Preprint FIAN, 34, Moscow, (1983) 50 p.

[8] N.A. Konoplev, A.A. Stepanov, and V.A. Shcheglov, Calculation of rate constants and energy distribution among products in the two-channel reaction between hydrogen atoms and bromine monochloride by the method of classical trajectories, Khim. Fizika, 3, (1984) 828.

[9] N.A. Konoplev, A.A. Stepanov, and V.A. Shcheglov, Influence of the isotopic effect on the energy distribution among products of the reaction between hydrogen atoms and bromine monochloride molecules, Krat. Soobshch. po Fizike, 2, (1985) 17.

[10] V.N. Kondrat´ev, E.E. Nikitin, A.I. Reznikov, and S.Ya. Umanskii, Thermal bimolecular reactions in gases, Nauka, Moscow, (1976).

[11] B.P. Demidovich and I.A. Maron, The foundations of computing mathematics, Fizmatgiz, Moscow, (1963).

[12] J.C. Polanyi and W.J. Skrlac, Energy distribution among reaction reaction products. 10. H + ICl, HCl + I, H + ClBr, HCl + Br, HBr + Cl, Chem. Phys. 23, (1977) 167.

[13] N.F. Chebotarev, L.I. Trakhtenberg, and S.Ya. Pshezhezkii, Determination of the rate constant of the elementary reaction H + ClF → HCl^* + F, Kvantovaya Elektron. 3, (1976) 2552.

[14] G.P. Zhitneva and S.Ya. Pshezhezkii, Influence of the chain interaction in the reaction of chlorine flouride with hydrogen, Kinetika i kataliz 19, (1978) 296.

[15]D. Brandt and J.C. Polanyi, Energy distribution among reaction products. 11. H + ClF → HF + Cl, HCL + F → HF + Cl, Chem. Phys. 35, (1978) 23.

[16] A.A. Stepanov and V.A. Shcheglov, Continuous chemical lasers on reaction products (review), Kvantovaya Elektron. 9, (1982) 1077.

[17] V.N. Kondrat´ev, Rate constants of reactions in the gase phase, Nauka, Moscow, (1970) 242 p.

APPLICATION OF THE QUANTUM THEORY OF SCATTERING TO THE CALCULATION OF VARIOUS PROCESSES IN NUCLEAR, ATOMIC, AND MOLECULAR PHYSICS

S.A. Pozdneev

INTRODUCTION

The practical demands for the development of new chemical lasers on the atomic and molecular electronic transitions [1-5], tools for plasma diagnostics in Tokamaks and laser nuclear fusion [6-9], the astrophysical investigations [8-10], progress in the plasmo-chemical technology [11, 12], the preparation of new chemical compounds with specified properties and a purposeful search for optimum methods of their synthesis [12], the development of new generations of computers, specifically, neuro-computers [13, 14], the investigations of a structure of atomic nucleus and elementary particles [15, 16], etc., call for new methods and means based on these methods for the calculations of the main characteristics of various elementary processes like the interaction of radiation with matter, collisions of electrons with atoms and molecules, between molecules and nucleons, etc.

For instance, molecular reactions, which produce the excited molecules, can be effectively used in chemical lasers if they have a number of the following salient features [1-5]: (1) the thermal effect and rate constants in these reactions have to be high enough and (2) a considerable part of the energy released must transfer to the energy of the reaction products. It is not surprising, then, that the practical demands for extending the spectral range and increasing the power of chemical lasers have stimulated an active search for such chemical reactions which can not be described accurately and completely on a basis of experimental investigations. Therefore, at present theoretical methods of investigations of elementary processes are attracting considerable attention [16-100].

The main difficulties encountered in calculating cross-sections of various elementary processes in atomic, molecular and nuclear physics are associated with a multidimensionality of the problem and with the fact that the perturbation theory approximations are invalid at the most typical conditions [17-28]. Thus,at present one can distinguish several lines of performing the calculations.

Classical calculations [17-24, 26-30] are based on the Hamilton-Jakobi equations in which the forces of interactions between atoms are determined from the potential energy surfaces (PES) of one kind or another.

The main advantage of classical methods lies in their clearness and simplicity when applied to calculations of averaged characteristics of chemical reactions. These methods, however, have limitations, specifically, pertaining to calculations of polyatomic molecules. In this case one needs to consider the motions of many atoms whereas the method of construction of PES for such molecules is practically unknown [28-33]. Limitations of classical calculations are also associated with nonadiabatic transitions between different PES (due to their quasiintersections), complications of integrations of equations of motion and labor-consuming character of obtaining statistics for calculations of trajectories to determine characteristics of chemical reactions.

In semiclassical calculations [17-35] the quantum problem of determination of nuclear wave-functions (WF) is reduced to semiclassical trajectory equations which have lesser dimension and have in some cases analytical solutions of Landau-Zinner, Nikitin, Demkov types etc.

In quasi-classical calculations [17-24] quantum problems are solved using solutions of equations for classical system and its geometrical and topological characteristics. Thus, a classical method is based on the principle of correspondence between quantun and classical problems. Mathematical basing of quasi-classical method is connected formally with the asymptotic expansions in quantum-mechanical equations in terms of parameter h/S, where h is Planck`s constant and S is an action. The main complexities of the derivation of the quasi-classical approximation in a case of the real three-dimensional scattering lay in the fact that it is impossible to determine the exact quantized classical trajectories and there are principle difficulties in calculations of multidimen- sional cofigurational integrals.Analogous difficulties also emerge in quasi-classical calculations of the energy levels for an arbit- rary potential with nonseparable variables. In this case the only methods to solve the problem are a procedure of construction of quantum tores, which can not be realized effectively with computers, and a method based on the "tracking" of energy on a classical trajectory when a perturbation is switched on adiabatically. The latter does not have a rigorous mathematical foundation, however it is effective at performing specific computer calculations.

Adiabatic approximation [17-35] is one of the most popular approximations in the collision theory and is based on the approximate separation of fast and slow variables in a dynamical system.

The region of the energies of collisions, where the adiabatic approximation is valid, depends on the specific process involved, its lower boundary being determined by the condition of feasibility of classical approximation to a description of nucleus motions while upper boundary is determined by the condition of feasibility of asymptotic expansion in terms of a small parameter which is a velocity of nuclei motion.

This approximation has been applied to the exactly solvable two-level model problems for transitions between bound states and the Demkov-Osherov problem for transitions from a bound state to a continious spectrum. These problems allowed to study various real physical processes and also to develop the foundations of more general asymptotic theory where they appear as an etalon problems (their set far exceeds the set of exactly solvable models).

At present the theory of two-level nonadiabatic transitions is generalized to the case of multilevel transitions and transitions to a complex spectrum [24], which allows to calculate processes for which the model representation is invalid. The main feature of this asymptotic theory of nonadiabatic transitions consists in using local characteristics of terms in the system investigated at complex values of nuclear distance R, which results from the approximate asymptotic solution of dynamical problem.

The system terms at complex values of R are connected with features of the adiabatic basis at real values of R for which the real collisional processes occure. For example, to calculate the probability of transition between two terms one needs to know a location of the complex branching point R_c and a difference between terms on the line connecting R_c with a real axes R, while the branching point always appears on the real axes R as a maximum of the matrix element of interaction between adiabatic states.

Thus, in [24] the most general theory of nonadiabatic transitions has been developed based on the Hamiltonian analyticity in terms of R. The theory allowed to consider various processes in the collision theory in the most natural and consistent way involving the transitons between bound states, the transitions between bound states and continuum spectrum states (ionization) and also to construct the modified adiabatic bases for calculations of isotopic effects. The latter are connected with a violation of the exact symmetry of the two-atomic molecules WF relative to the nuclei transposition due to their different masses and result from particularly quantum character of nuclei motion. The direct numerical calculations of these effects are very time consuming.

For example, at a numerical integration of the equations of tight coupling in an adiabatic presentation in the region of slow collisions, where the transition probabilities are small, considerable difficulties are associated with the incompleteness of basis states and elimination of nonphysical transitions between them. Besides, at a numerical integration of the system of equations of tight coupling in the finite difference approximation, the integration mistakes accumulate and can distort considerably the true value of the transition probability.

Method of strong coupling of channels [17-19, 25-39] is based on the fact that the wave functions of the system of several particles can be considered from the point of view of the functional analysis as a vector in the infinite-dimensional space and the Schrodinger equation can be treated as a vector equation with an infinite number of components, and if it is impossible to find an exact solution of this equation, one can set the problem of finding the approximate solution.

For example, if the initial Schrodinger equation is replaced by its projection to some finite-dimensional space, then the approximate WF in this space can be considered as some approximation to a true WF and the projection operator can be chosen in such a way that to reflect the most completely the substantial features of the processes involved, and, specifically, one can take into account the most essential open channels and also a part of closed ones. Then representing the approximate WF as a linear combination of known auxiliary functions, one obtaines the system of the ordinary differential equations for the coefficients in this expansion which can be solved numerically. The main advantage of this method is that the physical process is readily illustrated by a clear picture which allows to follow the details of the reactiom mechanism, and besides, the magnitude of the interaction which produces the reaction with a realigment of the particles, which can not be treated as a weak perturbation, does not limit an applicability of the method as in the case of the perturbation theory. This method allows to describe different types of reactions in the frames of a unified formalism involving direct reactions and reactions which occure with the formation of an intermediate complex. However, in spite of the above mentioned advantages of the method, a correct consideration of continuous spectrum in the expansion of a full WF poses great difficulties. For this reason there is no an unambiguous means of the constracti-on of WF describing the transition from one channel to another in terms of a full WF for the system of several particles. As a consequence difficulties emerge at the description of reactions with the realignment of the colliding complexes. In this case the modification of the method of tight coupling consists in presenting WF in the form of a finite mixed series comprised of the products of the eigenfunctions of clasters of every possible type, but the meaning of this expansion is not defined due to cramming of mixed basis.

Quantum-mechanical calculations [17-100] are based on the solution of the Schrodinger equation for the system of electrons and nuclei, e.g. a solution of the quantum-mechanical problem for many bodies. The principle difficulties at solving this problem are connected with the complexity of formulation of the asymptotic boundary conditions for the scattering states. This circumstance is more pronounced in calculations of cross-sections when the energies of the impacting particles exceed the break-down (dissociation) threshold. That is the reason why in solving such problems attention is focused [17-39] on the development of the various above-mentioned approximate methods and various modifications of the adiabatic and Born approximations, etc. The main feature of these methods is that the real many-particle problem is reduced to the two-paricle one for which the dynamical characteristics of the scattering processes (amplitudes, scattering phases, reaction rate constants, etc.) are determined.

Thus the next reasonable approximation for the consistent quantum-mechanical calculations of the cross-sections of reactions in nuclear, atomic, and molecular physics is the method of the quantum problem of scattering in the system of several particles. The most consistent and mathematically correct method of investigation of reactions in the region of low and medium energies is the method based on the integral equations formulated by L.D. Faddeev and O.Ja. Jakubovskii [36-41] which describe the nonrelativistic movement of several particles interacting by means of paired potentials. The correct formulation of the scattering problem for the system of several particles, developed in these works, has allowed to realize new numerical methods for calculations of various processes in nuclear, atomic, and molecular physics, to substantiate mathematically rigorously various above-mentioned approximate methods of solution of many-particle problems, to determine correctly the limits of applicability of these approximations and the interrelation betweem them.

The first attempts to apply the integral equations of the quantum theory of scattering to calculations of molecular reactions were made in [32, 33]. However, the main attention in these papers has been given to setting up a problem and the method of its solution. Numerical calculations in these papers are equivalent to calculations in Born approximation. This is connected with great computational difficulties of the numerical solution which are mainly determined by the nonlocal nature of these equations. For this reason it has been suggested in [41-45, 50, 52-54] to use modified differential equations which have the following main advantages:

the locality of differential equations reduces considerably computational resources at inversion of matrices appearing after quantization of the problem;

the asymptotics of every separate component of WF for the pair of particles is described in terms of corresponding Jakobian coordinates and contains only one member raised by the eigenfunction of a given coupled pair instead of a sum over all pairs as in a case of Schrodinger equation;

the differential formulation allows to derive correct equations for the problem of scattering in the system of charged particles on the basis of direct asymptotic methods, investigate low-energy and angular features of scattering amplitudes for many-particles coulomb systems, and obtain an extensive information about the peculiarities of elementary processes and the methods of their control;

the differential formulation allows to create methods and means for solution of the scattering problem in the system of several bodies in the model of boundary conditions.

Systems of singular integral equations (SSIE) and systems of integro-differential equations (SIDE) formulated in [36-41] provide the basis for the numerous theoretical and applied problems in classical and quantum mechanics, mathematical physics, hydrodynamics, control theory, theory of elasticity, etc.

In a general case an analytical solution of SSIE and SIDE is the exception rather than the rule. Hence, the methods of numerical solutions of these equations are of considerable interest, and they became especially actual in connection with the correct formulation of the quantum theory of scattering in the system of several bodies.

At present SSIE and SIDE are the only mathematically justified means to realize consistent calculations of various characteristics of elementary processes in nuclear, atomic and molecular physics. Therefore, a study and development of numerical methods of solutions of SSIE and SIDE and realization of these methods in the form of corresponding application packages (AP) [60-64] is an urgent problem at investigations and developments of new mathematical models of the collision theory in systems with several particles.

The software quality of a given AP acquires a great importance in the serial calculations of cross-sections of various elementary processes and operation of this AP in automated systems for realization of banks of models and main characteristics of elementary processes [71]. This means the effective use of the resources (CPU time, one-line and external memory of the input-output channels, etc.) of the one-processor and multi-processor computer centers, the possibility to use a given AP in any computers (mobility), reliability, etc.

To provide the effective use of resources, the methods and corresponding software for statical and dynamical analysis of AP were developed [61], which allow to perform the statical and dynamical optimization of resources of the computer systems involved, including special ones consisting from matrix and vector processors [60, 62]. The realability of operation of this AP was predicted by means of the specially developed system of realability models [60] taking into account groups of arguments, while the mobility was provided by consecutive application of the modular principle which determines the localization of dependence on a particular computer inside of individual modules only and allows to develop well-structured and documented programming systems.

In this paper, based on the method of quantum theory of scattering in the system of several particles both in integral and differential forms using numerical methods developed and AP realized on their basis, various processes (an elastic scattering, excitation, realignment, and dissociation) were calculated which occure at collisions between electrons and two-atomic molecules of hydrogen, hydrogen halides and their isotopically substituted modifications, halogens, nitrogen, and alkali halides both in the ground and excited vibration-rotation states [46-70]. Various processes were calculated which occure at collisions of hydrogen atoms with carbon, lithium, and hydrogen ions, of protons with hydrogen atoms and ions, of sodium atoms with iodine atoms. In addition, processes of scattering of atoms in high-energy excited Rydberg states, collisions of atoms with two- atomic molecules ($H + H_2$, $O + CS$, $S + O_2$, $O + O_2$, $O + N_2$) and three-atomic molecules ($O + CS_2$, $O + OCS$), collisions of three atoms, which are initially in a free state ($H + H + H$ $H (\nu = n) +$

H), and also collisions between two-atomic molecules, in particular, an exchange by the vibrational energy at collisions between HCl molecules, were calculated.

The reaction [59]

$$O + CF_3I \longrightarrow IO + CF_3$$

was calculated based on the cluster model in the framework of the quantum problem of scattering in the system of three bodies. Based on the method suggested, calculation were made of various processes in nuclear physics going on in collisions of neutrons and protons with deutons, α-particles with ^{8}Be nuclei, and also bound states of positronium ion, tritium, and helium-3 nuclei were calculated [52-54, 65, 67].

The material in the paper is grouped as follows. In the first part a formulation of the integral equations of the quantun theory of scattering in the system of several particles is presented and the conditions are considered imposed on the paired interaction potentials for which these equations can be solved unambiguosly (p. 1.1). In p. 1.2 the special cases are presented arising from the solution of a set of integral equations in a system of three bodies, and the approximations are considered which allow to obtain analytical solutions. The relationship between adiabatic, impulsive, and multiple scattering approximationsis investigated.

In p. 1.3 the method of separation of angular variables in the integral equations is presented. This method allows to reduce a set of multidimensional integral equations to a set of singular integral equations (SSIE) of lesser dimension which can be solved numerically by the most general and universal method designed both for calculations of bound states and scattering states of many-particle systems interacting by means of arbitrary paired short-range potentials. A brief description is presented of a software support realized on the basis of the method suggested.

In a second part a formulation is presented of the differential equations of the quantun theory of scattering in the system of several particles interacting by means of arbitrary paired potentials, and corresponding boundary conditions are considered (par. 2.1). In par. 2.2 one of the methods of separation of angular variables in the differential equations of quantum theory of scattering in the system of of several particles is considered, which allows to reduce multidimensional partial differential equations to a set of partial integro-differential equations for a function of two variables. In 2.3 a method of numerical solution of SIDE is presented based on the difference approximation on the nonuniform network, a description of the application package realized on the basis of the method suggested is given, and pecularities of the software support are discussed. In 2.4 a theory is developed for calculation of bound states and scattering states in the system of three particles based on the model of boundary conditions.

In a third part the results of calculations of main characteristics of various processes in atomic (par. 3.1), molecular (par. 3.2), and nuclear physics (par. 3.3) are presented, and main approximations used in calculations are discussed. In par. 3.4 the results of calculations of the threshold peculiarities in the systems of three particles are considered.

In Conclusion the main results of the paper are formulated.

1. INTEGRAL EQUATIONS OF THE QUANTUM THEORY OF SCATTERING

1.1. Formulation of the integral equations

The Faddeev integral equations in the quantun theory of scattering in the system of three particles [40] are formulated for the three parts into which the total wave function of three particles is broken

$$\Psi = \sum_{i=1}^{3} \Psi_i. \tag{1.1}$$

each corresponding to all possible divisions of the system in the noninteracting subgroups. These equations in the impulse space in a case of scattering of a particle 1 on a coupled pair (2, 3) have a form [40, 41]:

$$\begin{gathered}\Psi_i = \Phi_1\delta_{i1} - G_0(Z)\,T_i\,(\Psi_j + \Psi_k),\\ i,\,j,\,k = 1,\,2,\,3;\;\;3,\,1,\,2;\;\;2.\,1,\,3;\end{gathered} \tag{1.2}$$

where Φ_1 describes the initial state of the system of three bodies: a free motion of the particle 1 and a bound state of the pair (2,3); G_o and T_i are the integral operators. $G_o(Z) = (H_o - Z)^{-1}$, $Z = E + i0$, H_o is an operator of a free motion of three particles; E is a total energy of asystem of three bodies equal to the sum of the kinetic energy of theimpacting particle1 and bonding energy of the pair (2,3); T_i is a paired T-matrix which is defined in terms of the paired interaction potentials by means of Lipman-Schwinger equations [18]:

$$T_i = V_i + V_i G_i T_i. \tag{1.3}$$

To describe three particles in the system of the inversion center we will use conventional Jakobi coordinates which are defined as follows:

$$\begin{gathered}k_1 = (m_3 q_2 - m_2 q_3)/(m_2 + m_3),\\ p_1 = m_3(q_1 + q_2) - (m_1 + m_2)\,q_3\Big/\sum_{i=1}^{3} m_i.\end{gathered} \tag{1.4}$$

where m_i and q_i are masses and impulses of particles. Coordinates k_2, p_2, k_3, and p_3 are defined in a similar manner. In these variables

$$\Phi_1(k_1,\; p_1) = \varphi(k_1)\delta(p_1 - p_1^{\,0}),$$

where φ is the wave function of the initial state of the system (2,3) with the bonding energy $\varkappa_1^2$ and p_1^0 is an impulse of the impacting particle 1,

$$Z = p_1^{02}/2n_1 + \varkappa_1^2/2m_{23} + i0,\quad n_1 = m_1(m_2 + m_3)\Big/\sum_{i=1}^{3} m_i,$$
$$m_{23} = m_2 m_3/(m_2 + m_3).$$

G_0 and T_i are the integral operators with kernels of the type:

$$T_i\left(k_i k_i',\ p_i,\ p_i';\ Z\right) = t_i\left(k_i,\ k_i';\ Z - p_i^2/2n_i\right)\delta\left(p_i - p_i'\right),$$

$$G_0\left(k_i,\ k_i',\ p_i,\ p_i';\ Z\right) = \frac{\delta\left(k_i - k_i'\right)\delta\left(p_i - p_i'\right)}{k_i^2/2m_{jk} + p_i^2/2n_i - Z}.$$

One needs to keep in mind that as integration variables in (1.2) it is necessary to use a single system of variables which is the most convenient. For example, it is convenient to use as integration variables k_2 and p_2 in the integral corresponding to the expression $G_0T_1\Psi_2$. In this case variables k_1 and p_2, which define the kernel of operator T_1, must be expressed in trems of variables k_2 and p_2. Sometimes in the same case it is convenient to chose p_1 and p_2 as integration variables.

Paired T-matrices $t_i(k_i,\ k_i;\ Z)$ in the kernels of Eq. (1.2) have singularities over variable Z: poles, corresponding to the discrete spectrum of paired subsystems, and a cut in a positive part of the real axes raised by the continous spectrum of the problem of two bodies. The explicit form of these singularities is given by so called spectral representation of T-matrix [18, 41]:

$$t(k,\ k';\ Z) = V(k - k') + \sum_i \psi_i(k)\,\psi_i(k')/(\varkappa^2 + Z) + \int t(k,\ x;\ \varkappa^2 \pm i0)\ \times$$

$$\times\ t(k',\ \varkappa,\ \varkappa^2 \mp i0)\,(x^2/2m - Z)^{-1}\,dx,$$

where $\psi_i(k) = (\varkappa_i^2 + k^2)\varphi_i(k)$.

Poles of T-matrix corresponding to the discrete spectrum give rise to the singularities in the components of wave functions Ψ_i which after separating give the following representation:

$$\Psi_i\left(k_i,\ p_i;\ p_i^{(0)}\right) = \varphi_i(k_i)\,\delta\left(p_i - p_i^0\right) - \frac{B_i\left(k_i,\ p_i;\ p_i^0;\ Z\right)}{p_i^2/2n_i + k_i^2/2m_{jk} - Z}. \qquad (1.5)$$

$$B_i\left(k_i,\ p_i,\ p_i^0;\ Z\right) = -\sum_j\Big[Q_j\left(k_i,\ p_i;\ p_i^0;\ Z\right) -$$

$$-\sum_h \varphi_j^h(k_j)\,R_{ij}^h(p_j,\ p_i^0;\ Z)/\left(p_i^2/2n_i - \varkappa_h^2 - Z\right),$$

where Q_j and R_{ij} are the differentiable functions of their variables. This separation of singularities occures naturally by itself at the numerical solution of the integral equations (1.2). To determine functions B_i, Q_i, and R_{ij} uniquely, one can proceeds as follows: to insert Ψ_i in the form (1.5) in Eq. (1.2) and to equate coefficients at identical singularities. Thus, one obtains the equations for these functions which express explicitly all main characteristics of the three-body problem: wave functions, elements of scattering S-matrix, amplitudes and cross-sections of various processes going on in thethree-particlesystem[41-45]:

$$1 + (2,3) \rightarrow \begin{cases} 1 + (2,3) & \text{- elastic scattering} \\ 2 + (1,3) & \text{- processes of realignment} \\ 3 + (1,2) & \\ 1 + 2 + 3 & \text{- processes of decay into three particles} \end{cases} \quad (1.6)$$

The cross-section of elastic scattering is:

$$d\sigma_{11}/d\Omega = (2\pi)^4 n_1 \, |R_{11}|^2, \quad (1.7)$$

that of realignment is:

$$\frac{d\sigma_{1i}}{d\Omega} = (2\pi)^4 \, n_i \, \frac{p_f}{p_1^0} \, |R_{1i}|^2, \quad (1.8)$$

and that of decay is:

$$\frac{d\sigma_{1\rightarrow 3}}{d\Omega \, dp} = (2\pi)^4 \, n_i \, \frac{p_j}{p_1^0} \, | B_{0i} |^2. \quad (1.9)$$

$$p_f = 2n_i \left(p_i^{02}/2n_i - \varkappa_1^2 - \varkappa_i^2 \right).$$

Analogous formulas can be obtained for cross-sections of scattering in processes when there are three free particles in the initial state:

$$1 + 2 + 3 \rightarrow \begin{cases} 1 + (1, 3), \\ 2 + (1, 3), \\ 3 + (1, 2), \\ 1 + 2 + 3. \end{cases} \quad (1.10)$$

It is necessary to point out that the potentials do not present explicitly in Eq.(1.2), whereas these equations contain more general characteristic - T-matrices which are connected with potentials by Eq.(1.3). Therefore, although this method uses formally potentials, in fact, T-matrices are simulated which are constructed by the Beitman method [18, 41, 72]. According to this method, a partial harmonic of the local potential in the impulse space

$$V_e(k, k') = \frac{\pi}{2} \int r^2 j_e(kr)\, V(r)\, j_e(k'r)\, dr$$

is approximated by the expression

$$\widetilde{V}_e(k, k') = \sum_{i,j}^{n} V_e(k, s_i)\, V_e(k', s_j)/V_e(s_i, s_j), \qquad (1.11)$$

while parameters s_i and s_j are determined from the expression for the minimum of a functional

$$\chi(s_1, s_2, \ldots, s_n) = \frac{\int\int |V_e(k, k') - \widetilde{V}_e(k, k', s_1, \ldots, s_n)|\, dk\, dk'}{\int\int |V_e(k, k')|\, dk\, dk'}.$$

The solution of Eq.(1.3) with a potential (1.11) has the form

$$\widetilde{t}(k, k'; Z) = \sum_{i,j}^{n} C_{ij}(Z)\, V(k, s_i)\, V(k', s_j), \qquad (1.12)$$

where

$$C_{ij}(Z) = [V(s_i, s_j) + 8\pi m_{ij} I_{ij}(Z)]^{-1}.$$

$$I_{ij}(Z) = \int_0^\infty \frac{V(k, s_i)\, V(k', s_j)\, k^2\, dk\, dk'}{k^2/2m_{ij} - Z}.$$

It is necessary to point out that a given method of construction of the separable T-matrix (1.12) is applied for any local potential. In addition, this method simplifies considerably a numerical solution of a set of integral equations (1.2) and in some cases allows to obtain an analytical solution.

The integral equations (1.2) have good properties from the mathematical point of view (being Fredholmian, unambiguosly solvable, etc.) only at certain conditions for two-particles data [40, 41]:

(1) paired potentials $V_i(k, k')$, which are in general non-local, are differentiable functions of k, k′ and meet the condition $|V_i(k, k')| \leqslant C(1 - |k - k'|)^{-1-\varepsilon}$, $\varepsilon > 0$;

(2) a point Z = 0 is not a singular one for Eq.(1.3), i.e. all three scattering distances in paired channels are finite;

(3) the positive two-particle spectrum is continuous. This condition is essential for nonlocal potentials, because positive eigenvalues can appear only in this case, and it is valid practically for all real physical potentials. Colomb potentials and potentials of solid core do not satisfy the first condition with colomb potentials giving a singularity of type $|k - k'|^{-2}$ in T-matrices and potentials of solid core resulting in a slow decrease of T-matrix at large impulses. If the second condition is violated the equations (1.2) loose their Fredholmity at Z = 0, which results in the Efimov effect [72-82], i.e. an emergence of an infinite discrete spectrum in the system of three particles at certain conditions at critical value of the coupling constant when the scattering distance tends to infinity for the first time.

The integral Faddeev equations (1.2) are the exact equations for the description of the dynamics of three nonstructured particles ineracting in pairs. The assumption about the interaction in pairs between particles in the three-particle system is a natural one, because all characteristics of the processes in this system are first of all will be determined by the interaction in pairs. This is confirmed completely by the experimental results in direct reactions in nuclear and atomic physics where collisions in close proximity and paired potentials determine all reaction dynamics [17-34].

Here we emphasize that the method of classical trajectories does not apply in a general case to the calculation of collisions of an atom with a two-atomic molecule on a basis of

PES, i.e. PES can only emerge when a chemical reaction occures with a formation of intermediate complex. On the other hand, the method of quantum problem of scattering can be applied to any chemical reactions for which the scattering equations are Fredholmian.

In a case of molecular reactions, particles involved in the reaction are composite complexes (atoms with an electron shell), and therefore, their inner structure may play an essential role. As a result, the interaction potential in such a system, which is nonlocal in a general case, will contain many-particles potentials along with paired potentials. In [32, 33, 58, 59] it is shown how many-particle potentials emerge in a three-patricle system consisting of three identical atoms interacting by means of potentials of the type [32]

$$V_i(r_i) = {}^1V_i^1(r_i)^1\chi_i + {}^3V_i^3(r_i)^3\chi_i,$$

where

$${}^1\chi_i = {}^1/_4 + s_j s_k, \qquad {}^3\chi_i = {}^3/_4 - s_j s_k,$$

s_i is a spin of an atom i, whereas a total interaction is of the form

$$V = \sum_i V_i. \qquad (1.13)$$

It is necessary to point out a paper [33] where the analogous considerations were used for the construction of PES for polyatomic molecules LiF and LiBr. At a direct solution of Eq.(1.2), many-particle nonlocal potential [86-90] emerges after a separation of angular variables and an expansion in terms of partial waves in the impulse space. The first members of the Fourier-transform of this potential in the Born-Oppenheimer approximation coincide with (1.13). However, contrary to (1.13), this potential has an asymptotic $\sim C/\rho^2$ at $\rho \to \infty$ where ρ is a distance between the impacting particle and the center of mass of a pair. This property of the interaction potential results in threshold peculiarities of amplitudes and scattering cross-sections which show up as oscillations in the threshold of the decay into three particles [72-82]. We point out that these peculiarities do not depend on the concrete type of paired forces, but are determined by the presence of resonances in them and, therefore,they may exist in any system of three particles.

The integral equations of the quantum theory of scattering in the system of many bodies are rather complex for specific calculations, although presently some results are known which were obtained on a basis of these equations for the system of four particles [36-39, 70]. These equations, however, allow to develop various mathematically correct approximate approaches to the solution of concrete problems of the scattering theory in the system of many particles.

For example, in [36-39] one of these approaches is suggested to the constraction of approximate equations of the type of the strong coupling equations for the scattering problem in the system of three and four particles. These equations are formulated for functions of a single vector variable which can be represented as wave functions of the relative motion of the coupled complexes of particles. After aseparation of angular variables these equations are reduced to the one-dimensional equations. Components of the wave function in these equations containing information about scattering in some two-particle channel are expanded in terms of the total system of eigenfunctions of the channel Hamiltonians. In this case one obtaines for the wave function itself a known expansion which acquires an exact meaning due to only exact integral equations (1.2). For example, from Eq.(1.2) formulas are obtained for the contributions from discrete and continuous spectra of complexes expressed through the total wave function. In the problem of three particles this expansion appears using an identity $G_0T_i = G_iV_i$ and a spectral expansion for the Green function of the operator $G(Z) = (Z - H_i)^{-1}$, $H_i = h_0 + V_i$, $V_i = \sum_{a^-\alpha} V_\alpha$ [13, 36 - 39]. Analogous to the case of three particles the wave function of the system of four particles can be represented as a sum of components [41]

$$\Psi = \sum_{a,\alpha} \Psi^{a\alpha},$$

which satisfy to a set of equations

$$\Psi^{a\alpha} = \Psi^{a}_{\alpha_0}\delta_{\alpha\alpha_0} + \sum_{\substack{(b\neq c)\subset\alpha \\ \beta\neq\alpha}} G^{ab}(Z)\, V_b \Psi^{c\beta}. \tag{1.14}$$

Here Latin letters a, b,... number pairs of particles, Greek letters α, β,...number divisions of the system of four particles into two groups: $\Psi^a_{\alpha_0}$ is a component of the wave function in the initial state and

{Eq. 14, p. 71}

$$G^{ab}_{\alpha} = G_0\delta_{ab} + G_0 V_\alpha G_\alpha, \qquad V_\alpha = \sum_{a\subset\alpha} V_a.$$

For kernels G^{ab}_{α} one can write the following spectral representation

$$\begin{aligned} G^{ab}_{\alpha}(k, p, q, k', p', q'; Z) &= \delta(q-q')\Bigg\{\frac{\psi^a_\alpha(k, p)\,\psi_\alpha(k', p')}{Z - q^2/2\mu - \varkappa_i^2} + \\ &+ \sum_c \int dp_0 \frac{\psi^a_{p_0 c}(k, p)\,\psi_{p_0 c}(k', p')}{Z - q^2/2\mu + b_c - p_0^2/2n} + \int dk_0\, dp_0 \frac{\psi^{ab}_{k_0 p_0}(k, p)\,\psi_{k_0 p_0}(k, p)}{Z - q^2/2\mu - k_0^2/2m - p_0^2/2n} + \\ &+ (Z - q^2/2\mu - k^2/2m - p^2/2n)^{-1}\Big[\big(\psi_\alpha(k, p)\,\delta_{ab} - \psi^a_\alpha(k, p)\big)\psi_\alpha(k, p)\times \\ &\times \sum_c \int dp_0 \big(\psi_{p_0 c}(k, p)\,\delta_{ab} - \psi^a_{p_0 c}(k, p)\big)\psi_{p_0 c}(k', p') + \\ &+ \int dp_0\, dk_0 \big(\psi_{k_0 p_0}(k, p)\,\delta_{ab} - \psi^{ab}_{k_0 p_0}\big)\psi_{k_0 p_0}(k', p')\Big]\Bigg\}, \end{aligned} \tag{1.15}$$

$$\psi_\alpha = \sum_a \psi^a_\alpha, \quad \psi_{p_0 a} = \sum_b \psi^b_{p_0 a},$$

$$\psi^a_\alpha = G_a(-\varkappa^2 + i\varepsilon)\, V_\alpha \sum \psi^b_a,$$

$$\psi^b_{p_0 a} = \Phi_{p_0 a}\delta_{ab} + G_в\big(-b_a + p_0^2/2n + i\varepsilon\big)\, V_b \sum \psi^c_{p_0 a},$$

$$\psi^{ab}_{k_0 p_0} = \lim i\varepsilon G^{ab}_\alpha\big(k_0^2/2m + p_0^2/2n + i\varepsilon\big)\,\Phi_{k_0 p_0},$$

where m, n, and μ are reduced masses corresponding to the Jakobi variables k, p, q; μ^2_α is a bond energy of two and three particles, respectively; ψ_α is the wave function of the bound state of three particles; $\psi_{p_0\alpha}$ is the wave function describing scattering of a particle on the bound pair; $\psi_{k_0 p_0}$ is the wave function correspondibg to the scattering process when three particles are free in the initial state. Inserting (1.15) to (1.14), one obtains the following expansion in terms of a total set of the channel Hamiltonian functions

$$\psi^{\alpha}(k, p, q) = \psi_\alpha(k, p)\chi_\alpha(q) + \sum \int \psi_{p_0\alpha}(k, p)\chi_{\alpha p_0}(q)\,dp_0 + \int \psi_{k_0 p_0}(k, p) \times \times \chi_{\alpha k_0 p_0}(q)\,dk_0\,dp_0,$$

where functions $\xi_\alpha(q)$ satisfy to equations

$$\chi_\alpha(q) = \delta(q - q_0)\,\delta_{\alpha\alpha_0} + \sum_\beta \int B_{\alpha\beta}(q, q')\,\chi_\beta(q')\,dq' \tag{1.16}$$

and have a meaning of the wave functions in the problem of scattering of two complexes of particles

$$\chi_\alpha(q) = \delta_{\alpha\alpha_0}\delta(q - q_0) + A_\alpha(q)/(q_0^2/2\mu - q^2/2\mu + i\varepsilon),$$

with $A_\alpha(q)$ at $q^2 = q_0^2$ corresponding to the scattering amplitude in two-particle channel, whereas kernels of the integral Eq.(1.16) have the form

$$B_{\alpha\beta}(q_\alpha, q'_\beta) = (1 - \delta_{\alpha\beta})(Z - q'^2_\alpha/2\mu_\alpha + \varkappa_\alpha^2)^{-1} \sum_{(\alpha\neq\beta)\subset\alpha} \int \Phi_\alpha^a(k'_0, p'_{a\alpha}) \times$$
$$\times\, \delta(q_\alpha - q'_\alpha)\,\psi_\beta^b(k'_\beta, p'_{b\beta})\,dk'_b\,dp'_{b\beta} + \sum_{\substack{(a\neq b)\subset\alpha\\(c\neq\alpha)\subset\gamma}} (1 - \delta_{\beta\gamma}) \int \Phi_\alpha^a(k'_a, p'_{a\alpha}) \times$$
$$\times\, \delta(q_\alpha - q'_\alpha)\,\psi_\beta^\alpha(k'_\alpha, p'_{\alpha\beta})\,\widetilde{\mathcal{G}}_\alpha^{bc}(k'_b, p'_{b\gamma}, q'_\alpha, k'_c, p'_{c\gamma}, q'_\gamma)\,dk'_\alpha\,dp'_{\alpha\beta}\,dk'_b\,dp'_{b\gamma}\,dq'_\gamma,$$

$$\Phi_\alpha^a(k, p) = \int V_\alpha(k, k')\,\delta(p - p')\,\psi_\alpha(k', p')\,dp'\,dk',$$

$$\widetilde{\mathcal{G}}_\gamma^{bc} = \int \widetilde{G}_\gamma^{bc}(k, p, q; k'', p'', q''; Z)\,\delta(p'' - p')\,\delta(q'' - q')\,V(k', k'')\,dk''\,dp''\,dq'',$$

$$\widetilde{G}_\alpha^{ab} = \widehat{G}_\alpha^{ab} + G_\alpha^{ab},$$

$$G_\alpha^{ab}(k, p, q; k', p', q'; Z) = \delta(q - q')\,\psi_\alpha^a(k, p)\,\psi_\alpha(k', p')/(Z - q^2/2\mu + \varkappa_\alpha^2),$$

where k_α, $p_{a\alpha}$, q_α are Jakobi coordinates for the case of four particles which are determined similarly to Eq. (1.4).

To take into account a contribution of the continuous spectrum in kernels of Eq.(1.16) it is necessary to know the Green function for the problem of two and three particles. This hampers significally the application of these equations for the calculation of the scattering characteristics in the many-body system, therefore there is a need to perform further simplifications analogous to ones made in [39]. In this paper, keeping in the spectral expansion (1.15) only a contribution from a single bound state and neglecting contributions from other states, the analytical solutions of a set of integral equations (1.16) were obtained in a case of scattering of three particles (an electron scattering on a hydrogen atom) and four particles (an electron scattering on a helium atom). As a result, simple analytical formulas for the amplitudes and phases of the elastic scattering were obtained which are valid when the impacting particle energy is less than the decay threshold. The calculations based on these formulas are in a good agreement with experimental data and calculations of other authors. With increasing energy of the impacting particle it is necessary to take into account the excited states and the continuous spectrum states in the expansion (1.15) which results in Eqs.(1.16) with nonseparable kernels which by analogy with (1.11) and (1.12) may be inerpreted as nonlocal many-body interaction potentials in the impulse space in the system of many bodies.

This method allows to calculate scattering in the systems of arbitrary number of bodies by dividing a system into bound complexes [18, 41, 59]. The total wave function of a system in this approximation is presented in a form analogous to (1.14) whereas each component is expanded in terms of the eigenfunctions of the channel Hamiltonian with coefficients of the expansion proportional to amplitudes of all possible processes. The set of equations for these coefficients is obtined from the requirement for the the wave function that it satisfies to all two-particle Lipman-Schwinger equations. These equations unambiguously determine the solution of the scattering problem in this system. Notice that a natural approximation sequence arises in this exact set of equations, the first of which is equivalent to the method of the resonated groups and corresponds to considering only bound states in subsystems, the second approximation corresponds to the process of decay into three clusters and can be constructed knowing the wave functions of the scattering problems in subsystems, etc.

1.2. Analytical solutions of the set of integral equations

The equations of motion of three and more particles (1.2) and (1.14) interacting by means of arbitrary paired potentials are in a general case nonintegrable. Therefore, neither bound states nor scattering states in such systems can be reduced to quadrature. It is connected with the fact that a number of integrals of motion available is less than a number of dynamical variables necessary for the system description [41-47, 72]. Thus the methods of construction of analytical solutions of a set of integral equations (1.2) and (1.14) assume a great significance, which is possible in the cases when systems of many bodies have either additional integrals of motion connected with the peculiarities of the paired interaction or the system is constrained [41-70, 72].

The importance of construction of analytical solutions of integral equations (1.2) and (1.14) describing various dynamical processes in the systems of three and more particles is associated with the fact that, for example, in a case of three particles new qualitative features appear which are not inherent in the least in two-particle model: Efimov effect [73-81] considered above, using in the equations T-matrices instead of potentials which include some new characteristics of the interaction not observable in a problem of two bodies [41-86].

Therefore we will consider a number of specific cases when sets of integral equations (1.2) and (1.14) can be solved analytically [38, 39, 47, 51, 55, 79, 80].

Consider a case of a single light particle $m_1 = m$ and two identical heavy ones $m_2 = m_3 = M$. This situation emerges in problems of proton scattering on hydrogen atoms or collisions of an electron with a two-atomic molecule consisting of identical atoms. In a latter case the requirement that the ratio of mass of light particle to the reduced masses of heavy ones tends to zero forms a basis for the adiabatic approximation [18-21].

Let

$$\Psi_2 = -G_0F_2, \quad \Psi_3 = -G_0F_3. \tag{1.17}$$

Then from Eq.(1.2) we get the following set of equations [47, 51] for F_2 and F_3:

$$\begin{aligned}(1 - T_2G_0T_1G_0)F_2 + (T_2G_0 - T_2G_0T_1G_0)F_3 = T_2\Phi,\\ (1 - T_3G_0T_1G_0)F_3 + (T_3G_0 - T_3G_0T_1G_0)F_2 = T_3\Phi,\end{aligned} \tag{1.18}$$

anf for the total WF the following expression:

$$\begin{aligned}\Psi = \Phi - G_0T_1G_0(F_2 + F_3) - G_0T_3(\Phi + G_0F_2) - G_0T_2(\Phi + G_0F_3) -\\ - G_0(T_3 + T_2)G_0T_1(F_2 + F_3).\end{aligned} \tag{1.19}$$

The meaning of various members of the expansion becomes clear if one takes into account that residues of the expressions of the type $G_0T_iF_i$ in poles T_i are within a constant amplitudes of one of the possible excitation or redistribution processes of particles at collisions, while a residue in a pole G_0 gives the decay amplitude.

Thus, the second member in Eq.(1.19) gives the amplitude of the elastic scattering or excitation while third and fourth ones correspond to amplitudes of the realignment reaction in the system of three bodies. The last term gives contributions to all amplitudes and thus takes into account the mutual influence of these processes to each other. One can show that an action of the operator $G_0T_1G_0$ in Eq.(1.19) on an arbitrary sufficiently differentiable function F dissapears in the limit $m/M \to 0$ provided $æ_n$ are fixed. Thus, Eqs. (1.18) in the limit $m/M \to 0$ take the following form:

$$F_2 + T_2G_0F_3 = T_2\Phi, \qquad F_3 + T_3G_0F_2 = T_3\Phi. \tag{1.20}$$

This approximation is made to determine $F_2$and F_3 at finite small values of m/M from the system (1.20) in which m/M → 0. Here we notice that at æ_n fixed and M → ∞ all energy levels of the bound system (2, 3) tend to zero. However at M finite an energy of the bound system (2, 3) is not zero.

Therefore Eq. (1.20) will be a reasonable approximation only in the case when a presence of the bond in the system (2, 3) at the elastic scattering does not play an essential role. But that can be only when the impacting particle energy far exceeds the pair bond energy.

The operational solution of the system (1.20) can be presented in the form

$$F_2 = \frac{T_2(1 - G_0T_3)\Phi}{1 - T_2G_0T_3G_0}, \qquad F_3 = \frac{T_3(1 - G_0T_2)\Phi}{1 - T_3G_0T_2G_0}, \tag{1.21}$$

which coincides with expressions obtained for multiple scattering on two overlapping force centers [18, 26], i.e. in the adiabatic approximation all peculiarities of amplitudes and cross-sections of processes in the system of three particles are determined in terms of the multiple scattering.

Taking into account Eqs. (1.19) and (1.21), we get the following representation for the total wave function of the system of three bodies:

{Eq. 18, p. 74}

$$\Psi \simeq (1 - G_0T_3 - G_0T_2)\Phi - G_0T_3G_0T_2\Phi - G_0T_2G_0T_3\Phi - \ldots$$

The second and third terms in this approximate expression correspond to the single scattering approximation, while fourth and fifth ones correspond to the two-scattering approximation. Thus, terms $G_0T_3G_0F_2$ and $G_0T_2G_0F_3$ in Eq. (1.19) in this approximation allow for the multiple scattering on potentials V_2 and V_3, i.e. all peculiarities of amplitudes and cross-sections are determined by multiple scattering. Therefore the scattering in the system of three bodies (an electron and atoms in a molecule) can be treated in this approximation as a result of the consecutive two-particle collisions. Neglect of the operator $G_0T_1G_0$ is equivalent to the change of Ψ_1 by Φ in Eqs. (1.2). This means that the set of equations (1.20) describes the scattering of a particle 1 as if the distribution of the impulses of particles 1 and 2 conserves in the scattering process its initial value φ_0 . This approximate description of the scattering process forms the basis of the well-known impulse approximation based on the following assumptions.

(1) The effective radius of interaction of the impacting particle with the particles from a complex system is far less than the average distance between the particles in the system, i.e. the interaction between the impacting particle and the particle from the system is vastly stronger than that between particles in a bound system, and therefore the interaction potential can be considered as a paired one.

(2) The time of the paired interaction is much less than the characteristic time for the system of scatteres, i.e. the interaction forces between the particles in the system have no time to manifest themselves during paired interaction.

Thus, it is shown that the adiabatic and impulse approximations are related as Fourier transform and its original in the framework of the three-body problem of scattering of a light particle on the system of two heavy particles.

Analogous result appears at the quasi-classical description of the collision of a particle with a bound pair which also contains the representation of the scattering amplitude in the terms of paired amplitudes [18, 19, 43]. As is seen from the above examples, the equations of the quantum theory of scattering in the systems of several particles can serve as the universal mathematical basis for the deduction of such approximations. The iterations of these equations give a representation for the scattering amplitudes in the form of a series (1.21) which contains a product of paired amplitudes, while a number of cofactors in the components determines the rescattering order. The approximations of multiple scattering can be obtained using in the iterations of Eqs. (1.2) corresponding approximations for T-matrices (for example, a separable approximation). Besides, paired one-fold and two-fold scatterings of colliding particles result in the appearance of noninegrable δ-functions and pole singularities in kernels of the scattering matrix for the decay processe to which correspond free terms and the first iterations of Eqs. (1.2).

Similar expressions can be also obtained for the scatering in the systems of charged particles [42, 43]. In this case singularities of the scattering matrix are determined by the resolvent

$$R_{00} = F_c + \sum_{\alpha} F_\alpha + \sum_{\alpha \neq \beta} F_{\alpha\beta} + \widetilde{F}_0,$$

where

$$F_c = A_c / |p - p'|^{5+2i\eta},$$

which singularities are determined by:

the residual Coulomb interaction

$$V_\alpha^0 = (1 - \chi_\alpha)\, n_\alpha / |\, x_\alpha \,|,$$

one-fold scattering

$$F_\alpha = f_\alpha\left(k_\alpha, k'_\alpha\right) A_\alpha / |\, p - p'_\alpha \,|^{3-ia_\alpha},$$

and two-fold scattering

$$F_{\alpha\beta} = A_{\alpha\beta} \frac{f_\alpha\left(k_\alpha, k'_{\alpha\beta}\right) f_\beta\left(k'_\beta, k'_{\alpha\beta}\right)}{\left(k'^2_{\beta\alpha} - k'^2_\beta - i0\right)^{1+ia_{\alpha\beta}}},$$

where f_α are two-particle scattering amplitudes for operators h_α. The form of functions A_α, $A_{\alpha\beta}$ and parameters a_α, $a_{\alpha\beta}$ is presented in [41-43].

It should be stressed here that singularities of S-matrix for the decay processes, which are the most typical of the system of three bodies, are grouped on the same directions of the configurational space as in a case of neutral particles. However contrary to the decay processes in the systems of neutral particles, which are characterized by singularities of the type $\delta(p - p')\, \delta\left(p_\alpha - p'_\alpha\right)$, in the case of charged particles generalized functions $|p - p'|^{-5-2i\eta}$ и $|\, p_\alpha - p'_\alpha \,|^{-3-ia_\alpha}$, appear and the distorted pole $\left(k'^2_{\alpha\beta} - k'^2_\beta - i0\right)^{-1-ia_{\alpha\beta}}$, emerges instead of the pole $\left(k'^2_{\alpha\beta} - k'^2_\beta - i0\right)^{-1}$

These singularities appear because the region where paired rescatterings of three particles can occure, is infinite [18-45]. All singularities in the processes of the elastic scattering, realignment, and decay are determined by two-particle collisions of two coulomb centers. In a case of the elastic scattering one can explicitly obtain senior amplitude members, considering $R_{\alpha\beta} = R^{\alpha}$ or R^{β} and using instead of Green functions for the intermediate states their eikonal approximation [43].

In this case we have

$$f_{AA} = f_c + \widetilde{f}_{AA}, \quad \widetilde{f}_{AA} \sim U_A^1 V_1 (\widehat{R}_2\widehat{V}_2 + \widehat{R}_3\widehat{V}_3 + \widehat{R}_2\widehat{V}_2\widehat{R}_2\widehat{V}_3 + \widehat{R}_3\widehat{V}_3\widehat{R}_3\widehat{R}_2\widehat{V}_2)\, U_A^1.$$

and for the decay processes we obtain

$$f_{0A} \sim U_0^1\widehat{V}_1 (\widehat{R}_2\widehat{V}_2 + \widehat{R}_3\widehat{V}_3 + \widehat{R}_2\widehat{V}_2\widehat{R}_3\widehat{V}_3 + \widehat{R}_3\widehat{V}_3 R_2\widehat{V}_2)\, U_A^1,$$

where

$U_A^{\alpha}(X, p_\alpha) = \psi_A(x_\alpha)\,\psi_c^{\alpha}(y_\alpha, p_\alpha)$ is WF for the discrete spectrum; ψ_α are eigenfunctions of h_α; $A = \{\alpha, i\}$, $i = 1, 2, 3, \ldots, N$; ψ_c^{α} is WF describing scattering in coulomb field $V_c^{\alpha} = n_{\alpha\alpha'}/|y_\alpha|$; $\psi_\alpha(x_\alpha k_\alpha)\,\psi_c^{\alpha}(y_\alpha, p_\alpha)$ — is WF for the continuous spectrum.

Thus, the adiabatic approximation in the impulse space is shown in a general case to be equivalent to the impulse approximation and can be expressed in terms of multiple scattering. This approximation in the configurational space corresponds to the division of a system into fast and slow subsystems in some or other regions of the configurational space. This division is not universal and has a dynamical character. Hence attempts were made in a number of papers [87-90] to apply methods of the differential geometry and algebraic topology to solve a problem of such adivision. This approach allows to express nontrivial covariant derivative and differential topological invariants in terms of the spectral geometry, i.e. in terms of spectral properties of fast Hamiltonian.

Notice that the geometrical analysis of the problem of several bodies in the configurational space is complicated due to noncompactness of the space and in a general case different spectrum divisions result in different effective dynamical equations for slow subsystems. Therefore the properties of the adiabatic representation are studied by constracting unifirm asymptotics of the wave functions on the basis of methods of the homotopy theory. For example, in [95] the procedure is suggested of deriving effective dynamical levels for Faddeev equations. In this paper the relation between the spectral features of S-matrix and topological invariants of a variety has been studied based on the method of spectral projecting giving rise to nontrivial Abelian commutativities. Some geometrical properties resulting from the spectral characteristics of effective Hamiltonians have been demonstrated by simple examples.

The set of equations (1.20) in the impulse space has the form:

$$F_3(p, k) + 2m_1 \int dk' \frac{t_3(p, k - k'; Z)\, F_3(k - k', k')}{(k - k')^2 - p_0^2 - i0} = t_3(p, p_0; Z) \times \tag{1.22}$$
$$\times \varphi_0 (k - p_0 m_3/(m_2 + m_3)),$$
$$F_2(p, k) + 2m_1 \int dk' \frac{t_2(p, k - k'; Z)\, F_2(k - k', k')}{(k - k')^2 - p_0^2 - i0} = t_2(p, p_0; Z) \times$$
$$\times \varphi_0 (k + m_2 p_0/(m_2 + m_3)),$$

and allows to get the analytical solution in cases when T-matrices can be presented in a separable form (1.11), (1.12).

In a case of the approximation of the zero-radius potentials (ZRP), which are a specific case of the separable potentials, T-matrices do not depend on k and have the form [18,20]:

$$t_i(Z) = [(2\pi)^2 m_{jk} (\alpha_i + i\sqrt{2m_{jk} Z})]^{-1},$$

where α_i are the reciprocal scattering distances $a_i^{-1} = \alpha_i$. Hence functions F_2 and F_3 in the ZPR approximation satisfying to the set of equations (1.22) will depend not on two but on a single vector variable. The limiting expressions of t_i at $m/M \to 0$ which must be presented in Eq. (1.22) have the form

$$t_3 = [(2\pi)^2 m_1(\alpha_3 + ip_0)]^{-1}, \qquad t_2 = [(2\pi)^2 m_1(\alpha_2 + ip_0)]^{-1}.$$

Then the following set of equations for F_2 and F_3 is obtained:

$$F_3(k) + 2t_3 m_1 \int \frac{F_2(k')\,dk'}{(k-k')^2 - p_0^2 - i0} = t_3\varphi_0'(k - \gamma p_0),$$

$$F_2(k) + 2t_2 m_1 \int \frac{F_3(k')\,dk'}{(k-k')^2 - p_0^2 - i0} = t_2\varphi_0(k + \varkappa p_0),$$

$$\gamma = m_3/(m_2 + m_3), \quad \varkappa = m_2/(m_2 + m_3).$$

Notice that it is impossible to apply ZRP beyond the framework of the adiabatic approximation, because in the system of three particles 1, 2, and 3 of which the two, say, 2 and 3 interact with 1 via ZRP, there appears an atraction between particles 2 and 3 proportional to $-C/r^2$, where the coefficient C depends on the ratio of masses. This interaction leads to peculiarities in the behaviour of the real physical system of three particles, namely, to a fall to the centre. This causes the ambiguity of the solution of the integral equation describing scattering in such a system, because the corresponding uniform equation appears to have a nontrivial solution [41, 73, 75].

However, uniform equations appearing after dropping right parts in the set (1.22) have no non-zero solutions because t_2 and t_3 are complex numbers.

Thus, singularities described above do not appear in the adiabatic approximation. This proves formally the possibility of application of ZRP in the framework of approximation considered.

The set of integral equations (1.22) is solvable exactly by means of the Fourier transform, and the solution has the form:

$$F_3(r) = \frac{(\alpha_2 + ip_0)\exp(-i\gamma p_0 r) - r^{-1}\exp[i(p_0 r + \varkappa p_0 r)]}{(2\pi)^2 m_1\left[(\alpha_3 + ip_0)(\alpha_2 + ip_0) - r^{-2}\exp(2ip_0 r)\right]}\varphi_0(r),$$

$$F_2(r) = \frac{(\alpha_3 + ip_0)\exp(i\varkappa p_0 r) - r^{-1}\exp[i(p_0 r - \gamma p_0 r)]}{(2\pi)^2 m_1\left[(\alpha_3 + ip_0)(\alpha_2 + ip_0) - r^{-2}\exp(2ip_0 r)\right]}\varphi_0(r).$$

In a similar manner, analytical solutions of the equations can be obtained using approximations (1.11) with form-factors of the type

$$V_i(k, \beta_i) = (k^2 + \beta_i^2)^{-1},$$

to which corresponds the local potential in the coordinate space:

$$V_i(r) = \lambda e^{-\beta_i r}/r.$$

In this case the solution of the set has the followig form [47, 51, 55]:

$$F_i(r, \rho) = \Lambda_i(\rho, \beta_i) C_i(r),$$
$$\Lambda_i = \lambda_i / m_{jk} d_i(Z),$$
$$C_2(r) = \frac{V_2(p_0)\, e^{i\varkappa p_0 r} - 2m_1 \Lambda_2 V_3(p_0)\, e^{i\gamma p_0 r}}{1 - \Lambda_2\Lambda_3 [2m_1 \Phi(r)]^2} \varphi_0(r),$$
$$C_3(r) = \frac{V_3(p_0)\, e^{-i\gamma p_0 r} - 2m_1 \Lambda_3 V_2(p_0)\, e^{i\varkappa p_0 r}}{1 - \Lambda_2\Lambda_3 [2m_1 \Phi(r)]^2} \varphi_0(r),$$
$$\Phi(r) = \frac{2\pi}{r}\left[\frac{e^{ip_0 r}}{(p_0^2 + \beta^2)^2} - \left(1 - \frac{r}{2\beta}(p_0^2 + \beta^2)\right)\frac{e^{-i\gamma p_0 r}}{(p_0^2 + \beta^2)^2}\right].$$

Amplitudes of various processes (1.6) going on in the system of three particles are determined from Eq. (1.7) - (1.9) or from Eq. (1.19) for the total WF. In the latter case the amplitude is determined as follows [17, 46-49, 51]. Near the pole corresponding to the system bound state we have

$$t_i\left(k, k'; Z - \tilde{p}^2\right) = \frac{\left(k^2/2m_{jk} - \varepsilon_n\right)\left(k'^2/2m_{jk} - \varepsilon_n\right)}{Z - \tilde{p}_i^2 - \varepsilon_n} +$$

+ smooth part of the function.

where φ_n are WF for the bound state of particles (j, k) and ε_n is the system bond energy.

Therefore, the expression $G_0 T_i G_0 F_i$ is factorized in the form of the product of three factors:

$$G_0 T_i G_0 F = \varphi_n(k)[p^2/2n + \varepsilon_n - Z]^{-1} M.$$

The medium factor describes free motion of theparticle relative the bound system (j, k). In the coordinate space the divergent wave corresponds to this factor. It is clear from this that factor M, expressed by the integral of the product of functions φ_n and F_i, is the scattering amplitude. Thus, amplitudes and cross-sections of the elastic scattering and excitation processes have the form:

$$M_{if} = (2\pi)^2 m_1 \int \varphi_i^* (r) \left[e^{i\gamma pr} F_3 (r, \rho) + e^{-i\varkappa pr} F_2 (r, \rho)\right] \varphi_f (r)\, dr d\rho,$$
$$d\sigma_{if}/d\Omega = |M_{if}|^2,$$

where φ_i and φ_f are the vibration-rotation WF which in the adiabatic representation have the form [17-21]:

$$\varphi_i (r) = \psi_v (r)\, Y_{JM} (\Theta, \zeta), \qquad \varphi_f (r) = \psi_v (r)\, Y_{J'M'} (\Theta, \zeta),$$

where $Y_{JM}(\Theta, \zeta)$ are spherical functions [91], while vibrational wave functions ψ_v of the Morse potential have the form

$$\psi_v (r) = A_v F_1 (-v, 1 + 2x_v, y)\, e^{-y/2} y^{x_v} r^{-1},$$
$$y = \mathscr{L} e^{-\alpha(r-r_0)}, \quad x_v = (\mathscr{L} - 1)/2 - v,$$
$$\mathscr{L} = 2 \sqrt{m_{23} D}/\alpha,$$
$$A_v = \sqrt{\frac{\alpha\,(2x_v + 1) \ldots (2x_v + v)}{\Gamma\,(2x_v)\, v!}},$$

where $F_1(a, b, x)$ is a degenerate hypergeometric function [91].

The amplitude of the rearrangement process has the form

$$M_R = (2\pi)^2 m_{23} \int \varphi_{12}^* (r) \exp\left[i \sqrt{m_{23}/m_1}\, pr\right] F_2 (r, \rho)\, dr\, d\rho,$$

where φ_{12} is WF of the bound system produced in the reaction process. The effective cross-section of the rearrangement process is

$$d\sigma/d\Omega = \sqrt{m_{21}/m_{23}}\, |M_R|^2.$$

For the dissociation processes of molecules by electrons, the final state is the state of the continuous spectrum, and therefore one has to use for ψ_v the function of the following form {17-22, 93]

$$\psi_\varepsilon (r) = A_\varepsilon e^{-y/2} r^{-1} \{\exp (i\delta (x))\, F_1 (1/2 + x - L/2,\ 1 + 2x,\ y)\, y^x +$$
$$+ \exp (i\delta (x))\, F_1 (1/2 - x - L/2,\ 1 - 2x,\ y)\, y^{-x}\},$$
$$x = i\,(2\varepsilon m_{23})^{1/2}/\alpha,$$
$$\delta (x) = \arg \Gamma (-2x)/\Gamma (1/2 - x - L/2) = -\delta (-x),$$
$$A_\varepsilon = (m_{23}/8\pi^2\varepsilon)^{1/4},$$

where ε is the kinetic energy of the nuclei flying apart, and at calculations of the transitions into continuous spectrum from high mlecular vibrational states one can use the following asymptotics [93]: $\psi_\varepsilon(r) \sim 2A_\varepsilon \sin[\sqrt{2m_{23}}(r - r_0 - \ln L/\alpha)]/r.$

In the case of identical particles $m_2 = m_3$ the elastic scattering and excitation amplitudes have the form

$$M_{if} = 2\int \varphi_i^*(r)\left[\frac{\cos(pr/2)\cos(p_0r/2)}{\alpha + ip_0 - r^{-1}e^{ip_0r}} + \frac{\sin(pr/2)\sin(p_0r/2)}{\alpha + ip_0 - r^{-1}e^{ip_0r}}\right]\varphi_f(r)\,dr. \quad (1.23)$$

Summary cross-sections of all processes including dissociation and various reactions occuring in the system of three particles can be obtained using the optical theorem[17,18]. As a result we have

$$\sigma_{tot} = \frac{4\pi}{p_0^2}\int |\varphi_0|^2\left\{\left[1 + \left(\frac{\alpha r + \cos pr}{p_0 r + \sin p_0 r}\right)^2\right]^{-1} + \left[1 + \left(\frac{\alpha r + \cos pr}{p_0 r - \sin p_0 r}\right)^2\right]^{-1}\right\} dr, \quad (1.24)$$

which coincides with the expressions obtained in papers [17-21]. The expressions (1.23) and (1.24) reflect the main features of the processes cross-sections, particular,those occuring at collisions of electrons with diatomic molecules. For instance, at collisions of electrons with homonucleous ($m_2 = m_3$) molecules the cases are possible in the denominators of functions F_i at fixed r values and changing impulse p_0, when the real part of the expression $J_\pm(p_0) = \alpha + ip_0 \pm e^{ip_0 r}/r$

is equal to zero. These points correspond to the resonance maxima of functions F_i which results in the appearance of maxima in the cross-sections of all processes and also in σ_{tot}. In addition, the presence of the ratio $(m_{23}/m_1)^{1/2}$ in the exponent power for the reaction amplitudes results in the appearance of the noticible isotopic effect which is observed distinctly in the experiment [97]. As was mentioned above, all pecularities of the process cross-sections (1.6) are expressed in the terms of multiple scattering which action is described by the term $e^{ip_0 r}/r$ in functions F_i.

However, the traditional interpretation of the resonance phenomena at the interaction of electrons or atoms with molecules is based on the assumption that transitional complexes are produced in the collision process. The connection between the multiple scattering approximation and the transitional state theory is seen directly, if one takes into account that imaginary zeroes of $J_{\pm}(p_0)$ determine the quasi-stationary terms of the transitional state [17-21], while complex zeroes correspond to the prolongation of terms to the quasi-stationary states region.

Consider one more method of the analytical solution of the set of integral equations (1.2) in which we first proceed to the limit at $m/M \to 0$ directly in equations (1.2) and then use the separable potentaial approximation. In this case it is convenient to chose p_2 and p_3 as independent varaibles. Then

$$k_1 = (p_3 - p_2)/2, \quad k_2 = -p_3 - p_2, \quad k_3 = p_2 + p_3. \tag{1.25}$$

Similary one can obtain analytical solutions for equations (1.2) in the case when one heavy particle $m_3 = M$ and two identical light ones $m_1 = m_2 = m$ are present, for instance, in the case of the electron scattering on a hydrogen atom. In this case it is convenient to chose momemta k_1 and k_2 of light particles as independent variables similary to Eq. (1.25). Then we shall have [98, 41]

$$\begin{aligned}
&k_3 = k_1 - k_2/2, \ p_3 = k_1 + k_2, \\
&\Psi^{\pm}(k_1, k_2) = \Psi(k_1, k_2) \pm \Psi(k_2, k_1) = \\
&= A^{\pm}(k_1, k_2) \pm A^{\pm}(k_2, k_1) + B^{\pm}(k_1, k_2), \\
&G_0(Z) = 2\left(k_1^2 + k_2^2 - p^2 + 1\right)^{-1}, \\
&T_{ep} = t_{ep}\left(k_2, k_2';\ Z - k_1^2/2\right)\delta\left(k_1 - k_1'\right),
\end{aligned}$$

$$T_{ee} = t_{ee}\left(\frac{k_1 - k_2}{2}, \frac{k_1' - k_2'}{2};\ Z - \frac{(k_1 + k_2)^2}{4}\right)\delta\left(k_1 + k_2 - k_1' - k_2'\right).$$

In this approximation the set of equations (1.2) has the form:

$$\begin{aligned}
A^{\pm}(k_1, k_2) = \Phi_1(k_1, k_2) - G_0(Z)\int t_{ep}\left(k_1, k_1';\ Z - k_1^2/2\right) \times \\
\times \left[\pm A^{\pm}\left(k_1', k_1\right) + B^{\pm}\left(k_1, k_1'\right)\right] dk_1'.
\end{aligned} \tag{1.26}$$

$$B^{\pm}(k_1, k_2) = -G_0(Z)\int t_{ee}\left(\frac{k_1 - k_2'}{2}, \frac{k_1 + k_2 + k_2'}{2}, Z - \frac{(k_1 + k_2')^2}{2}\right) \times$$
$$\times \left[A^{\pm}(k_1 + k_2 - k_2') \pm A^{\pm}(k_2', k_1 + k_2 + k_2')\right] dk_2',$$

where

$$\Phi_1(k_1, k_2) = 4\sqrt{2}\,\delta(k_1^2 - p^2)/p\,(k_2^2 - 1)^2,$$

t_{ee} and t_{ep} are coulomb T-matrices. To solve the set (1.26) we use the spectral representation $G_0 T_i$ of the form:

$$G_0(Z)\,T_i = G_{0i}V_i = \sum_i \psi_i^k \psi_i/(Z - \alpha_i^{n2}/2m_{ij}) + \int \frac{\psi^i(k^0)\,\varphi_i(k^0)\,dk^0}{Z - k^{02}/2m_{ij}},$$

which is based on the spectral representation of T-matrix. We shall only use in this representation the contribution from one term corresponding to discrete spectrum and neglect contributions from other terms. The main advantage of this approximation is that the kernels of the set of equations (1.26) become degenerate and the set is reduced to a one-dimension one. In our case we have

$$G_0 t_{ee} \simeq \frac{8\pi}{1 - p^2 + k_1^2 + k_2^2}\,\frac{1}{|k - k_1'|^2} \simeq 0,$$
$$G_0 t_{ep} \simeq \frac{\psi(k)\,\varphi(k)}{Z - k_2^2/2 - {}^1/_2} + \frac{8\pi}{1 - p^2 + k_1^2 + k_2^2}\,\frac{1}{|k_1 - k_1'|^2} \simeq$$
$$\simeq \frac{\psi_0(k)\,\varphi_0(k)}{Z - k_2^2/2 - i\varepsilon}, \qquad \varphi_0(k) = \sqrt{2}/\pi\,(k^2 + 1), \quad \psi = V\varphi.$$

In this approximation

$$B^{\pm}(k_1, k_2) \simeq 0,$$

while for the function $A^{\pm}(k_1, k_2)$ we get the expression

$$A^{\pm}(k_1, k_2) = \psi(k_1)\left[(2\pi)^3\,\delta(k_2 - p) + \frac{4\pi f^{\pm}(k_2)}{k_2^2 - p^2 - i\varepsilon}\right],$$

where the function $f^{\pm}$ satisfies to the equation

$$f^{\pm}(k) = \mp\frac{\psi(k)\,\varphi(k)}{2\pi} \mp \frac{2\psi(k)}{(2\pi)^3}\int \frac{\varphi(k)\,f^{\pm}(k')\,dk'}{k_2'^2 - i\varepsilon - p^2},$$

having the solution of the type

$$f^{\pm}(k) = \mp\frac{\psi(k)\,\varphi(p)}{2\pi\,[1 \pm 2\lambda(p)]}, \quad \lambda(p) = \frac{1}{2\pi^2}\int \frac{\varphi(k)\,\psi(k)}{k^2 - p^2 - i\varepsilon}\,dk.$$

Upon increasing energy of the impacting electron it is necessary to take into account the contribution from continuous spectrum. In this case we have

$$G_1(k, k';\ Z) \simeq \frac{\psi(k)\,\psi(k')}{Z - k_2^2/2 - {}^1/_2} + G_0(k, k';\ Z)$$

while kernels of the set (1.26) take the form

$$G_0 t_{ee} \simeq \frac{8\pi}{\left(1 - p^2 + k_1^2 + k_2^2\right)} \frac{1}{|k - k_1'|^2}, \tag{1.27}$$

$$G_0 t_{ep} \simeq \frac{\psi(k)\,\varphi(k')}{Z - k_2^2/2 + {}^1/_2} + \frac{8\pi}{1 - p^2 + k_1^2 + k_2^2} \frac{1}{|k_1 - k_1'|^2}.$$

Inserting Eq. (1.27) into the set (1.26), we get the following equation for functions $f^{\pm}(k)$ [39]:

$$f^{\pm}(k) = -\frac{R^{\pm}(k, p)}{2\pi} - \frac{1}{4\pi^3}\int \frac{R^{\pm}(k, k')\, f^{\pm}(k')\, dk'}{k^2 - p^2 - i\varepsilon},$$

where

$$R^{\pm}(k, k') = \sum_{i=1}^{3} R_i^{\pm}(k, k'),$$

$$R_1^{\pm}(k, k') = \pm 2\psi(k)\,\varphi(k'),$$

$$R_2^{\pm}(k, k') = \frac{1}{\pi^2 |k - k'|^2}\int \frac{\varphi(q)\,[\psi(q) - \psi(k + q - k')]}{(2\pi)^3\,[1 - p^2 + k^2 + q^2]}\, dq,$$

$$R_3^{\pm}(k, k') = \mp \frac{1}{\pi^2}\int \frac{\varphi(q)\,\psi(k + q - k')\, dq}{|q - k|\,(1 - p^2 + k^2 + q^2)}.$$

For large enegies of the impacting particle we have

$$f^{\pm}(p_f) \simeq -\frac{1}{2\pi} R^{\pm}(p_f, p),$$

where p_f is the momentum of the scattered particle.

Using this approximation and also definitions of direct and reverse amplitudes [17, 18]

$$f = (f^+ + f^-)/2, \quad g = (f^+ - f^-)/2,$$

we get the expressions

$$f(0) = \left[1 - (1 + p^2)^{-2}\right]/P, \tag{1.28}$$

$$g(0) = \frac{16}{(1 + p^2)^2} - \frac{4}{p^2(1 + p)^3}\left[\frac{p^4 - (P/2)(3 + p^4 - p^2)}{(1 - p^2)^2} - \frac{(p^2 - P^2)\,\mathrm{arctg}\,P}{P}\right],$$

where $P = p \sin \theta/2$. Formulae (1.28) correspond to Born approximation [18].

The approximation given above can be shown [38, 39] to be valid in the case when the Green function G_i has the pole corresponding to the bound state of the pair which is located well away from the cut and, besides, this state must be tightly bound, i.e. $\gamma = \alpha/\beta \gg 1$, where α is a bond energy and β is a radius of the potential (1.11) action. To take into account a continuous spectrum it is sufficient to take into account free Green function G_0. Notice that this statement does not depend on the specific potential form which allows to use this approximation to obtain also analytical solutions for a set of integral equations (1.14) for four particles.

Consider the case of three light identical particles and a heavy one, for instance, the electron scattering on a helium atom [38, 70]. In the approximation (1.15) we have

$$G_\alpha^{ab}(k, p, q, k', p', q'; Z) = \delta(q - q') \frac{\Psi_\alpha^a(k, q)\,\psi_\alpha(k, p)}{Z - q^2/2\mu + \varkappa_\alpha^2},$$

where ψ_α is WF of the helium atom, $\varkappa_\alpha^2$ is the helium atom bond energy, $\Psi_\alpha = \sum_\alpha \Psi_\alpha^a$ is WF for a given four-particle system which is factorized and may be presented in the form

$$\Psi_\alpha^{a\alpha}(k_1, k_2, k_3) = \Psi_\alpha^a(k_1, k_2)\,\chi_\alpha(k_3).$$

Because there is no bound state of two and three electrons, α takes only three values α = (1, 234), (2,341), and (3,412) (where 1, 2, 3 are electrons and 4 is the helium atom nucleus). Thus, the set of equations (1.14) is transformed into the set of three equations for functions χ_α of the form:

$$\chi(k) = \delta(k - p)\,\delta_{i1} + [2\,2\pi^3(p^2 - k^2 + i\varepsilon)]^{-1} \sum_{j \neq i} \int R_{ij}(k, k')\,\chi_j(k')\,dk'.$$

Choosing WF for helium atom in the form $\Psi_\alpha(k_1, k_2) = \psi(k_1)\psi(k_2)$, we get the set of integral equations for χ with degenerated kernels of the form

$$R_{ij}(k, k') = \frac{1}{2\pi^3} \int \Phi_{jk4}^{i4}(k', q)\,\psi_{k4i}^{4k}(q, k)\,dq.$$

Here $\Phi^{j4}_{ik4}(k_1, k_2)$ are components of the form-factor of WF for the bound state of the helium atom:

$$\Phi^{j4}_{j4k}(k_1, k_2) = \frac{1}{(2\pi)^6} \int V_{j4}(k_1, k')\, \delta\left(k_2 - k_2'\right) \psi_{j4k}\left(k_1', k_2'\right) dk_1'\, dk_2'.$$

Notice that the set of equations for χ is obtained from the set (1.14) which does not take into account an identity of electrons. Consideration of the electrons identity leads to the expression for the total WF of the system of four particles of the form [18-20, 39]:

$$\Psi_{1/2}(k_1, k_2, k_3) = \Psi(k_1, k_2, k_3) + \Psi(k_2, k_1, k_3) - \Psi(k_1, k_3, k_2) - \Psi(k_3, k_1, k_2),$$

and as a result we get the analytical solution for the set of equations:

$$f(p) = -\frac{\alpha(p)\,\psi(p)\,\varphi(p)}{\pi\,[1 + 4\lambda_1(p)]},$$

$$\alpha(p) = \frac{1}{2\pi^2}\int_0^\infty \frac{\varphi(k)\,\psi(k)}{\varkappa^2 + k^2 + p^2}\, dk,$$

$$\lambda_1(p) = \frac{1}{2\pi^2}\int_0^\infty \frac{\alpha(k)\,\psi(k)}{k^2 - p^2 - i\varepsilon}\, dk.$$

Similarl., one can consider the case of two identical heavy particles and two light ones (a collision between hydrogen atoms, a collision between proton and a negative hydrogen ion etc.) and the case of three identical heavy particles and a light one (proton scattering on a positive molecular hydrogen ion).

Notice that to obtain analytical solutions of the set of equations (1.2) and (1.14) a small mass ratio has been used and some or other models of T-matrices and Green functions. Undoubtedly it is the main advantage of the formulation of these equations, which allows to separate out explicitly the objects for which it is necessary to constract approximate expressions.

In addition, when one considers a number of limiting cases such as a light particle and two identical heavy ones or a heavy particle and two identical light ones, a separated coordinate system appears which express characteristics of all dynamical processes going on in such systems.

1.3. Method of numerical solution of sets of integral equations

The set of integral equations (1.2) is nine-dimensional because the system of three particles is described by nine spatial variables (by the particles coordinates and impulses). However, due to translational symmetry of the isolated system of three particles the WF dependence on three variables of the total impulse is separated out explicitly, while rotational symmetry allows to separate out the dependence on additional three variables. As a result the set of integral equations (1.2) is reduced to the system of coupled integral equations depending on three independed variables which, as has been pointed out above [36-70], are nonintegrable in a general case of arbitrary pair interaction potentials. The numerical solution of these equations can not even be obtained using modern high capacity computers. Hence, to find an approximate numerical solution further simplifications are required. For this purpose, we expand components of WF

$$\Psi(k, p) = \sum_{l,\lambda,L,M} \psi_{l\lambda L}(k, p, p_0)\, \mathcal{Y}_{l\lambda}^{LM}(n_k, n_p)\, Y_{lL}\left(n_{p_0}\right) \tag{1.29}$$

(where n_k and n_p are unit vectors directed along vectors k and p) and T-matrix

$$t(k, k';\ Z) = \sum_l (2L+1)\, t_l(k, k';\ Z)\, P_l(\cos\theta) \tag{1.30}$$

over eigenstates $E_{L\lambda}^{lM}$ of the total angular momentum l ($l = L + \lambda$) where L is an orbital moment of a pair of particles and λ is an orbital moment of a third particle relative the center of inertia of the pair) and its projection M on a fixed axes

$$\mathcal{Y}_{l\lambda}^{LM}(n_k, n_p) = \sum_{m,\mu} (lm\lambda\mu \,|\, LM)\, Y_{lm}(n_k)\, Y_{\lambda\mu}(n_p),$$

where $\cdot lm\lambda\mu | LM$ are Klebsh-Gordan coefficients [91]. Inserting expansions (1.29) and (1.30) in Eqs.(1.2), we get for $\psi_{l\lambda L}$ the infinite set of two-dimensional integral equations. In the case of paired interactions with a finite radius of action manifested in a finite number of states of the relative orbital momentum, it is possible to approximate the infinite set of integral equations by a finite one. Using then the separable approximation for T-matrices (1.11), (1.12), we obtain a finite set of integral equations for the function of one variable of the type [41, 51, 55-59]:

$$\Psi_i(x) = \Phi_i(x) + \sum_j \int_0^\infty Q_{ij}(x, y)\, \tau_i(y)\, \Psi_j(y)\, y^2\, dy, \tag{1.31}$$

where

$$Q(x, y) = \int_{-1}^{1} \frac{F(x, y, u)\, du}{E - x^2/2m_1 - y^2/2m_2 - (x^2 + y^2 + 2xyn)/2m_3}, \tag{1.32}$$

F(x, y, u) is the differentiable function at real positive values of. After the integration of (1.32) we get

$$Q(x, y) = B_0(x, y) + B_1(x, y) \ln A(x, y),$$

where B_0 and B_1 do not have singularities at real x and y and

$$\begin{aligned} A(x, y) &= (y - \nu)(\mu + y)/[(y + \nu)\quad(y - \mu)], \\ \gamma(x) &= -m_i x/(m_i + m_j) + a_i \sqrt{b_i^2 - x^2}\Big/b_i, \\ \mu(x) &= m_i x/(m_i + m_j) + a_i \sqrt{b_i^2 - x^2}\Big/b_i, \\ a_i^2 &= m_j m_k E/(m_j + m_k), \\ b_i^2 &= m_i(m_j + m_k)\, E\Big/\sum_i m_i, \end{aligned} \tag{1.33}$$

where τ_i are paired T-matrices, Φ_i is the wave function describing the initial state of the three-body system. Later on to simplify calculations, we shall consider an equation (1.31). E is the total energy of the three-body system and m_i are masses of particles.

For instance, in the case of three identical particles $m_1 = m_2 = m_3 = m$ (scattering of a neutron on a deuteron) interacting in s-states via separable potentials (1.11) and (1.12), Eqs.(1.31) have the form:

$$\Psi_{iL}^J(k) = \chi_{i1}^J K_{i1L}(k) + \sum_{j=1}^{N_J} \chi_{ij}^J \int_0^\infty \frac{K_{ijL}(k, x)\, \tau_j \Psi_{jL}^J(x)\, x^2\, dx}{E - 3x/4m - i\varepsilon},$$

$$K_{ijL}(k,k') = \frac{\pi}{3}\int_{-1}^{1} \frac{P_L(x)\,V_i(|k'+k/2|)\,V_j(|k+k'/2|)\,dx}{k^2+k'^2+kk'x-mE-i\varepsilon} =$$

$$= \frac{2\pi}{3(kk')^3}\left[\frac{Q_L(-b_3)}{(b_1-b_2)(b_1-b_3)} + \frac{Q_L(-b_2)}{(b_2-b_1)(b_2-b_3)} + \right.$$

$$\left. + \frac{Q_L(-b_3)}{(b_2-b_1)(b_3-b_2)}\right],\ b_1 = \frac{k^2+k'^2-mE}{kk'},$$

$$b_2 = \frac{k^2/4+k'^2+\beta_i^2}{kk'},\ b_3 = \frac{k^2+k'^2/4+\beta_i^2}{kk'}, \tag{1.34}$$

$$\tau_i(y) = \frac{6\beta_j(\alpha_i-\beta_j)^2\left(\beta_j - i\sqrt{m}\,y\right)^2}{\pi^2\left(2\beta_j+\alpha_j-i\sqrt{m}\,y\right)\left(\alpha_i-i\sqrt{m}\,y\right)},$$

$$N_J = 3 \text{ при } S = {}^1\!/_2, \qquad \chi_{ij}^{1/2} = \begin{pmatrix} {}^1\!/_2 & -{}^1\!/_2 & -1 \\ -{}^3\!/_2 & -{}^1\!/_2 & -1 \\ -{}^3\!/_2 & -{}^1\!/_2 & 0 \end{pmatrix},$$

$$N_J = 1 \text{ при } S = 3/2, \qquad \chi_{ij}^{3/2} = \begin{pmatrix} -1 & 0 & 0 \\ 0 & 0 & 0 \\ 0 & 0 & 0 \end{pmatrix}.$$

The numerical solution of the integral Eqs.(1.31) is connected with some difficulties because kernels of the integral equations (1.31) have a number of singularities (Fig. 1):

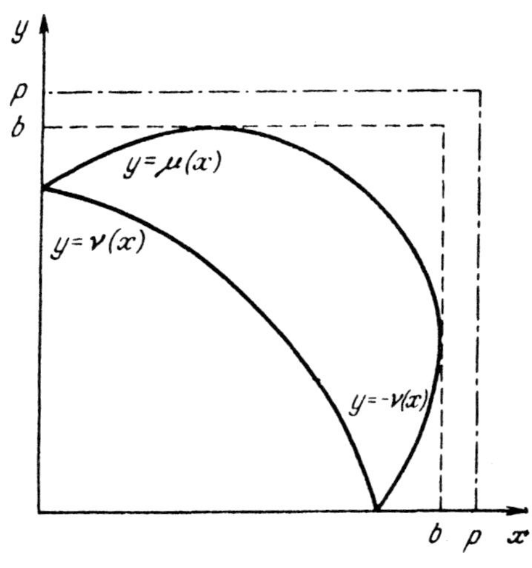

Fig. 1. Representation of the singularities of the integral equations kernels (1.2)

(1) Branch points of the square root type corresponding to thresholds of two-paricle processes, and poles determining bound states of two-particle subsystems in paired T-matrices $\tau(y)$.

(2) Logarithmic singularities $Q(x, y)$ which layon a closed curve and are determined by functions $\gamma(x)$ and $\mu(x)$.

Location of these singularities and their type depends on the form of the paired interaction potentials, masses of colliding particles and the total energy of the three-particle system, as one can see from Eqs.(1.33).

Therefore, different methods of solution of the set (1.31) take into account the following circumstances.

1. It is necessary to get around these singularities deforming the integration contour into a complex plane of a variable y . It is possible only in the case when paired separable potentials are presented in an analytical form and do not introduce additional singularities in kernels of Eqs. (1.31), which happens, for instance, with potentials of the square well or hard core type [72, 99].

2. When integrating along a real y axes, it is necessary to approximate a solution in the vicinity of these singularities: summing up Born series by means of pade-approximant, approximation of the solution with interpolation polynomials, moment methods, variation principles of different types, collocation methods, methods of the artificial separation of singularities [60, 41, 86, 72, 99, 100].

All the above-mentioned methods have one common disadvantage - they are only applicable for some single specific three-particle system. Mainly these are calculations of bound states and scattering states of three identical particles [72, 99].

Therefore, a method suggested [60, 100] of solution of a set of integral equations (1.31) for systems of three particles with arbitrary masses interacting via paired separable potentials of the type (1.11), (1.12). In this method the domain of an unknown function $\Psi(x)$ is divided into a number of intervals. Inside each of these intervals the function $\Psi(x)$ is defined as

$$\Psi(x) \simeq \sum_i L_i(x) \Psi(x_i), \tag{1.35}$$

where L_i are the interpolation functions which are chosen in the form of Legendre polynomials convenient for the description of the function Ψ behaviour in the region of the singularity of the square root type. Inserting (1.35) in (1.31), we get

$$\Psi(x_i) = \Phi(x_i) + \sum_j K_{ij} \Psi(x_j), \tag{1.36}$$

where

$$K_{ij} = \int Q(x_i y)\, \tau(y)\, y^2 L_j(y)\, dy. \tag{1.37}$$

Notice that the integration in (1.37) is only performed in the interpolation region, which contains x_i, rather than from 0 to ∞. Thus Eqs. (1.36) and (1.37) allow to separate the following problems: finding of the reciprocal operator in (1.36) which in a given case is equivalent to the finding of the reciprocal matrix, and a calculation of integrals with singularities by means of corresponding quadrature formulae [91, 92]. A given method of the solution of the set (1.31) is a modification of the standart method of the solution of integral equations [92], where it is suggested for solving (1.31) to change the integral in the right part for some quadrature formula. As a result we obtain a set of algebraic equations for the values of the sought-for function in the nodes of quadrature formula. Solving this set, we find approximations $\tilde{\Psi}(x_i)$ for the values of the sought-for function $\Phi(x_i)$, and then the values of the function itself can be found by means of one or other interpolation method [60, 100]. In the method suggested, the domain of the sought-for function is first divided into a number of intervals where it is defined by means of interpolation polynomials reproducing right behaviour of the function in the vicinity of singularities mentioned above, and then the integration is performed by quadrature formulae.

Consider how the region must be divided in the presence of the singularity of the square root type at a point y = b, which has a form $(b^2 - y^2 + i\varepsilon)^{1/2}$. In this case we have two interpolation regions divided by a point b . Choosing new variables ν and ω

$$y^2 = b^2 - \nu^2, \ y \leqslant b,$$
$$y^2 = b^2 + \omega^2, \ y > b,$$

one can use for L_j Lagrange's interpolation formula over five points:

$$L_j(y) = \begin{cases} \prod\limits_{k \neq j} \nu - \nu_k / (\nu_j - \nu_k), & y \leqslant b, \\ \prod\limits_{k \neq j} \omega - \omega_k / (\omega_j - \omega_k), & y > b. \end{cases}$$

To the right of this region one can use Lagrange's interpolation formula over three points

$$L_j(y) = \prod_{k \neq j} y - y_k / (y_j - y_k).$$

To the left of this region, choosing a new variable $y = y_0 + Z/(1 - Z)$, the integration is performed by means of Gauss quadrature formula [92], in which a point Z is included in the system of nodal points.

Notice that the root square singularuty can be separated explicitly, taking into account analytical properties of Eqs.(1.31). In this case Ψ is presented as a sum of two functions, one of which is continuous Ψ_0 while another one has the square root singularity at the point x = b:

$$\Psi(x) = \Psi_0(x) + C \exp\,[-(x - b)^2]\sqrt{x - b}\,\Psi(b/2).$$

Therefore after transition to a new function Ψ_0 the solution of Eq. (1.31) is considerably simplified. Other singularities can be also separated by a similar method [60, 100].

Consider now calculations of integrals (1.36) using standart techniques. If singularities in the integrand (1.37) are absent, the nodes of the quadrature formula can be used both for integration and finding an approximate solution [60]. In the case of logarithmic singularities one can use the representation (1.34). Then

$$K_{ij} = K_{1ij} + K_{2ij},$$

where K_{1ij} are calculated as in the absence of singularities, and

$$K_{2ij} = \int_a^b B_1(x, y)\,\tau(y)\,L_i(y)\ln A(x, y)\,y^2\,dy.$$

Suppose that y = S determines a location of the logarithmic singularity. Then, the following cases may appear.

1. S is a point at the end of the integration contour S = b or S = a and does not belong to the region of the square root singularity. Then Lagrange's interpolation formula over three points can be used for L_j, and the integral can be calculated by means of the quadrature formula for the integration of functions with logarithmic singularity which have a degree of accuracy 2n - 1, and their weights and nodes can be found [91, 92]. For concrete calculations it is usually taken n = 4, which is sufficient for a comparision with experimental data available.

2. a < S < b. Using a representation

$$\int_A^b = \int_A^S + \int_S^b,$$

one can perform calculations as in p.1.

3. S = a and y belongs to the region of the square root singularity. In this case calculations are performed as in p.1 after a change of variable $y^2 = b^2 - y'^2$. Lagrange's interpolation polynomials of fourth degree are used as $L_j(y')$.

4. a < S < b and y belongs to the region of the square root singularity. Calculations are performed as in pp.2, 3.

In the case of singularity of the pole type

$$\tau(y) = \bar{\tau}(y)/(y - p - i\varepsilon),$$

kernels of (1.37) at $y_j < p$ can be presented in the form:

$$K_{ij} = \int_0^{\sqrt{p^2-b^2}} Q(x_i, y') \frac{\bar{\tau}(y') L_j(y') y'y\, dy}{y' - p - i\varepsilon}, \quad (1.38)$$

where $y'^2 = b^2 - y^2$ and Lagrange's interpolation formula over five points is used as L_j. Notice that the integrand in Eq.(1.38) is limited because zeroes of Lagrange's polynomial cancel with zeroes in the denominator. Hence forcalculation K_{ij} Gauss quadrature formula is used. This is also valid for $y_j > p$. At $y = p$ we have

$$K_{ij} = i\pi Q(x_i, p)\,\bar{\tau}(p)\,p + \int_0^{\sqrt{p^2-b^2}} Q(x_i, y') \frac{\bar{\tau}(y') L_j^1(y') y'y\, dy}{y' - p - i\varepsilon} + \quad (1.39)$$

$$+ \int_{\sqrt{p^2-b^2}}^{2\sqrt{p^2-b^2}} Q(x_i, y') \frac{\tau(y') L_j^2(y') y'y\, dy}{y' - p - i\varepsilon}.$$

Integrals in Eq.(1.39) at $y_j = p$ are limited and can be calculated using common Gauss quadrature formulas. Notice that Simpson quadrature formula or other quadrature formulas, which require knowledge of integrands on the ends of the integration contour, are not applied here, because the integrand in Eq.(1.39) has a singularity of the type 0/0 on the end of the integration contour.

At the point $a(y = 0)$ an argument of logarithm has the form

$$A(x, y) = (y + \mu(x))(y + 0)/[(y - \mu(x))(y + 0)]$$

and singularities are absent.

At the point a at $x = 0$ two logarithms of the kernel degenerate into a pole at $y = a$, and calculations are performed as in the case of the pole. In regions, where all three types of singularitied are present, Eq.(1.37) may be calculated using various combinations of methods described above.

The main advantage of the method suggested above consists in its universality. It is applicable to calculations of bound states and scattering states of the system of three particles having different masses and interacting via arbitrary paired short-ranged potentials, which can be defined numerically as well. This advantage distinguishes the method among others which can be only applied either to calculations in the system of three bodies with paired potentials allowing only an analytical representation (as in the method of deformation of the integration contour into a complex plane) or only to the system of three particles with identical masses.

To realize the considered numerical method of solution of the set of integral equations (1.31) the software package in FORTRAN [60, 100] has been used. The operating sequence of the package is as follows: INTRO; PLOTK; GEOMET; CHECK; GRID; MOON; SEACH; MTRX; BLOCK1-BLOCK9; CARRY; TAUPL; NOPOLE; QEQO; LOGG; STRAIGT; FINAL; SOLVE; BREAK.

Subroutine INTRO is intended to check and analyse correctness of input data and organization of calculations of bound states in three-body system or scattering states. In a latter case a number of calculations can be performed both for different energies of the impacting particle and an orbital moment fixed, and for different orbital moments and the impacting particle energy fixed. A detailed description of input data and their formats is presented in Ref.[60]. Notice here that the initial data entry is organized in such a way that data, which are entering in an incorrect format or have an improper order number, are discarded automatically, counting is terminated and a correspoding error diagnostics appears.

PLOTK is intended for determination and graphic display of singularities of kernels in the set of integral equations (1.31) in the plane of variables y, x depending on the impacting particle energy, particles masses and paired potentials operating in the three-particle system. This is necessary in any numerical method of solution of the set (1.2), and if a series of calculations is performed with paired potentials of different forms or with different parameters, one can obtain a spatial image of these singularities using subroutine PLOTDM [63, 64]. These subroutines can run in computers with a minimum volume of computer graphics.

GEOMET fills an array containing numerical values of the integrand in Eq.(1.32) consisting of Klebsh-Gordan coefficients, 6-j and 9-j symbols, Legendre polynomials etc.

CHECK performs control and analyses of the entry information for the GEOMET subroutine.

CRID is intended to choose nodes and weights of the quadrature formula according to the kernels singularities in Eq.(1.32) for transforming the set of integral equations to the set of algebraic equations (1.36). If a user is not satisfied by the standart set of nodal points (as may happen at calculations of three-body systems with different masses or with paired potentials defined in a numerical form), then using MOON subroutine, one can constract any set of nodes and weights of corresponding quadrature formula.

MTRX controls calculations of the value TRIX = I - K_{ij} where κ_{ij} is defined in Eq.(1.37). Subroutines BLOCK1 - BLOCK9 perform calculations of TRIX in the regions where the integrand singularities can be separated explicitly, for instance, in the case of analytical representation of paired interaction potentials.

In this case the following variations are possible: (1) an absence of singularities in the integration region; (2) singularities may be accounted exactly using standart quadrature formulae as in the case of singularities of the square root type and logarithmic singularities.

CARRY performs calculation of the TRIX value in those regions where singularities can not be separated explicitly, for instance, when paired interaction potentials in the three-body system are specified numerically. This subroutine activates the following subroutines:

TAUPL calculates TRIX in the region of pole τ;≃

NOPOLE and LOGG calculate TRIX in the region of logarithmic kernel singularities.

QEQO calculates TRIX in the region of singularities of the square root type;

STRAIGT calculates in the region where logarithmic singularities are absent.

FINAL calculates phases of the elastic scattering and absorption coefficients, and activates the following subroutines:

SOLVE is intended to find a solution of a set of algebraic equations;

BREAK calculates cross-sections of the three-particle decay, e.g. of processes of the type 1 + (2, 3) → 1 + 2 + 3.

SEACH subroutine is used for calculations of bound states of three-particle systems , which by means of subroutines MTRX and SOLVE calculates TRIX determinant with a prescribed accuracy.

POT, TAUQ, and NORM subroutines are used for calculations of various characteristics of paired subsystems:

POT calculates form-factors $V_i(k, \beta_i)$ for paired separable potentials (1.11) and (1.12);

TAUQ calculates τ by solving Lipmann-Schwinger equations (1.3) for the potential (1.11);

NORM calculates a normalization constant for the wave function of the bound state with the potential (1.11).

An accuracy of solution of the set of integral equations (1.31) by the method described above is connected with an uncertainty in values of the elastic scattering phases in paired subsystems, by which paired potentials (1.11) are restored, and with dimension of the set of algebraic equations (1.36), and is defined by the following expression [60, 101]:

$$\Delta\Psi/\Psi = \max\,(2/\Delta,\ 4N\delta/\delta_s),$$

where N is the number of equations (1.36), Δ is the number of divisions of the integration interval, δ/δ_s is the ratio of the rejected phase to S-phase. As is seen from the formula, to increase an accuracy of solution, one needs to increase a number of divisions of the integration interval and dimension of the set of algebraic equations, which, in turn, requires increasing a number of partial amplitudes considered and results again in an increase of the set (1.36) dimension. This principally restricts an accuracy of solution of the three-particle problem by this method. In reality an error can not be less than 10% for characteristics of the elastic scattering both for separable and local interaction potentials, and is equal to 20 - 30% in the case of the inelastic scattering [60, 101].

However, the accuracy of solution of the set of integral equations can be increased up to 5 - 10% by increasing a dimension of the set of equations (1.36) and decreasing a time of solution, using special-purpose multiprocessing systems, in particular, two-processing system, consisting of a EC-1045 computer with a EC-2345 matrix processor [60 - 62].

2. DIFFERENTIAL EQUATIONS OF THE QUANTUM THEORY OF SCATTERING

2.1. Formulation of differential equations

For a description of the system of three particles with masses m_α, α = 1, 2, 3 in the coordinate space,« relative coordinates x_α, y_α are used, which are connected with particle coordinates as follows [41-54]:

$$x_\alpha = (2\mu\alpha)^{1/2}(r_\beta - r_\gamma), \qquad (2.1)$$
$$y_\alpha = (2\nu_\alpha)^{1/2}\left(\frac{m_\beta r_\beta - m_\gamma r_\gamma}{m_\beta + m_\gamma} - r_\alpha\right),$$

where $M_\alpha = m_\beta m_\gamma/(m_\beta + m_\gamma)$ is a reduced mass of a pair α, ν_α = $m_\alpha(m_\beta + m_\gamma)\sum_{\alpha=1} m_\alpha$ the reduced mass of the particle α and pair γ. Further on, presenting this part, we shall use results from papers [41-45].

A transition from x_α, y_α to x_β, y_β is performed by the transformation

$$x_\beta = c_{\beta\alpha}x_\alpha + s_{\beta\alpha}y_\alpha, \quad y_\beta = c_{\beta\alpha}y_\alpha - s_{\beta\alpha}x_\alpha, \tag{2.2}$$

where

$$c_{\beta\alpha}^2 + s_{\beta\alpha}^2 = 1.$$

In the momentum space the relative momenta p_α and q_α and corresponding to them coordinates x_α and $y_\alpha$of particles are combined into six-dimensional vectors $X = x_\alpha \oplus y_\alpha$, $K = p_\alpha \oplus q_\alpha$ so that a choice of the definite pair x_α, y_α or p_α, q_α fixes the certain coordinate system and

$$(X, X') = (x_\alpha, x'_\alpha) + (y_\alpha, y'_\alpha),$$
$$(K, K') = (p_\alpha, p'_\alpha) + (q_\alpha, q'_\alpha).$$

Denote unit vectors in the direction X, K, x_α, y_α by $\hat{X}$, $\hat{K}$, $\hat{x}_\alpha$, $\hat{y}_\alpha$. In these designations the kinetic energy of three particles is

$$E = p_\alpha^2 + q_\alpha^2,$$

while the kinetic energy of the pair is $\alpha \mp \varepsilon_\alpha = p_\alpha^2$ and the kinetic energy operator is $H_0 = -\Delta_{x_\alpha} - \Delta_{y_\alpha}$.

Suppose that paired potentials of interaction between particles have the form

$$V_\alpha(x) = n_\alpha / |x| + V_\alpha^S(x), \quad H = H_0 + \sum_\alpha V_\alpha(x_\alpha),$$

where $n_\alpha = e_\beta e_\gamma$, e_α is a charge of a particle α, $V_\alpha^s(x)$ are short-range fast decreasing potentials at $|x| \to \infty$.

Denote a self-conjugate operator h_α of the two-particle subsystem by

$$(h_\alpha f_\alpha)(x_\alpha) = -\Delta_{x_\alpha} f(x_\alpha) + V_\alpha(x_\alpha) f_\alpha(x_\alpha) = (h_0 + V_\alpha) f_\alpha$$

and suppose that:

(1) a positive spectrum of h_α is absolutely continuous;

(2) an invariant part of h_α acting in the subspace corresponding to the absolutely continuous spectrum, is unitary equivalent to the operator of multiplying on p_α in $L^2(R^3)$;

(3) a positive spectrum of h_α consists of finite-fold eigenvalues.

Denote eigenstates of h_α by $\varphi_{\alpha_i}(x_\alpha)$ and their corresponding energies by $\varepsilon_{\alpha,i}$, $\varepsilon_{\alpha,i} > 0$, $i = 1,2,...N$. If Coulomb forces correspond to a repulsion $n_\alpha > 0$ (to an attraction $n_\alpha < 0$), then the number N_α of bound states is finite (infinite).

Denote a two-particle Hamiltonian in the three-particle space by H_α^S

$$\left(H_\alpha^S f\right)(X) = -\Delta_X f + V_\alpha^S(x_\alpha) f.$$

Consider a number of regions in R^6 which will be used for a construction of three-particle Coulomb potentials necessary in the following for description of the coordinate asymptotics of wave functions:

$\tilde{\Omega}_\alpha(a_o)$ is the region in R^6 where the scattering between particles of the pair α is limited by the constant a_o: $\tilde{\Omega}_\alpha(a_o) = \{X;\ |x_\alpha| < a_o\}$;

$\tilde{\Omega}_{0\alpha}(\nu) \in$ is the region where inequalities

$$a_0 \leqslant |x_\alpha| < y_\alpha|^\nu,\ 0 < \nu < {}^1/_2;$$

are valid.

$\Omega_\alpha(\nu) \ni$ is a combination of regions $\tilde{\Omega}_\alpha(a_o)$ and $\Omega_{0\alpha}(\nu)$. In this region a distance between particles of the pair α is far less than a distance to a third particle, if $|X| \to \infty$.

$\Omega_{0\alpha}$ is the rest of the space R^6, where $|x_\alpha| > \max\{|y_\alpha|, a_o\}$.

Denote by Ω_o the intersection of regions $\bigcap_\alpha \Omega_{0\alpha}$.

Denote

$$\chi(|x_\alpha|, |y_\alpha|) = \begin{cases} 1 \text{ at } x_\alpha, y_\alpha \subset \Omega_\alpha(\nu) \\ 0 \text{ in the opposite case} \end{cases} \qquad (2.3)$$

and present

$$V_\alpha(x) = \widehat{V}_\alpha(x) + V^0_\alpha(x),$$

where $\widetilde{V}_\alpha(x) = V_\alpha(x_\alpha) + n_\alpha \chi_\alpha / |x_\alpha|$ is the short-range part of V_α; $V^0_\alpha(x_\alpha) = n_\alpha(1 - \chi_\alpha)/|x_\alpha|$ is a long-range part of V_α.

Introduce

$$H^{as} = -\Delta_X + \sum_\alpha V^0_\alpha = \Delta_X + V^0.$$

Here it should be noted that the perturbation value V^0 can be made as small as one likes by choosing a cutoff constant a_0, and the continuous spectrum H^{as} consists only of a single branch $[0, \infty]$ corresponding to the scattering of particles moving far apart in the presence of weak Coulomb potentials:

$$H_\alpha = H^{as} + \widehat{V}_\alpha = -\Delta_X + V_\alpha + \sum_\beta V^0_\beta.$$

The continuous spectrum H_α consists of two types of branches: $[-æ^2_{\alpha,i}, \infty)$ and $[0, \infty)$ with branches $[-æ^2_{\alpha,i}, \infty)$ corresponding to the scattering of the bound pair α (with the bond energy $\varepsilon_{\alpha,i} = æ^2_{\alpha,i}$, $i = 1,2,\ldots N_\alpha$) in the presence of the long-range potentials V^0_β, $\beta = \alpha$ created by other pairs.

Denote by $R^{as}(Z)$, $R_\alpha(Z)$, $R(Z)$ resolvents of operators H^{as}, H_α, H:

$$R(Z) = (H - Z)^{-1}.$$

Define $M_{\alpha\beta} = V\delta_{\alpha\beta} - V_\alpha R V_\beta$ (where V_α is the multiplying operator)

and $R_{\alpha\beta} = -R^{as} M_{\alpha\beta} R^{as}$. It can be shown [41, 45] that $R_{\alpha\beta}$ satisfies the set of modified equations

$$R_{\alpha\beta} = \delta_{\alpha\beta}\left(R_\alpha - R^{as}\right) - R_\alpha \widehat{V}_\alpha \sum_{\gamma \neq \alpha} R_{\gamma\beta}, \tag{2.4}$$

and if the charges are equal to zero ($n_\alpha = 0$), the set coincides with a common set of integral equations (1.2).

Using (2.4) one can show that the continuous spectrum of H operator consists of $1 + \sum_\alpha N_\alpha$ branches $[-æ^2_{\alpha,i}, \infty)$, $[0, \infty)$

with wave functions $\psi_{\alpha,i}(X, q_\alpha)$ describing scattering of a particle on a bound pair, e.g. processes of the type (1.6) and scattering of free moving particles to which corresponds functions, satisfying to equations

$$(H - E)\psi = 0, \tag{2.5}$$

where

$$E = \begin{cases} q_\alpha^2 - \varkappa_{\alpha,i}^2 & \text{for processes (1.6) described by wave functions } \psi_{\alpha,i}(X, q_\alpha) \\ K^2 & \text{for processes described by wave functions } \psi_o(X, K). \end{cases}$$

It is convenient to write equations (2.5) in the form

$$\left(H^{as} - E\right)\psi_i = -\widehat{V}\psi_i,\ \widehat{V} = \sum_\alpha \widehat{V}_\alpha. \tag{2.6}$$

Then

$$\psi = -R^{as}\widehat{V}\psi, \tag{2.7}$$

and defining

$$\Phi_{\alpha,i} = -R^{as}V_\alpha\psi_i \tag{2.8}$$

we get

$$\sum_\alpha \Phi_{\alpha,i} = -R^{as}\sum_\alpha V_\alpha\psi_i = -R^{as}\widehat{V}\psi_i = \psi, \tag{2.9}$$

$$\Phi_{\alpha,i} = -R^{as}\widehat{V}_\alpha\sum_\beta \Phi_{\beta\alpha}.$$

Acting on the right and left parts with an operator (H^{as} - E), we obtain a set of equations for components $\Phi_{\alpha i}$ (hereafter we shall omit index i)

$$\left(H^{as} - E\right)\Phi_\alpha = -V_\alpha\left(\Phi_\alpha + \sum_{\beta\neq\alpha}\Phi_\beta\right)$$

or

$$\left(H^{as} + \widehat{V}_\alpha - E\right)\Phi_\alpha = -\widehat{V}_\alpha\sum_{\beta\neq\alpha}\Phi_\beta. \tag{2.10}$$

The set of equations obtained is modified differential equations for the case of scattering of charged particles. Mathematically rigorous foundation of a given derivation (2.10) can be obtained by means of integral equations (2.4).

In calculations for likely charged particles, the equations are found to be convenient in which the interaction $n_\alpha/|x|$ is contained in Hamiltonian $H^c = -\Delta_x + \sum n_\alpha/|x|$, and for the total Hamiltonian we have

$$H = H^c + \sum_\alpha V_\alpha^S = H^c + V^S.$$

The equation for WF with this Hamiltonian has the form

$$\left(H^c + V_\alpha^S - E\right) U_\alpha = - V_\alpha^S \sum_{\beta \neq \alpha} U_\beta, \tag{2.11}$$

where components are defined by the relation

$$U_{\alpha,i} = - R^c\, V_\alpha^S \psi_i. \tag{2.12}$$

The wave functions $\psi_{\alpha,i}$, $\Phi_{\alpha,0}$ and their components satisfying to equations (2.10) have substantially different asymptotics in infinity. Notice that asymptotics of functions ψ_α and $\Phi_{\alpha,0}$ were studied in detail in the case of particles with like charges ($n_\alpha > 0$) for the short-range interaction potentials V_α^s each having a single bound state $\varphi_\alpha(x_\alpha)$ with bond energy $\varkappa_\alpha^2$ [41-45]; the asymptotics of $\Psi_\alpha(x_\alpha)$ is described by the formula

$$\Psi_\alpha = \sum_\beta U_{\alpha\beta}(X, q_\alpha). \tag{2.13}$$

Here $X \to \infty$, $X \subseteq \Omega_\alpha$, particles in a pair α are sufficiently close to each other, and the coordinate asymptotics of $\psi_\alpha(X, q_\alpha)$ contains only terms corresponding to particles elastically scattered relative y_α coordinate which is nonparallel to the momentum q_α:

$$U_{\alpha\alpha} = \varphi_\alpha(x_\alpha)\Big\{\exp\left[i(q_\alpha y_\alpha) + iW_\alpha^0\right](1 + O(1)) + \\ + A_{\alpha\alpha}\left(\widehat{y}_\alpha, q_\alpha\right)\frac{\exp\left[i\,|q_\alpha|\,|y_\alpha| + iW_\alpha\right]}{|y_\alpha|}[1 + O(1)]\Big\}. \tag{2.14}$$

where

$$W_\alpha^0 = n_{\alpha\alpha}\ln(|q_\alpha|\zeta_\alpha),\ n_{\alpha\alpha} = \sum_{\beta\neq\alpha} n_\beta/|S_{\alpha\beta}|,$$
$$W_\alpha = -\widetilde{n}_{\alpha\alpha}\ln(2\,|q_\alpha|\,|y_\alpha|),$$
$$\widetilde{n}_{\alpha\alpha} = n_{\alpha\alpha}/2\,|q_\alpha|,\ \zeta_\alpha = |y_\alpha| - (q_\alpha, y_\alpha),$$
$$S_{\alpha\beta}^2 = m_\gamma \sum_\alpha m_\alpha/[(m_\alpha + m_\gamma)(m_\beta + m_\gamma)], \tag{2.15}$$

where $A_{\alpha\alpha}$ is the amplitude of the elastic scattering cross-section. If y_α is close to $\hat{q}_\alpha$ ($\zeta \to 0$) then the formula (2.14) is invalid because W^o_α and $A_{\alpha\alpha}$ become infinity. In this case the asymptotics has the form

$$U_{\alpha\alpha} = \varphi_\alpha(x_\alpha) \exp[i(q_\alpha x_\alpha)] - F_\alpha(\zeta_\alpha) + \\ + \varphi_\alpha(x_\alpha) \widetilde{A}_{\alpha\alpha}(\widehat{y}_\alpha, q_\alpha) \frac{\exp[i|q_\alpha||y_\alpha| + iW_\alpha]}{|y_\alpha|}(1 + O(1)), \tag{2.16}$$

where the function $F_\alpha(\zeta_\alpha)$ is expressed via the degenerate hypergeometric function $\Phi(a, b, t)$ [91]:

$$F_\alpha(\zeta_\alpha) = e^{\frac{-\pi \widetilde{n}_{\alpha\alpha}}{2}} \Gamma(1 + in_{\alpha\alpha}) \Phi(-\widetilde{i n}_{\alpha\alpha}, 1; i/q_\alpha|\zeta_\alpha|). \tag{2.17}$$

In Eq.(2.16) $\widetilde{A}_{\alpha\alpha}$ is a smooth function for all directions of y_α, and if $\zeta_\alpha \to \infty$, then $\Phi(a, b, t)$ can be changed for the asymptotics at $t \to \infty$. Then Eq.(2.16) can be transformed to Eq.(2.14) where the incident and scattered flows are separated. In this case the amplitude $A_{\alpha\alpha}$ is presented in the form

$$A_{\alpha\alpha} = \widetilde{A}_{\alpha\alpha} + A^S_{\alpha\alpha},$$

where the singular part of $A^S_{\alpha\alpha}$ is presented in the form

$$A^S_{\alpha\alpha}(q_\alpha, y_\alpha) = \frac{n_{\alpha\alpha} \exp\left[-i\widetilde{n}_{\alpha\alpha} \ln \sin(\Theta_\alpha/2) + 2i\eta_\alpha\right]}{2|q_\alpha| \sin^2 \Theta_\alpha/2}, \tag{2.18}$$

In the region Ω_o, where all three particles are far apart, the asymptotics has the form

$$U_{\alpha\alpha}(X, q_\alpha) = \frac{\exp\left[i\sqrt{E}|X| + iW_0\right]}{|X|^{5/2}}\left[A_{\alpha\alpha}(X, q) + O(1)\right]- \tag{2.19}$$

where

$$W_0 = -(2\sqrt{E})^{-1}|X| \sum_\alpha n_\alpha \ln(2\sqrt{E}|X|)/|x_\alpha|. \tag{2.20}$$

A square of modulus $A_{\alpha\alpha}$ is proportional to a differential cross-section of decay into three particles and is a smooth limited function for all x directions.

If Coulomb forces between particles in a pair α are attractive, then the amplitude $A_{\alpha\alpha}$ has singularities of the type $|X/|x_\alpha||^{1/2}$ and formulas (2.13)-(2.20) are only valid in the region $\tilde{\Omega}_{o,\alpha}, \nu > 2/3$. Then the WF asymptotics has the form

$$\Psi_\alpha(X, q_\alpha) = \tilde{U}_{\alpha\alpha}(X, q_\alpha) + \tilde{U}^0_\alpha(X, q_\alpha), \tag{2.21}$$

where $\tilde{U}_{\alpha\alpha}$ are defined by formulas(2.13)-(2.17) depending on the direction of scattering, while a component $\tilde{U}^o_\alpha$ at $|x_\alpha| \to \infty$ transforms into a distorted spherical wave and is described by the formula:

$$\tilde{U}^0_\alpha(X, q_\alpha) = \left(2\pi \sqrt{E}\right)^{-3/2} \int dp A^0_\alpha\left(q_\alpha, \frac{|p|}{\sqrt{E}}, p, y_\alpha\right) F(p, x_\alpha) \times$$

$$\times \exp\left\{i \sqrt{\varepsilon(p)}\, |y_\alpha| - i\tilde{n}_{\alpha\alpha} \ln 2\varepsilon(p)\,|y_\alpha|/2\varepsilon(p) + i n_\alpha \ln \frac{|p|}{\sqrt{E}} \Big/ |p|\right\},$$

$$F(p, x_\alpha) = \sum_{l=0}^{\infty} \frac{2l+1}{2} \exp\left(i\frac{\pi l}{2} - i\eta_l + i\frac{3\pi}{4}\right) P_l(\cos\Theta) \times$$

$$\times (2|p||x_\alpha|)^{l+1} \exp\left(-\frac{n_\alpha \pi}{4|p|} + i|p||x_\alpha|\right) \Phi_l(l+1+in_\alpha/2|p|, 2l+2,$$

$$-2i|p|x_\alpha|) \frac{\Gamma(l+1+in_\alpha/2|p|)}{(2l+1)!},$$

$$\varepsilon(p) = \sqrt{E - p^2}, \; \cos\Theta_\alpha = (x_\alpha, \hat{p}), \; \eta_l = \arg\Gamma(l+1+in_\alpha/2|p|). \tag{2.22}$$

An amplitude A^o_α in Eq.(2.22) are defined by the formula (2.19). Asymptotics of integrals defining $\tilde{U}^o_\alpha$, are investigated in detail in Ref.[41-45], where it has been found that the asymptotics (2.22) has the form of the distorted spherical wave if $|x_a|^2 \gg |y_\alpha|^2$. In the region $\Omega_{\alpha_o}(a, \nu)$, $\nu > 2/3$ the function $\tilde{U}^o_\alpha$ decreases faster than the spherical wave, i.e. $\tilde{U}^o_\alpha \simeq o\,(|X|^{5/2})$, and does not contribute to the WF asymptotics in the $\tilde{\Omega}_\alpha$ region, where the distance between particles in a pair is limited.

If $|X| \to \infty$, $X \subseteq \Omega_\beta$, $\beta \neq \alpha$, the asymptotics of $\Psi_\alpha(X, q_\alpha)$ is determined by the terms describing rearrangement processes:

$$\psi_\alpha(X, q_\alpha) \sim U_{\alpha\beta}(X, q_\beta),$$

$$U_{\alpha\beta}(X, q_\alpha) = \varphi_\beta(x_\beta) \frac{\exp\{i\sqrt{E_\beta}\,|y_\beta| + iW_{\alpha\beta}\}}{|y_\beta|} A_{\alpha\beta}(\widehat{y}_\beta, q_\alpha)\,[1 + O(1)], \tag{2.23}$$

$$E_\beta = q_\alpha^2 - \varkappa_\alpha^2 + \varkappa_\beta^2, \tag{2.24}$$

where $A_{\alpha\beta}$ are smooth functions of their arguments proportional to the amplitudes of the rearrangement process:

$$W_{\alpha\beta} = -\sum_{\gamma\neq\beta} n_\gamma \ln(2\sqrt{E_\beta}\,|y_\beta|)/2\,|S_{\gamma\beta}|\sqrt{E_\beta}. \tag{2.25}$$

In the region Ω_{β_o} a summand $\overset{o}{U}_{\alpha\beta}$ is added to the term $\tilde{U}_{\alpha\beta}$ which is expressed explicitly through the amplitude $\overset{o}{A}_\alpha$, where instead of variables $|X|$, $|x_\alpha|$, $|y_\alpha|$, $|q_\alpha|$ it is necessary to put $|X|$, $|x_\beta|$, $|y_\beta|$, $(E_\beta)^{1/2}$.

In the region Ω_o the asymptotics of $U_{\alpha\beta}$, $b \neq \alpha$ has the form (2.19) and a sum of all amplitudes $\overset{o}{A}_{\alpha\beta}$ coincides with the decay amplitude

$$A_\alpha^0 = \sum_\beta A_{\alpha\beta}^0,$$

which square of modulus is proportional to the differential cross-section of the process of decay into three free particles.

The asymptotic behaviour of WF components $U_{\alpha\beta}$, satisfying to modified equations (2.10), is as follows.

In the case of the elastic scattering in the system of three particles the asymptotics is determined by terms $U_{\alpha\alpha}$, and (1) in the region Ω_α this term is defined by formulas (2.13)-(2.17) depending on the mutual orientation of vectors y_α and q_α, (2) in the region $\Omega_o(a, \nu)$ at $\nu > 2/3$ $U_{\alpha\alpha}$ is described by the distorted spherical wave (2.20) with the amplitude $\overset{o}{A}_{\alpha\alpha}(X, q_\alpha)$ (2.18).

In the case of the inelastic scatttering processes (rearrangement processes) the asymptotics of nondiagonal elements $U_{\alpha\beta}$, $\alpha \neq \beta$ in Ω_β, where a β pair is poorly separated, coicides with that for WF $U_{\alpha\beta} \sim \tilde{U}_{\beta\alpha}$.

In transition regions $\tilde{\Omega}_{\gamma_o}$ (γ = 1, 2, 3) the asymptotics of components $U_{\alpha\beta}$ is more complex. The asymptotic form of $U_{\alpha\beta}$ in the $\Omega_{\beta o}$ region transforms smoothly from two-particle to three-particle one and has the form $U_{\alpha\beta} \sim \tilde{U}_{\alpha\beta} + \tilde{U}^{o}_{\alpha\beta}$, where $U^{o}_{\alpha\beta}$ is defined by Eq. (2.22), where $A^{o}_{\alpha\beta}$ have to be used instead of $A^{\cup}{}_{\alpha}$. In the region Ω_{γ}, where particles of a γ pair at $\gamma \neq \beta$ are close to each other the asymptotics of $U_{\alpha\beta}$ contrary to the total wave function does not contain in $\tilde{\Omega}_{\gamma}$ slowly decreasing summands $\tilde{U}_{\alpha\gamma}$ and is described by the expression of the type (2.22).

The detailed form of these asymptotics is presented in Ref.[41-45]. Notice that it is important to take into account in specific calculations that in these regions components decrease similar to Eq.(2.19) and the asymptotic behaviour of the function $|X|^{5/2}\exp\{(E)^{1/2}|X|\}U_{\alpha\beta}$ is sufficiently smooth.

In conclusion notice that the existence and uniqueness theorem for the above-considered equations proved in papres [41-45] defines a mathematically correct problem.

2.2. Separation of angular variables and asymptotics of radial components in the integro-differential equations

At present it is impossible to solve directly using computers the set of Eqs. (2.10), (2.13 - 2.20). Hence further simplifications are needed. In particular, such simplifications are appropriate at solving problems of low-energy scattering and also problem of bound states. In this case the equations dimension can be lowered by separating in some way or other the angular dependence, and resulting radial equations can be solved numerically with a computer. Consider a separation of angular variables for the case of three identical particles without spins [41 - 45]. In this case it is convenient to introduce the following notations: $(X_1, Y_1) \rightarrow (X, Y)$, $(X_2, Y_2) \rightarrow (X', Y')$, $(X_3, Y_3) \rightarrow (X, Y)$

Let S_3 is the group of permutations with elements

$$\pi = \begin{pmatrix} 1 & 2 & 3 \\ i_1, & i_2, & i_3 \end{pmatrix}, \quad \pi_0 = \begin{pmatrix} 1 & 2 & 3 \\ 1 & 2 & 3 \end{pmatrix}.$$

To construct the S_3 group representation in $L^2(R^\sigma)$ we introduce the operator $P\pi f(r_1, r_2, r_3) = f(r_{i1}, r_{i2}, r_{i3})$ which is a unitary one while the mapping $\pi \to P\pi$ realizes the S_3 group representation in $L^2(R^\sigma)$.

Using operators

the total WF for the system of three identical particles can be presented in the form [41]

$$\psi = (I + P^+ + P^-)U, \tag{2.26}$$

where $U(x, y) = U_1(x_1, y_1)$ and I is a unit operator of the unitary representation corresponding to π_0. By means of Eq. (2.26) modified differential equations can be written in the form

$$\left(-\Delta_x - \Delta_y + V + \sum_\beta V^0_\beta(x_\beta, y_\beta) - E\right) U(x, y) = \widehat{V}(x, y)\left(P^+ + P^-\right) U(x, y), \tag{2.27}$$

$$V = n/|x| + V^S, \; \widehat{V} = V^S + n\chi_S(|x|, |y|)/|x|,$$
$$V^0_2 = n(1 - \chi_s(|x|_s|y|)/|x'|),$$
$$V^0_3 = n(1 - \chi_s(|x|_s|y|)/|x''|).$$

$$P^+ = P\pi^+ = \begin{pmatrix} 1 & 2 & 3 \\ 2 & 3 & 1 \end{pmatrix}, \quad P^- = P\pi^- = \begin{pmatrix} 1 & 2 & 3 \\ 3 & 1 & 2 \end{pmatrix} \tag{2.28}$$

If one sets formally $\chi_s = 0$ in Eq. (2.27), the modified equations (2.27) transform into equations

$$\left(-\Delta_x - \Delta_y + V^S + \sum V^c_\beta - E\right) U(x, y) = V^S\left(P^+ + P^-\right) U(x, y);$$
$$V^c_\beta = n_\beta/|x_\beta|. \tag{2.29}$$

Denote by I, λ operators of the orbital moment of a pair (2,3) and particle 1 relative of the pair center of mass:

$$\mathbf{l} = -ix_\nabla, \; \boldsymbol{\lambda} = -iy_\nabla,$$

Their eigenfunctions are spherical functions $Y_{em}(\hat{x})$, $Y_{em}(\hat{y})$. In the representation of the total orbital moment $L = l \oplus \lambda$ we have [18 - 22],41 - 45]:

$$Y_{l\lambda}^{LM}(x, y) = \sum_{m_l m_\lambda} \langle l m_l \mid m_\lambda L M \rangle \, Y_{l m_l}(\widehat{x}) \, Y_{\lambda m_\lambda}(\widehat{y}), \tag{2.30}$$

where $Y_{l\lambda}^{LM}$ are eigenfunctions of operators L^2, L_z, l^2, λ^2. Therefore U(x, y) can be expanded over functions (2.30):

$$U(x, y) = \sum_{\alpha, L} U_\alpha^L(|x|, |y|) \, Y_\alpha^{LM}(\widehat{x}, \widehat{y}) / |x| \, |y|. \tag{2.31}$$

Here and further on $\alpha = \{l, \lambda\}$. To separate angular variables (x, y) in Eq. (2.27) or in Eq. (2.29) one needs to calculate matrix elements of operators $P^{\pm}$, V^S, V_β^o, V in a basis of Y_α^{LM}. Operators V^S, V and $\hat{V}$ do not depend on y and therefore they are diagonal on λ:

$$\langle \alpha L \mid V^S \mid \alpha' L' \rangle = \delta_{\lambda\lambda'} V_{ll'}^{LL'}(|x|),$$

$$\langle \alpha L \mid \widehat{V} \mid \alpha' L' \rangle = \delta_{\lambda\lambda'} \widehat{V}_{ll'}^{LL'}(|x|, |y|).$$

For the equations, matrix elements of operators $P^{\pm}$, V_β^o can be calculated analytically:

$$P^+ \frac{U_\alpha^L(x, y)}{xy} Y_\alpha^{LM}(x, y) = \frac{U_\alpha^L(x', y')}{x'y'} Y_\alpha^{LM}(x', y'),$$

where the relation between x′, y′ and x, y is determined by formulas (2.1) and (2.2) which in the case of identical particles have the form

$$x' = -x/2 + y\sqrt{3}/2, \qquad y' = -\sqrt{3}\,x/2 - y/2.$$

Expanding over Y_α^{LM}, we get

$$P^+ \frac{U_\alpha^L(x, y)}{xy} Y_\alpha^L(x, y) = \sum_{\alpha'} \frac{Y_{\alpha'}^{LM}(\widehat{x}, \widehat{y})}{2|x||y|} \int_{-1}^{1} h_{\alpha\alpha'}^L(x, y, u) \, U_\alpha^L(x', y') \, dx' \, dy', \tag{2.32}$$

where u is an angle between x and y.

Because $U_\alpha^L(x', y')$ depends only on an angle between x′ and y′, expanding in series over Legendre polynomials, we have

$$\frac{U_\alpha^L(x', y')}{|x'|^{l+1} |y'|^{l+1}} = \sum_{k=0}^{\infty} (-1)^k 4\pi \sqrt{2k+1} \, Y_{kk}^{00}(x', y') \times \tag{2.33}$$

$$\times \frac{1}{2} \int_{-1}^{1} du P_k(u) \, U_\alpha^L(x', y') / [|x'| \, |y'|]^{l+1}.$$

Using

$$|\alpha x+\beta y|^{l}\,Y_{l}(\alpha x+\beta y)=\sum_{l_1+l_2=l}(\alpha|x|)^{l_1}(\beta|x|)^{l_2}\times$$

$$\times Y^{l}_{l_1 l_2}(x,y)\,4\pi(2l+2)/(2l_1+1)!\,(2l_2+1)!,$$

we get

$$Y^{l}_{\alpha}(x',y')\,|x'|^{l}\,|y'|^{l}=\sum_{\substack{\lambda=\lambda_1+\lambda_2\\ l=l_1+l_2}} D^{l_1 l_2}_{\lambda_1\lambda_2}\,|y'|^{\lambda_1+\lambda_2}\times$$

$$\times\begin{Bmatrix}\lambda_1 & \lambda_2 & \lambda\\ l_1 & l_2 & \lambda\\ l' & l' & L\end{Bmatrix}\sum_{l'\lambda'}\sqrt{(2\lambda+1)(2l+1)(2l'+1)(2\lambda'+1}\,|x|^{l_2+\lambda_2}\times$$

$$\times\Big[Y^{\lambda_1}_{l_1\lambda_1}(x',y')\oplus Y^{l_1}_{\lambda_2 l_2}(x,y)\Big]_{L}, \qquad (2.34)$$

$$D^{l_1 l_2}_{\lambda_1\lambda_2}=4\pi(-1)^{l_2+\lambda}(\sqrt{3})^{\lambda_2+l_1}2^{-l-\lambda}\left[\frac{(2l+1)!\,(2\lambda+1)!}{(2\lambda_1+1)!\,(2\lambda_2+1)!\,(2l_1+1)\,(2l_2+1)}\right]^{1/2}.$$

From Eqs. (2.33) and (2.34) we obtain the representation

$$P^{+}\frac{U^{L}_{\alpha}(x,y)}{|x||y|}Y^{LM}_{\alpha}(x,y)=\frac{(-1)^{\lambda+l+L}}{4\pi}\sum_{k=0}^{\infty}(-1)^{k}\sqrt{2k+1}\,\frac{1}{2}\int_{-1}^{1}P_{k}(u)\times$$

$$\times\frac{U^{L}_{\alpha}(x',y')\,du}{|x'|^{l+1}\,|y'|^{l+1}}\sum_{\lambda=\lambda_1+\lambda_2}D^{l_1 l_2}_{\lambda_1\lambda_2}\times$$

$$\times\sum_{\lambda' l'}\sqrt{(2l+1)(2\lambda+1)(2\lambda_1+1)(2\lambda_2+1)(2l_1+1)(2l_2+1)}\,|y|^{\lambda_1+l_1}\times$$

$$\times|x|^{l_2+\lambda_2}\begin{Bmatrix}\lambda_1 & \lambda_2 & \lambda\\ l_1 & l_2 & l\\ \lambda' & l' & L\end{Bmatrix}\langle\lambda_1 0 l_1 0|\lambda' 0\rangle\,\langle\lambda_2 0 l_2 0|l' 0\rangle\times$$

$$\times\sum_{l''\lambda''}\sqrt{(2\lambda'+1)(2k+1)(2l'+1)}\,\langle k0\lambda'0|\lambda''0\rangle\,\langle k0l'0|l''0\rangle\times$$

$$\times\begin{Bmatrix}l'' & \lambda'' & L\\ \lambda' & l' & k\end{Bmatrix}Y^{LM}_{\lambda'' l''}(x,y).$$

Comparing with Eq. (2.32), which determines a kernel $h^{L}_{\alpha\alpha'}$, we obtain

$$h^L_{\alpha\alpha''}(x,y,u) = \frac{|x||y|(-1)^{l+\lambda}}{|x'||y'|}(2\lambda+1)(2l+1)2^{-\lambda-1}\times$$
$$\times\sqrt{(2\lambda''+1)(2l''+1)(2\lambda)!(2l)!}\sum_{k=0}^{k_{\max}}(-1)^k(2k+1)P_k(u)\times$$
$$\times\sum_{\lambda=\lambda_1+\lambda_2}(-1)^{l_2}3^{(l_1+l_2)/2}|y|^{\lambda_1+l_1}|x|^{\lambda_2+l_2}|y'|^{-\lambda}\times$$
$$\times|x'|^{-l}\sum_{\lambda' l'}(2l'+1)(2\lambda'+1)\begin{pmatrix}\lambda_1 & l_2 & \lambda'\\ 0 & 0 & 0\end{pmatrix}\begin{pmatrix}\lambda_2 & l_2 & l'\\ 0 & 0 & 0\end{pmatrix}\times$$
$$\times\begin{pmatrix}k & l' & l''\\ 0 & 0 & 0\end{pmatrix}\begin{Bmatrix}l'' & \lambda'' & L\\ \lambda' & l' & k\end{Bmatrix}\begin{Bmatrix}\lambda_1 & \lambda_2 & \lambda\\ l_1 & \lambda_2 & \lambda\\ \lambda' & l' & L\end{Bmatrix}[(2\lambda_1)!(2\lambda_2)!(2l_1)!(2l_2)!]^{-1},$$
$$k_{\max} = (\lambda''+l''+l+\lambda)/2. \qquad (2.35)$$

The action of P^- operator is determined by Eq. (2.32) in which the kernel $h^L_{\alpha\alpha'}$ should be changed by $(-1)^{\lambda+\lambda'}h^L_{\alpha\alpha'}$. Thus, the action of operators P^+ and P^- in the bispherical basis is given by the kernel $\varepsilon_{\lambda\lambda'}h^L_{\alpha\alpha'}$, where $\varepsilon_{\lambda\lambda'} = [1 + (-1)^{\lambda+\lambda'}]/2$.

Matrix elements of operators V^0_2, V^0_3 in the basis of $Y^{LM}_{\lambda l}$ have the form

$$V^{0LL'}_{\alpha\alpha'} = \langle Y^{LM}_\alpha | V^0_2 + V^0_3 | Y^{LM}_\alpha\rangle \equiv V^{2LL'}_{\alpha\alpha'} + V^{3LL'}_{\alpha\alpha'}.$$

For example, using Eq.(2.32), we obtain for V^0_2 at $L = M = \lambda = 0$ the following expression:

$$V^0_2(|x'||y'|) = \sum_k \frac{Y^{00}_{kk}(x,y)}{|x||y|}\frac{1}{2}\int_{-1}^{1} h^0_{kk00}(|x|',|y'|)V^0_2\,du, \qquad (2.36)$$

anf for the h_{kkoo} kernel using Eq. (2.35) we have

$$h^0_{kk00}(|x'|,|y'|) = |x||y|(-1)^k\sqrt{2k+1}\,P_k(u)/|x'||y'|.$$

Then

$$V^0_2(x',y') = \frac{1}{2}\sum_k 4\pi(-1)^k\sqrt{2k+1}\int_{-1}^{1}P_k(u)V^0_2\,du =$$
$$= \frac{1}{2}\sum_k 4\pi(-1)^k\sqrt{2k+1}\,Y^0_{kk}(x,y)\,n\,[N_k+F_k],$$
$$N_k = \int_{-1}^{1}[P_k(u)/|x'|]\,du, \qquad F_k = -\int_{-1}^{1}\chi_s(x',y')\frac{P_k(u)}{|x'|}\,du.$$

Thus

$$\left\langle Y_{\alpha_0}^{*LM} \middle| V_2^0 \middle| Y_{\alpha}^{L'M'} \right\rangle = \frac{1}{2} \sum_k 4\pi (-1)^k \sqrt{2k+1}\, n \tilde{J}_{\alpha\alpha' k}^{LL'} (N_k + F_k),$$

where

$$\tilde{J}_{\alpha\alpha' k}^{LL'} = \int dx\, dy Y_{kk}^{*LM}(x, y)\, Y_{\alpha'}^{L'M'}(x', y') =$$
$$= (-1)^{L+l+l'} [(2l+1)(2l'+1)(2\lambda+1)(2\lambda'+1)]^{1/2} (-1)^k \times \quad (2.37)$$
$$\times \sqrt{2k+1}\, [4\pi (2L'+1)]^{-1/2} \begin{pmatrix} k & l' & l \\ 0 & 0 & 0 \end{pmatrix} \begin{pmatrix} k & \lambda' & \lambda \\ 0 & 0 & 0 \end{pmatrix} \begin{Bmatrix} l' & l' & k \\ \lambda & \lambda' & L \end{Bmatrix} \delta_{LL'},$$

$$V_{\alpha\alpha}^{2LL'} = \sum_k (2k+1)\, n \left(N_k' + F_k'\right) J_{\alpha\alpha' k}^{L} \delta_{LL'},$$

$$J_{\alpha\alpha' k}^{L} = (-1)^k \tilde{J}_{\alpha\alpha' k}^{LL} \delta_{LL'} / \sqrt{2k+1}. \quad (2.38)$$

Similarly

$$V_{\alpha\alpha'}^{3LL'} = \left\langle Y_{\alpha}^{*LM} \middle| V_3^0 \middle| Y_{\alpha'}^{L'M'} \right\rangle =$$
$$= \frac{1}{2} \sum_k (-1)^k (2k+1)\, n \left(N_k'' + F_k''\right) J_{\alpha\alpha' k}^{L} \delta_{LL'},$$

where $N_k'' = N_k(x'')$ are determined by formulas (2.38), where x', y' are substituted by x'', y''. Therefore,

$$V_{\alpha\alpha'}^{0LL'} = n \sum_k (2k+1)\, J_{\alpha\alpha' h}^{L} (N_k + F_k)\, \delta_{LL'}, \quad (2.39)$$

and in the case of Eqs. (2.11) $F_k = 0$ for any k, while

$$N_k = \rho^{-1} 2^{-k-1} \sum_{m=0}^{[k/2]} (-1)^m \binom{k}{m} \binom{2k-2m}{k} J_{k-2m}.$$

$$J_q(a, b) = \int_{-1}^{1} u^q (a^2 - 2abu + b^2)^{-1/2}\, du =$$

$$= \begin{cases} J_q(b, a) \text{ при } a > b > 0,\ a = \cos\theta,\ b = \sin x \sqrt{3}, \\ \dfrac{a^2 + b^2}{2^q (ab)^{q+1}} \displaystyle\sum_{S=0}^{q} (-1)^S \frac{(q/S)\, b^{2S+1}}{(a^2+b^2)^S (2S+1)} \sum_{t=0}^{2S+1} \left(1 + (-1)^{t+1}\right) \times \end{cases}$$
$$\times \binom{2S+1}{t} \left(\frac{a}{b}\right)^t \text{ при } b > a > 0,\ \theta = \operatorname{arctg} |y|/|x|. \quad (2.40)$$

For N_k we get

$$N_k = \rho^{-1} \mu_k(\theta), \quad (2.41)$$

$$\mu_k(\theta) = 2^{-k-1} \sum_{m=0}^{[k/2]} (-1)^m \binom{k}{m} \binom{2k-2m}{k} J_{k-2m}. \quad (2.42)$$

This allows to present the matrix element in the form

$$V_{\alpha\alpha'}^{0LL'} = n\mu_{\alpha\alpha'}^{L}(\theta)\,\delta_{LL'}/\rho, \tag{2.43}$$

$$\mu_{\alpha\alpha'}^{L}(\theta) = 2^{-k-1}\sum_{k=0}^{[k/2]}(2k+1)\,J_{\alpha\alpha' k}^{L}\mu_k(\theta). \tag{2.44}$$

Notice here that functions $\mu_k(\theta)$ for any k according to Eq. (2.40) have a quite different behaviour at atransition through the value of $\theta = \pi/6$, because J_q are represented differently in sectors $\theta < \pi/6$ and $\theta \geq \pi/6$. This results in the disappearance of the smoothness of Coulomb operators $V_{\alpha\alpha'}^{0LL'}$ in R_+^2 along the ray $\theta = \pi/6$. Thus, separating angular variables in Eq.(2.27), we obtain the set of radial modified equations

$$\left(-\partial^2/\partial x^2 - \partial^2/\partial y^2 + V_\alpha^{cf} + n/|x| - E\right) U_\alpha^L + \\ + \sum_{l'\alpha'}\left(V_{ll'}^S + V_{\alpha\alpha'}^0\right) U_\alpha^L = \sum_{l'} V_{ll'} \sum_{\alpha''} \varepsilon_{\lambda\lambda'} \int_{-1}^{1} du h_{\alpha\alpha''}^L U_{\alpha''}^L(|x'|, |y'|), \tag{2.45}$$

where $V_{\alpha\alpha'}^{0}$ is a matrix element (2.43) without δ-factor, V_{ll}^{S} is the representation of the short-range potential in the three-particle basis $\langle\alpha|V^S|\alpha'\rangle = \delta_{\lambda\lambda'}\delta_{LL'}V_{ll'}^S$.

$$V_\alpha^{cf} = l(l+1)|x^2 + \lambda(\lambda+1)|y^2.$$

The following property of Eqs. (2.45) should be noted [40 - 45]. In contrast to the case of the neutral particles, rank of the set (2.45) is not finite, because the set equations are coupled both by means of $V_{ll'}^{S}$ (short-range interaction) and $V_{ll'}^{0}$ (Coulomb interaction) at any fixed value of the total angular moment L.

However if the short-range interaction is assumed to operate only in S-state, then neglecting nondiagonality of the Coulomb interaction, $V_{\alpha\alpha'}^{0}$ (the set (2.45)) can be approximated by yhe set of equations, in which equations corresponding to differen values of L are independent. Thus

$$\left(-\frac{\partial^2}{\partial x^2}-\frac{\partial^2}{\partial y^2}+\frac{L(L+1)}{y^2}+\frac{n}{|x|}+\frac{2n\mu_0(\theta)}{\rho}+V^S-E\right)U^L=$$
$$=-V^S\int_{-1}^{1}du h^L_{L0L0}(u)\,U^L(x',y'),$$
$$V^S=V^S_{00},\ U^L=U^L_{L0},\ V^0_{L0L0}=2n\mu_0(\theta)/\rho.$$

(2.46)

Notice, however, that asymptotics of solutions of the infinite set (2.45) and approximate set (2.46) are different. and asymptotics of (2.45) depend on masses, charges and a geometry of the collision in contrast to asymptotics of (2.46) [130].

Consider a behaviour of $V^0_{\alpha\alpha'}$ in different regions of space Ω_α.

1. The vicinity of the special direction $\theta = \pi/6$

$$J_q(Z,Z)=\frac{1}{Z}\sum_{S=0}^{q}(-1)^S\binom{q}{S}2^{2S+1}/(2S+1)\,J_q(1,1)/Z.$$

(2.47)

Then from Eqs. (2.43) and (2.44) we obtain that in the vicinity of $\theta = \pi/6$ contributions of $\mu_k(\theta)$ to the matrix element $V^0_{\alpha\alpha'}$ are of the same order. Therefore the Coulomb interaction in the bispherical representation is substantially nondiagonal in the vicinity of the special direction.

2. At $b \gg a$ in the region $\tilde{\Omega}_1$ the asymptotics is valid

$$J_q(a,\ b)\sim a^q b^{q+1}/2^q(2q+1), \qquad (2.48)$$

$$\mu_k(\theta)\sim 2^{-k-1}\sum_m(-1)^m\binom{k}{m}\binom{2k-2m}{k}\times$$
$$\times a^{k-2m}/\left[2^{k-2m}(2k-2m+1)\,b^{k-2m+1}\right].$$

(2.49)

It is easy to see from the function $\mu_k(\theta)$ behaviour that the main contribution is given by $\mu_0(\theta)$. In the case of $k \geq 1$, neglecting all $\mu_k(\theta)$, we obtain

$$J^L_{\lambda l\lambda' l'0}=(-1)^{\lambda+\lambda'}\delta_{\lambda\lambda'}\delta_{ll'}, \qquad (2.50)$$
$$V^0_{\lambda l\lambda' l'}=2n\rho^{-1}J^L_{\lambda l\lambda' l'0}\mu_0(\theta)=2n\rho^{-1}(-1)^{\lambda+\lambda'}\mu_0(\theta)\,\delta_{\lambda\lambda'}\delta_{ll'}.$$

Thus from Eqs. (2.48) - (2.50) it follows that the Coulomb interaction in $\Omega_1 \subseteq R^2_+$ is essentially nondiagonal.

To solve modified equations (2.45) uniquely, the boundary conditions should be added. Consider a simpliest system in which particles are only interacting in S-states. In this case we have

$$(-\partial^2/\partial x^2 - \partial^2/\partial y^2 + V^S(x) + 2n\mu_0(\theta)/\rho - E)U =$$
$$= -V^S \int_{-1}^{1} xyU(x', y')/x'y'\, du, \qquad (2.51)$$

where

$U(x, y) = U^0_{00}(x, y)$, $V^S_{ll'} = V^S_{00} = V^S$,
$V^\theta_{\lambda l\lambda' l'} = 2n\mu_0(\theta)/\rho$.

Denote $\rho^2 = x^2 + y^2$, $V = n/x + V^S(x)$, $h_S = -\partial^2/\partial x^2 - \partial^2/\partial y^2 + V$, $h_n = -\partial^2/\partial y^2 + 4n/\sqrt{3}y$ and consider processes of the elastic scattering, rearrangement and decay in which a potential V has a single bound state $\varphi_0(x)$ with the bond energy $\varkappa_0^2$.

The method of costruction of the coordinate asymptotics is as follows. In the case considered, during the movement of the particle 1 an additional interaction of the type $V^{eff} = 4n/\sqrt{3}y$ appears at the interaction with the target (2, 3). Because particles (2,3) are close, $\chi(x, y)$ is nonzero only in the region $\tilde{\Omega}_1 \subset R^2_+$, where $\mu_0(\theta) = 2\sin\theta/\sqrt{3}$ and the integral operator corresponding to the decay of the bound state (2, 3) can be neglected in Eq. (2.51). Then differential operator H in Ω_1 in the left part of Eq. (2.51) is an orthogonal sum $H = h_s \oplus h_n$ and hence we have

$$(h_S + \varkappa_0^2)\varphi_0 = 0, \qquad (2.52)$$

$$(h_n + q^2)\psi_0 = 0, \quad q^2 - \varkappa_0^2 = E, \qquad (2.53)$$

where regular boundary conditions should be added to Eq. (2.53):

$$\psi_0(|q|,\ 0) = 0. \qquad (2.54)$$

The boundary-value problem (2.53), (2.54) has a solution of the type

$$\psi_0(|q|,|y|) = |q||y|e^{-\pi n/\sqrt{3}|q|}\Gamma(1+2in/\sqrt{3}|q|)\times$$
$$\times e^{i|q||y'|}\Phi(1+2in/\sqrt{3}|q|, 2, -2i|q||y|). \quad (2.55)$$

If the nuclear interaction $V^S(x)$ at $|x| \to \infty$ decreases faster than $O(|x|^{-2})$, the boundary condition

$$\varphi_0(0, \varkappa^2) = 0 \quad (2.56)$$

and the requirement $\varphi_0 \in L^2(R_+)$ have also to be added to (2.52). Then the problem (2.52), (2.56) has a unique solution. Thus the initial state has the form

$$\chi(x, y) = \varphi_0(x)\psi_0(|q|,|y|). \quad (2.57)$$

The substitution $U \to \chi + \Phi$ leads (2.51) to the inhomogeneous equation for Φ:

$$(-\partial^2/\partial x^2 - \partial^2/\partial y^2 + V + 2n\mu_0(\theta)/\rho - E)\Phi = \quad (2.58)$$
$$= 16n\sin(\pi/6-\theta)\,\varepsilon(\pi/6-\theta)\,\chi(|x|,|y|)/\rho\sqrt{3}\cos\theta -$$
$$- V^S\int_{-1}^{1} du \frac{|x||y|}{|x'||y'|}[\chi(x',y') + \Phi(|x'|,|y'|)].$$

The equation (2.58) can be written in the form

$$(H-E)\Phi = \tilde{\chi} + \gamma\Phi,$$
$$\tilde{\chi} = \frac{16n\sin(\pi/6-\theta)}{\rho\sqrt{3}\cos\theta}\varepsilon(\pi/6-\theta)\,\chi(|x|,|y|) -$$
$$- V^S\int_{-1}^{1}\chi(|x'|,|y'|)|x||y|(|x'||y'|)^{-1}du,$$
$$\gamma\Phi = -V^S\int_{-1}^{1} du\,|x||y|\Phi(|x'|,|y'|)/|x'||y'|.$$

Using asymptotic properties of the Green function G(Z)and the representation [18, 41]

$$\Phi = \lim_{\alpha\to+0} G(E+i\alpha)(\tilde{\chi} + \gamma\Phi),$$

one can show that the coordinate asymptotics of the function @ has the form

$$\Phi \sim a_0\varphi_0(|x|)\exp\{i|q||y| + iW_3\} + A_0(\theta)\exp\{i\sqrt{E}\rho + iW_0\}/\sqrt{\rho},$$

where

$$W_3 = -2n\ln(2|q||y|)/\sqrt{3}|q|, \tag{2.59}$$

$$W_0 = -n[\sec\theta + 2\mu_0(\theta)]2\ln 2\sqrt{E}\rho/2\sqrt{E}. \tag{2.60}$$

This method becomes rather complex when generalized to the matrix equations. Therefore we consider more simple way to obtain asymptotics of (2.59) which can be simply generalized to the set (2.45). This method involves a construction of the basis functions which satisfy to the equation (2.58) in one part or another with an accuracy to the minor order terms. Then the linear combination of the basis functins gives the senior term of asymptotics, i.e. (2.59).

Consider a construction of the basis functions in R^2_+. In the region $|X| \to \infty$ it is appropriate to chose as basis functions the eikonal approximation distorted by the phase

$$\Phi^l = \varphi_0(|x|)f(|y|)\exp\{i|q|L_y + i\widetilde{W}/|q|\}, \tag{2.61}$$

where L_y is the eikonal and f(|y|) and are unknown functions.

Inserting Eq. (2.61) into Eqs. (2.57) and (2.58), equating coefficients at various powers of |q| to zero and retaining only senior terms, the following equations for L_y, f, $\widetilde{W}$ are obtained:

$$(\nabla_y L_y)^2 = 1, \tag{2.62}$$

$$2\nabla y f\nabla_y L_y + f\nabla_y L_y = 0, \tag{2.63}$$

$$2\nabla_y\widetilde{W}\,\nabla_y L_y = -4n/\sqrt{3}y. \tag{2.64}$$

We shall chose L_y = |y as the eikonal. Then Eq. (2.62) becomes an identity and the solution of Eq. (2.63) is

$f(|y|) = \text{const}$, while the solution of Eq. (2.64)

$-\partial\widetilde{W}/\partial|y| \to -2n/\sqrt{3}n$ has the form $\widetilde{W} = -2n\ln(2|q||y|)\sqrt{3} + \text{const}$.

Considering const = 0 and denoting $W_3 = \widetilde{W}/|q| = -2n \ln(2|q||y|)/\sqrt{3}q$, we get

$$\Phi^l = \varphi_0(|x|) \exp(i|q||y| + iW_3)\, a_0. \tag{2.65}$$

The function (2.65) determines a contribution of the elastic scatterinng into Eq. (2.59).

In the region $X \subseteq \Omega_{01}$, $X \to \infty$ we choose as a basis function the eikonal approximation distorted by the additional phase

$$\Phi^h = F(x) \exp\{i\sqrt{E}L_0 + i\widetilde{W}/\sqrt{E}\}, \tag{2.66}$$

In this case the integral term in Eq. (2.58) is small due to a presence of V^S in front of the integral, while the fast-oscillating function $\varphi_0(|x|)$ makes the inhomogeneous term infinitesimal, and therefore we have in Ω_{01} instead of Eq. (2.57) the following equation:

$$(-\Delta + V^S(|x|) + V^c(|x|, \ |y|) - E)\Phi = 0. \tag{2.67}$$

Inserting Eq. (2.66) in Eq. (2.67) we obtain the set of equations for L_0, F and $\widetilde{W}$

$$|\nabla L_0|^2 = 1, \tag{2.68}$$

$$2(\nabla F, \ \nabla L_0) + \nabla L_0 F_0 = 0, \tag{2.69}$$

$$2(\nabla \widetilde{W}, \ \nabla L_0) - V^c(|x|, |y|) = 0, \tag{2.70}$$

where

$$V^l = n/|x| + 2\mu_0(\theta)n/\rho. \tag{2.71}$$

Choosing as L_0 the spherical eikonal $L_0 = |X|$, we find from (2.69) $F(x) = |X|^{-1/2}$.

To solve Eq.(2.70), it is necessary to transfer to the new coordinate system (L_0, M), where L_0 is the corresponding eikonal and M is such transversal variable that (L_0, M) is an orthogonal coordinate system in R^2_+. In the case considered $L_0 = \rho$, while the transversal variable is a polar angle $\theta = \text{arctg}(|y|/|x|$.

Then Eq. (2.70) has the following form in new variables:

$$\partial\widetilde{W}/\partial\rho = -V^c(\rho, \ \theta)/2, \tag{2.72}$$

and, hence, $\widetilde{W} = [-n(\sec\theta + 2\mu_0(\theta)) \quad \ln \quad 2\sqrt{E}\rho + C(\theta)]/2$, where $C(\theta)$ is an arbitrary function of the transversal variable. Defining the Coulomb phase as

$$W_0 = \frac{\widetilde{W}}{\sqrt{E}} = -\frac{n}{2\sqrt{E}}(\sec\theta + 2\mu_0(\theta))\ln 2\sqrt{E}\,\rho, \tag{2.73}$$

we obtain the basis function of the form

$$\Phi^b = A_0(\theta)\ \exp\{i\sqrt{E}\rho + iW_0\}/\sqrt{\rho}, \tag{2.74}$$

which determines a contribution of the decay processes into three free particles into asymptotics of (2.59). Constructing a linear combination of basis functions Φ^l and Φ^b, we get

$$\Phi \sim \Phi^l + \Phi^b, \tag{2.75}$$

i.e. Eq. (2.59), while the senior term of asymptotics for Eqs. (2.51) has the form

$$U \sim \chi + \Phi^l + \Phi^b, \tag{2.76}$$

where χ, Φ^l, Φ^b are determined by Eqs. (2.57), (2.65) and (2.74), respectively.

To construct asymptotics of solutions of the set (2.45) we cut the set off at $\lambda \leq N$, $l \leq N$, so that

$$|\lambda - l| \leqslant L \leqslant \lambda + l. \tag{2.77}$$

We shall introduce the following definitions [41 - 45]: $\beta = \beta(N)$ is a space of sufficiently differentiable functions specified in R_+^2 with values in the additional space N. For N we choose a group of square nondegenerate matrices over field C. The elements of the space β will be called vectors. Denote by $M = M(N, N)$ a space of all limited functions of the variable $X \in R_+^2$ operating at each X fixed as operators from N to N, and the elements of this space will be called matrices. Denote unit operators in β and M by I_β and I_μ, and subscripts β and μ will be omitted for simplicity; hereafter instead of operators $\Delta \oplus I$, $\partial/\partial|y| \oplus I$, $\partial/\partial|x| \oplus I$ definitions Δ, ∂y, ∂x will be used and ∇ means a pair $\{\partial_x, \partial y\}$. An action of ∇ on the element $W \in \beta$

gives a vector $\nabla W = \{\partial_x W, \partial_y W\}$, i.e. an element in the space where the product is specified as follows: if $\omega = \{\omega_1, \omega_2\} \in \beta \oplus \beta$ and $\varphi = \{S_1, S_2\} \in \beta \oplus \beta$, then $\langle\omega, \varphi\rangle = W_1 o S_1 + W_2 o S_2$, where o is a group multiplication in GL(N, C). If one introduces definitions {Eq. 40, p. 103} for diagonal matrices containing in their diagonals quantum numbers l and λ, respectively, V_S for the short-range potentials matrix, {Eq. 41, p. 103} for the matrix of Coulomb potentials with elements $V^0_{\alpha\alpha'}$, $\tilde{h}^L$ for the integral operator with $h^L_{\alpha\alpha'}$, $\mathcal{E} = E\, I$, then the set (2.45) can be written in the form

$$\left[-\Delta + |x|^{-2}\mathscr{L}(\mathscr{L}+1) + |y|^{-2}\Lambda(\Lambda+1) + V^n + \right. \\ \left. + V^S + V^0 + \varepsilon\right]U^L = -V^S\tilde{h}^L U^L. \tag{2.78}$$

We define operators b_x and b_y by equalities

$$b_x = \text{diag}\ \{|x|^{-l-1}\}, \quad b_y = \text{diag}\ \{|y|^{-\lambda-1}\} \tag{2.79}$$

or

$$b_x = |x|^{-\mathscr{L}-I},\ b_y = |y|^{-\Lambda-I}.$$

Denote quantum numbers L, l, λ« q corresponding to the initial state by a subscript 0: $L_0 - \lambda_0$, l_0, q_0, $\alpha_0 = \{l_0, \lambda_0\}$. Then the initial state χ corresponding to the movement of a particle 1 with an orbital moment λ_0 and momentum q_0 in the effective potential produced by a target in the bound state with an orbital moment l_0, $L = \lambda_0 \oplus l_0$ is defined in the region $\tilde{\Omega}_1$ by the equation

$$\left[-\Delta_x - \Delta_y - \mathscr{L}(\mathscr{L}-1)/|x|^2 + \Lambda(\Lambda+1)/|y|^2 + V^S + V^n + V^0 - \varepsilon\right]\chi = 0 \tag{2.80}$$

with boundary conditions

$$\lim_{|x|\to 0} b_x\chi = I,\ \lim_{|y|\to 0} b_y\chi = I. \tag{2.81}$$

To solve the boundary-value problem (2.80), (2.81) we introduce the product $\overset{L}{\oplus}$ of vector-functions with a limitation on the addition of moments in the following way. Let $\varphi(|x|)$ is the vector-function with components $\varphi_l(|x|)$ and $\psi(|y|)$ is the vector-function with components $\psi\ (|y|)$.

Then we define $\varphi \overset{L}{\oplus} \varphi$ as an element of the form

$$\varphi \overset{L}{\oplus} \psi = \begin{pmatrix} \varphi_0\psi_0 & \varphi_0\psi_1 & \cdots & \varphi_0\psi_N \\ \varphi_1\psi_0 & \varphi_1\psi_1 & \cdots & \varphi_1\psi_N \\ \cdots & \cdots & \cdots & \cdots \\ \varphi_n\psi_0 & \varphi_n\psi_1 & \cdots & \varphi_n\psi_N \end{pmatrix}, \tag{2.82}$$

in which matrix elements with $\lambda \oplus 1 \neq L$ are equal to zero. For the diagonal matrix we define the expansion

$$\varepsilon = \theta_0^2 - \varkappa_i^2, \ \theta_0^2 = q_0^2 oI, \ \varkappa_i^2 = \mathrm{diag}\,\{\varkappa_i^2\}, \tag{2.83}$$

where i is a set of quantum numbers specifying the target state, and in the region Ω_1 the initial state χ has the form $\chi = \varphi \overset{L}{\oplus} \psi$, where φ and ψ satisfy to the following boundary-value problems:

$$\left[-\Delta_x + V^S + V^n + \mathscr{L}(\mathscr{L}+1)/|x|^2 + \varkappa_i^2\right]\varphi = 0, \tag{2.84}$$
$$\lim_{|x|\to 0} b_x\varphi = \mathrm{I},$$
$$\left[-\Delta_y + V^0 + \Lambda(\Lambda+1)/|y|^2 - \theta_0^2\right]\psi = 0,$$
$$\lim_{|y|\to 0} b_y\psi = \mathrm{I}. \tag{2.85}$$

The solution of Eq. (2.85) has the form

$$\psi_\lambda(|y|,|q|) = \frac{1}{2}(2|q_0||y|)^{\lambda+1} e^{-\pi n/\sqrt{3}|q_0|} \frac{\Gamma\left(\lambda+1+i2n/\sqrt{3}\,|q_0|\right)}{(2\lambda+1)} \times$$
$$\times \Phi\left(\lambda+1+2ni/\sqrt{3}\,|q_0|,\ 2\lambda+2,\ -2i|q_0||y|\right). \tag{2.86}$$

The basis functions determining senior terms of asymptotics of solutions of Eqs. (2.78) are constructed similarly to the scalar case.

1. In the region $\tilde{\Omega}_1(X \in \Omega_1,\ X \to \infty)$ Hamiltonian H is a diagonal one, and hence the basis function has the form

$$\Phi^l = \varphi \overset{L}{\oplus} \psi, \tag{2.87}$$

where φ and ψ are solutions of Eqs. (2.84) and (2.85) which satisfy to boundary conditions with c $\varkappa_f^2 = \mathrm{diag}\,\{\varkappa_f^2\}$, and in Eq. (2.84) (f is a set of quantum numbers characterizing the target in the final state), ψ is an irregular solution of Eq. (2.85) with the following asymptotics:

$$\psi_\lambda(|q|, \ |y|) \sim \exp\{i|q||y| + iW_3 - i\pi\lambda/2\}, \quad q = E - \varkappa_j^2. \tag{2.88}$$

2. In the region $\Omega_{01}(X \in \Omega_{01}, \ X \to \infty)$ we shall construct a basis function in the eikonal approximation

$$\Phi^b = B(X)\exp\{i\sqrt{E}Z + \widetilde{W}/\sqrt{E}\}, \tag{2.89}$$

where $Z = L_0 I$ is a matrix eikonal, $\widetilde{W}$ is a matrix phase, $B(X) = b(X)$ is a matrix which is a multiple of a unit matrix. Neglecting in Ω_{01} the integral operator in Eq. (2.78), we get (as in the scalar case) the following equations for unknowns Z, $\widetilde{W}$, B:

$$\langle\nabla Z, \ \nabla Z\rangle = I, \tag{2.90}$$

$$2(\nabla B, \ \nabla Z) + \Delta Z B = 0, \tag{2.91}$$

$$2\sqrt{E}\langle\nabla\widetilde{W}, \ \nabla Z\rangle - \varepsilon(X) = -V, \tag{2.92}$$

where

$$V = \mathscr{L}(\mathscr{L}+1)/|x|^2 + \Lambda(\Lambda+1)/|y|^2 + V^S + V^c,$$
$$V^c = V^0 + V^n, \ \varepsilon(X) = -\langle\nabla\widetilde{W}\cdot\nabla\widetilde{W}\rangle + i\Delta\widetilde{W} + 2i\langle\nabla\widetilde{W}, \ \nabla B \oplus B^{-1}\rangle. \tag{2.93}$$

One can seek the formal solution of Eq. (2.92) in the form of the following asymptotic series:

$$\widetilde{W} = \sum_{k=0}^{\infty} W_k(X)/E^{k/2+1/2}. \tag{2.94}$$

Inserting Eq. (2.94) into Eq. (2.92) we obtain

$$2\langle\nabla\widetilde{W}_0, \ \nabla Z) = -V + \nabla BoB^{-1}, \tag{2.95}$$

$$2\langle\nabla\widetilde{W}_k, \ \nabla Z) = F_k(\widetilde{W}_k), \quad k \geqslant 0, \tag{2.96}$$

$$F_k(\widetilde{W}_k) = i\Delta W_k - \langle\nabla\widetilde{W}_k, \ \nabla\widetilde{W}_k\rangle + 2i\langle\nabla\widetilde{W}_k, \ \nabla BoB^{-1}\rangle, \quad k \geqslant 0. \tag{2.97}$$

To solve Eqs. (2.95) - (2.96) we transfer to the new "action-angle" ($L_0 M$) coordinate system. Then (2.96) has the form

$$\partial L_0 \widetilde{W}_k = F_k(L_0, \ M)/2. \tag{2.98}$$

Considering the scattering processes, we take

$$L_0 = |x|, \ M = \theta = \mathrm{arctg}\,(|y|/|x|).$$

In this case Eq. (2.90) transforms to the identity, while Eq. (2.91) has a solution

$$b(X) = \sqrt{|X|}, \quad B(X) = b(X)\mathrm{I}.$$

The estimations

$$V - V^c \underset{L_0 \to \infty}{=} O\left(L_0^{-1}\right),$$

$$V^c \underset{L_0 \to \infty}{=} O\left(L_0^{-1}\right) \qquad (2.99)$$

(2.100)

allow to seek the solution of Eqs. (2.95) in the form

$$\widetilde{W}_0 = \widetilde{W}^0 + \delta\widetilde{W}_0, \qquad (2.101)$$

where

$$2\langle \nabla \widetilde{W}^0, \ \nabla Z \rangle = V^c \qquad (2.102)$$

$$2\langle \nabla \delta W_0, \ \nabla Z \rangle = -(V - V^c) + \nabla B o B^{-1}. \qquad (2.103)$$

From Eqs. (2.101) - (2.103) we have

$$\widetilde{W} \underset{L_0 \to \infty}{\sim} O(\ln L_0), \ \delta\widetilde{W}_0 \underset{L_0 \to \infty}{\sim} o(1), \ F_k \sim o\left(L_0^{-1}\right),$$

from which it follows that $\widetilde{W}_k = o(1)$, $k \gg 1$ and the senior term of asymptoptics is determined by Eq. (2.102) which has in the coordinates $(L_0, M) = \rho, \theta$ the form:

$$\partial_\rho \widetilde{W}_0 = -V^c\,(\rho, \ \theta)/2, \qquad (2.104)$$

$$\widetilde{W}^0 = -\int V^c\,(\rho, \ \theta) d\rho/2,$$

while matrix elements are equal to

$$W^0_{\alpha\alpha'} = -n\,(\sec\theta\delta_{\alpha\alpha'} + 2\mu_{\alpha\alpha'}(\theta))\ln\left(2\sqrt{E}\,\rho/2\right). \qquad (2.105)$$

Defining a phase as $\widetilde{W}_0 = \widetilde{W}^0/\sqrt{E}$, we obtain the basis function

$$\Phi^b = \exp\{i\sqrt{E}Z + iW_0\}/\rho^2. \qquad (2.106)$$

Thus the senior term of asymptotics (2.78) has the form

$$U \sim \chi + a\Phi^l + A(\theta)\Phi, \qquad (2.107)$$

where matrix elements a and A correspond to amplitudes of processes of the elastic scattering, rearangement and decay, respectively, and depend on the impacting particle energy.

In conclusion of this section we present an explicit form of equations with asymptotic boundary conditions in the case of scattering of proton on deuton [41 - 45, 60]. In this case at S = 3/2 Eqs. (2.45) have the form

$$\left(H_L + V_{3_{V_1}} + n\mu(\theta)/\rho - E\right)\Phi_L(x, y) = \left(\frac{2n}{\sqrt{3}\,|y|} - \frac{2n}{|x|}\right)\times$$
$$\times\psi_d^L(x, y)\,\varepsilon(\pi/6 - \theta) - \frac{1}{2}V_{3_{S_1}}\int du\,h^L(u)\left[\psi_d^L(x', y') + \Phi_L(x', y')\right],$$
$$\theta = \mathrm{arctg}\,|y|/|x|,\ \psi_d^L = \varphi_d\varphi^L,$$
$$H_L = -\partial^2/\partial x^2 - \partial^2/\partial y^2 + L(L+1)/y^2,$$
$$h^L(u) = \frac{|x||y|}{|y'||x'|}\frac{4}{\sqrt{3}}\left(-\frac{1}{2\sin\theta}\right)^L\sum_{k=0}^{L}\frac{L!}{k!\,(L-k)!}P_k(u)\times$$
$$\times\left(\sqrt{3}\cos\theta\right)^k(\sin\theta)^{L-k},\ \varepsilon(t) = \begin{cases}1, & t > 0,\\ 0, & t < 0,\end{cases}$$
$$u = -(\cos 2\theta + 2\cos\theta')/\sqrt{3}\sin\theta,$$
$$\varphi^L(|y|, |q|) = \frac{1}{2}(2qy)^{L+1}\exp\left(-\pi n/2\sqrt{3}q\right)\times$$
$$\times\frac{\Gamma(L+1+in/\sqrt{3}q)}{(2L+1)!}\Phi\left(L+1+in/\sqrt{3}q,\ 2L+2,\ -2iqy\right),$$
(2.108)

where φ_d is the deuton WF corresponding to the bond energy $\varkappa^2$.

To solve Eqs. (2.108) uniquely, the boundary conditions are needed which after expansion over angular harmonics have the form [41, 60]:

$$\Phi_L(x, y)|_{x=0} = \Phi_L(x, y)|_{y=0} = 0,$$
$$\Phi_L(x, y) \underset{\rho\to\infty}{\sim} a_L\varphi_d(x)\exp\{i|q||y| + iW_3 - i\pi L/2\} +$$
$$+ A_L\exp\{i\sqrt{E}\rho + iW_0 - i\pi L/2\}/\rho^{1/2},$$
$$W_3 = -n\ln(2|q||y|)/\sqrt{3}|q|,\ W_0 = -n\mu(\theta)\ln\left(2\sqrt{E}\rho\right)/2\sqrt{E}.$$
(2.109)

Experimentally observed cross-sections of processes which are determined by amplitudes $\tilde{A}_L$ are related with A_L by equations

$$\widetilde{A}_L = A_L - \frac{1}{2}\int_{-1}^{1} h^L A_L \, du,$$

similar to the case of neutral particles. The amplitude of the elastic scattering process can be presented in the form

$$a = \tilde{a} + a^s,$$

where

$$a^s = \frac{2n}{\sqrt{3}\, q^2 \sin^2\theta/2} \exp\left\{- i \frac{2n}{\sqrt{3}\,|q|} \ln(\sin^2\theta/2) + i \arg\Gamma\left(1 + i2n/\sqrt{3}\,|q|\right)\right\},$$

$$\tilde{a} = \frac{1}{|q|}\sum_{L=0}^{\infty}(2L+1)\, a_L P_L(\cos\theta),$$

$$a_L = \left(\eta_L e^{2i\Delta_L} - 1\right)\Big/ 2i,$$

$$\Delta_L = \delta_L + \arg\Gamma\left(L + 1 + in/\sqrt{3}\,|q|\right),$$

here Δ_L is the phase shift corresponding to the short-range potential. In the case of $S = 1/2$ similarly to $S = 3/2$ we have

$$\left(H_L + V_{{}^3S_1} + n\mu(\theta)/\rho - E\right)\widetilde{\Phi}_1 = -V_{{}^3S_1}F_1,$$

$$\left(H_L + V_{{}^1S_0} + n/|x| - E\right)\widetilde{\Phi}_2 = -V_{{}^1S_0}F_2, \qquad (2.110)$$

$$\left(H_L + V_{{}^1S_0} + n\mu(\theta)/\rho - E\right)\widetilde{\Phi}_3 = -V_{{}^1S_0}F_3,$$

$$\widetilde{\Phi} = \mathcal{U}\Phi, \qquad \mathcal{U} = \begin{pmatrix} 1 & 0 & 0 \\ 0 & \sqrt{2}/3 & -1\sqrt{3} \\ 0 & 1/\sqrt{3} & \sqrt{2}/3 \end{pmatrix},$$

$$F = \widetilde{\Phi} + \int_{-1}^{1} h^L(u)\, \widetilde{\mathcal{G}}\Phi(x', y')\, du,$$

$$\widetilde{\mathcal{G}} = \mathcal{U}\mathcal{G}\mathcal{U}^{-1} = \begin{pmatrix} 1/4 & -\sqrt{6}/4 & -\sqrt{3}/4 \\ -\sqrt{6}/4 & 0 & \sqrt{2}/4 \\ -\sqrt{3}/4 & \sqrt{2}/4 & -1/4 \end{pmatrix},$$

$$\widetilde{\Phi}(x, y)\,|_{x=0} = \widetilde{\Phi}(x, y)\,|_{y=0} = 0,$$

$$\widetilde{\Phi}_1(x, y) \underset{\rho\to\infty}{\sim} \left(a_L \varphi_d(x) \exp\{i\,|q|\,|y| + iW_1 - i\pi L/2\} + \varphi^L + \right.$$

$$+ \widetilde{A}_L^1 \exp\{i\sqrt{E}\,\rho + iW_0^1 - i\pi L/2\}/\sqrt{\rho},$$
$$\widetilde{\Phi}_2(x, y) \underset{\rho\to\infty}{\sim} \widetilde{A}_L^2 \exp\{i\sqrt{E}\,\rho + iW_0^2 - i\pi L/2\}/\sqrt{\rho},$$
$$\widetilde{\Phi}_3(x, y) \underset{\rho\to\infty}{\sim} \widetilde{A}_L^3 \exp\{i\sqrt{E}\,\rho + iW_0^3 - i\pi L/2\}/\sqrt{\rho},$$
$$W_0^1 = W_0^3 = -n\mu(\theta)\ln(2\sqrt{E}\,\rho)/2\sqrt{E}, \qquad (2.111)$$
$$W_0^2 = -n\sec\ln(2\sqrt{E}\,\rho)/2\sqrt{E},$$
$$W_1 = -n\ln(2|q||y|)/\sqrt{3}|q|,$$

$$A_L = \widetilde{A}_L + \int_{-1}^{1} h^L(u)\,\mathcal{G}\,\widetilde{A}_L\,du, \quad A_L = \mathcal{U}^{-1}\widetilde{A}_L.$$

2.3. Methods of numerical solution of differential equations of quantum theory of scattering

To solve equations (2.108), (2.110) with corresponding boundary conditions (2.109), (2.111) the method is used which is a generalization of the common scheme of solution of Schrodinger equations.

In this method a finite-difference approximation of equations [41 - 45, 50, 52 - 54, 60] in polar coordinates $\rho^2 = x^2 + y^2$, $\theta = \operatorname{arctg} y/x$ is used. By means of the net with N_θ knots on the arc ρ = const and $N_{\rho+1}$ knots on the beam θ = const the finite-difference approximation of equations (2.108) and (2.110) is performed.

Denote by $\chi = \{\chi_1, \chi_2, \ldots, \chi_{N_m}\}$ the vector corresponding to the function values in knots of the net $x \in R_{N_m}$, where

$$N_m = (N_{\rho+1})N_\theta. \qquad (2.112)$$

Eqs. (2.108), (2.110) after transfer to the finite-difference ones can be written in the matrix form

$$(B - E)\chi = \Phi, \qquad (2.113)$$

where B is a linear operator in the finite-dimension vector space R^{N_m} corresponding to the Laplacian and the integral in (2.108), (2.110), Φ is a vector $\Phi \in R_{N_m}$ corresponding to the values of the inhomogeneous term in knots of the net. The problem (2.108), (2.110) is reduced to the determination of vector χ from a given Φ, while on the square boundary $\theta = 0$, $\theta = \pi/2$ the components values are determined by boundary conditions (2.109), (2.111).

Vector χ is conveniently described using the following numbering of the net knots and components χ_k, respectively. Let $\tilde{\chi}_j$ is a vector from the space $R^{N_{\theta_j}}$ corresponding to the values of the solution Φ_L on j-th arc $\rho = \rho_j$. Components of this vector are defined by relations:

$$\tilde{\chi}_j = \Phi_L(\rho_j \cos\theta_k, \rho_j \sin\theta_k),$$

$$j = 1, 2, \ldots, N_{\theta_j}, \ 0 < \theta < \pi/2. \tag{2.114}$$

The number N_{θ_j} of divisions of j-th arc can be different depending on the distance ρ_j to the coordinate origin, while the component numbering is performed in order of increasing angle θ.

The vector χ can be written as an orthogonal sum

$$\chi = \sum_{i=1}^{N_\theta} \oplus \tilde{\chi}^i, \tilde{\chi}^i \in R^{N_\theta}, R^{N_m} = \sum_{i=1}^{N\theta_i} \oplus R^N. \tag{2.115}$$

In the representation (2.115) the difference equation can be transformed to the form

$$L_i\tilde{\chi}^{i-1} + M_i\tilde{\chi}^i + R_i\tilde{\chi}^{i+1} = \mathrm{I}_i, \tag{2.116}$$

where L_i, M_i, R_i are operators acting from subspaces $R^{N_{\theta_{i-1}}}$, $R^{N_{\theta_i}}$, $R^{N_{\theta+1}}$ in R^{N_θ} and having the form of real square matrices of rank N_{θ_i}, matrices L_i, R_i being diagonal ones corresponding to the radial part of the Laplace operator

$$\partial^2/\partial\rho^2 + \partial/\rho\partial\rho, \tag{2.117}$$

After transition to the finite-difference approximation differential operators (2.117) connect the function Φ_L value on near arcs. The matrix M_i corresponding to the contribution of the angular part of the Laplace operator (2.115) (differentiation with respect to angular variables $\partial^2/\rho^2\partial\theta^2$ and the integration over an arc $\rho = \rho_i$) is nondiagonal due to integrating in the right part of Eqs. (2.108), (2.110). I_i is a vector in the space $R^{N_{\theta_i}}$ corresponding to the inhomogeneous term Φ^o in Eq. (2.113)

$$\Phi = \sum_{i=1}^{N_\rho} \oplus \, I_i .$$

Thus the matrix B in Eq. (2.113) is rarefied and has in the representation (2.115) the tape-like structure.

The number of unknowns in Eqs. (2.113) is equal to $(N_\rho+1)N_\theta$, and there are $N_\rho N_\theta$ relations for their determination. To find a unique solution corresponding to the physical wave function, Eqs.(2.113) must be supplemented with a boundary condition (2.109), (2.111). Neglecting terms on the order of $O(\rho^{-3/2})$, we can obtain

$$\tilde{\chi}^{N_\rho} = a^L \chi^{N_\rho} + A^L_{N_\rho} \frac{\exp\{iK\rho_{N_\rho} + i\pi L/2 + iW_0(\rho_{N_\rho})\}}{\sqrt{\rho_{N_\rho}}},$$

$$\tilde{\chi}^{N_\rho+1} = a^L \chi^{N_\rho+1} + A^L_{N_\rho+1} \frac{\exp\{iK\rho_{N_\rho+1} + i\pi L/2 + iW_0(\rho_{N_\rho+1})\}}{\sqrt{\rho_{N_\rho+1}}}, \tag{2.118}$$

where $A^L_{N_\rho}$ is a vector $A^N_{N_\rho} \in R^{N_\rho}$ corresponding to the amplitude $A^L(\theta)$ values in the net knots (2.112). Eliminating the unknown vector $A^L_{N_\rho}$, we obtain the relation between vectors $\tilde{\chi}^{N_\rho+1}$ and $\ddot{\chi}^{N_\rho}$ which allows to find the vector $\ddot{\chi}$:

$$\tilde{\chi}^{N_\rho+1} = C_{N_\rho}\chi^{N_\rho} + a\hat{\chi}^{N_\rho},$$

where

$$C_{N_\rho} = \sqrt{\rho_{N_\rho}/\rho_{N_\rho+1}}\exp\{iK(\rho_{N_\rho+1} - \rho_{N_\rho}) + i[W_0(\rho_{N_\rho+1} - \rho_{N_\rho})]\},$$

$$\hat{\chi}^{N_\rho} = C_{N_\rho} I_{N_\rho} - I_{N_\rho+1}.$$

The equation (2.116) supplemented with boundary conditions (2.118) takes the form

$$\widetilde{M}\chi = \Phi + a^{L}\widehat{\chi}. \tag{2.119}$$

Here all vectors are specified in the representation (2.115). The vector $\widetilde{\chi}$ has only one nonzero component $\widehat{\chi}^{N_\rho}$ from the space R^{N_ρ}. $\widetilde{M}_{N_\rho}$ denotes a linear operator in R^{N_ρ} :

$$\widetilde{M}_{N_\rho} = M_{N_\rho} + C_{N_\rho}R_{N_\rho}. \tag{2.120}$$

Notice that there is an unknown quantity a^{L} in the right part of Eq. (2.118) corresponding physically to the elastic scattering amplitude. Due to its linearity, the solution of Eq. (2.119) can be presented in the form

$$\chi = \alpha + a^{L}\beta, \tag{2.121}$$

where vectors α and β are determined from equations

$$\widetilde{M}\alpha = \Phi, \quad \widetilde{M}_{\beta} = \widehat{\chi},$$

while inhomogeneous terms are known.

The solution of Eqs. (2.119) is similar to the reciprocation of a matrix of the order of $N_\rho N_\theta$, which is in a general case a rather complex problem. However, in our case the matrix has the tape-like structure and for its reciprocation the Gauss algorithm is the most effective. Using a scheme of thesubsequent elimination of unknowns, one can transform the matrix to the triangular form, while vectors α and β and solution χcan be found in the form (2.121) by means of a reciprocal procedure.

To find an unknown quantity a^{L}, either asymptotic formulas (2.109), (2.111) can be used or common integral representations of the scattering amplitude [18 - 21, 41 - 45]. Consider determination of a^{L} by means of asymptotic formulas (2.109), (2.111). For this purpose, compare equalities (2.118) and (2.121) at $x \ll y$. In this case at large values of ρ_{N_ρ} the second term in Eq. (2.118) is far less than the firstone (at energies below the decay threshold @ it is exponentially small and is of order @. Therefore, neglecting the second term in Eq. (2.118), we obtain the relation for determining a^L:

$$a^L \widehat{\chi}^{N_\rho} = [\alpha]_i^{N_\rho} + a^L [\beta]_i^{N_\rho}.$$

It follows from that

$$a^L = [\alpha]_i^{N_\rho} \Big/ \left\{ \widehat{\chi}_i^{N_\rho} - [\beta]_i^{N_\rho} \right\}. \qquad (2.122)$$

Here $\hat{\chi}_i^{N_\rho}$ denotes vector components $\hat{\chi}^{N_\rho}$ corresponding to the small values of angles θ_i. In the case i = 1 when a^L is known, one can find a vector $\tilde{\chi}^{N_\rho}$ corresponding to the function Φ^L values on the external arc $\rho = \rho_{N_\rho}$.

Then the decay amplitude A^L is defined by the equality

$$A^L_{N_\rho} = \left(\tilde{\chi}^{N_\rho} - a^L \widehat{\chi}^{N_\rho} \right) \frac{\exp\left[- iK\rho_{N_\rho} - i\pi L/2 - iW_0\left(\rho_{N_\rho}\right)\right]}{\sqrt{\rho_{N_\rho}}}. \qquad (2.123)$$

Notice that to find values of a^L and A^L, it is not necessary to calculate the function Φ^L in the inner points x_i, y_i of the net. These values are determined from the function Φ^L values on the boundary $\rho = \rho_{N_\rho}$.

At the energies above the threshold of the three-particle decay E > 0, neglect of the second term in Eq. (2.118) may result in the inadequate accuracy of the amplitude a^L determination due to a slow decrease of the second term at $\rho \to \infty$. Hence in this case a method based on the integral representations for the scattering amplitude was found to be preferable [17 - 22]:

$$f = -\frac{m_\alpha}{|\rho_\alpha|} \int dx_\alpha \, dy_\alpha \left[\chi^*_{\alpha_i} \right] [V(x_2) + V(x_3)] \, \psi_\alpha(x_\alpha, y_\alpha). \qquad \textbf{(2.124)}$$

A discrete analog of Eq. (2.124) for s-component of the scattering amplitude f can be written in the form:

$$a = \left(\theta, \varphi + \chi_i^m \right) + a\,(\theta, \psi_0), \qquad (2.125)$$

where a vector θ, $\theta \in R^{N_m}$ is determined by the integrand values in the net knots. From Eq. (2.125) it is possible to find the a^L value corresponding to the physical wave function. The decay amplitude A_L and wave function can be obtained by the method considered above.

It is necessary to dwell on the interpretation of the scheme described above from the point of view of projection methods [41 - 45]. Consider a class B of smooth functions defined in the region $x \geq 0$, $y \gg 0$ which have asymptotics of the form (2.109), (2.111) at $\rho \to \infty$. This class can be connected with the Banach space $\tilde{B}_s$ of functions which are obtained after a separation of fast-oscillating exponents. Then Eqs. (2.108) and (2.110) can be related to the linear equation in the space $\tilde{B}_s$. This equation, supplemented with proper boundary conditions (2.109) and (2.111), has a unique solution in $\tilde{B}_s$.

The transition to the difference equation can be considered as the B narrowing to the subspace of functions in $\tilde{B}_s$ which take constant values in the vicinity of the net knots.

Such functions can be presented as a linear combination

$$F(x, y) = \sum_{i=1}^{N} C_i g_i(x, y), \tag{2.126}$$

where $g_i(x, y)$ $i = 1, 2, \ldots N$ is a division of unity corresponding to the net introduced (the function g_i is equal to unity in the vicinity of the j-th knot of the net).

The boundary condition (2.118) on the arc with the number N_ρ means that the dependence of coefficients C_i at $\rho_i > \rho_{N_\rho}$ is fixed. Coefficients C_i are described by recursive relations (2.118). As a result differential equations (2.108), (2.111) are reduced to the finite-dimension space specified by equalities (2.109) and (2.126).

A set of functions of the type (2.118) is dense in $\tilde{B}_s$ [41 - 45, 60]. In the case considered an accuracy of the solution of equations (2.108) and (2.110) is defined in a sense of the norm of the subspace $\tilde{B}_s$ which is a projection of the space B_N. This accuracy is determined by the net knots density and value of ρ_{N_ρ}. In this case an error due to fixating the asymptoptical functions behaviour at $\rho > \rho_{N_\rho}$ is of order $O(\rho^{3/2} N_\rho)$. The analysis of the character of errors resulting from the quantization is rather involved due to the net inhomogeneity and has been investigated in detail in Ref. [24 - 34, 41 - 62].

Notice that to increase the calculation accuracy and to provide an effective operation of the corresponding software support, the net knots must be located nonuniformly. The net must be dense in the regions adjacent to an axes $x = 0$, while at $x \to \infty$ it becomes less dense. Such an arrangement of the net knots is necessary for sufficiently accurate description of the complex behaviour of the wave function in the region, where $x \ll y$. In the region $x \to \infty$ the wave function takes an asymptotical form (2.108), (2.111) and depends smoothly on the angular variable, hence in this region the net must be chosen to be less dense [41 - 60].

At calculations on the inhomogeneous net, it is very important to comply with the condition of self-consistence, which implies that the error due to introducing the cut off radius must be far less than the quantizing error in the vicinity of the boundary $\rho = \rho_{N_\rho}$, i.e.

$$|\Delta h|^2 \gg \rho_{N_\rho}^{-3/2}.$$

where h is a minimum step in the boundary vicinity.

The stability and self-consistence of the calculations may be checked by various methods: determinations of the net, the cut off radius and the scattering amplitude. The analysis of calculations based on the method developed and software packages realized has shown that at calculations of the scattering states in the system of three particles the results are stabilized at $\rho_{N_\rho} = 20R$, $N_\rho \sim N_\theta \sim 2\rho_N$, where R is a characteristic distance which in the case of nuclear forces is proportional to 1 fm, for atomic forces to $1a_0$, and for molecular forces to the characteristic distance between clusters. For example, the accuracy of calculation of the amplitude of the proton elastic scattering on a deuton with the net parameters $\rho_{N_\rho} = 20$ fm, $N_\rho = 42$, and $N_\theta = 40$ is equal to 15%.

The method of numerical solution of equations in the system of three particles considered in this section can be extended without the loss of generality to the systems of four and more particles. However the numerical solution of the equations obtained presents computational difficulties which can be overcome using rather powerfull computers [60 - 64].

Notice that the approach to the numerical solution of the scattering problem discussed above combines the advantages of the integral equations (the solution existence and uniqueness) with the computational scheme simplicity. Also it has the advantage that as unknown two-particle quantities are specified not T-matrices but potentials and eigenfunctions corresponding to the bound states. The simplicity of the computational algorithm allows to consider scattering problems in real systems.

To realize the numerical method of solution of differential equations (2.108) and (2.110) with asymptotic conditions (2.109) and (2.111) discussed above the software package in FORTRAN has been developed [41, 60 - 62]. We list here the main subroutines.

CASCHA is a control subroutine which determines arrays that are necessary for the operation of other subroutines and activates them. POT is intended for the calculation of paired short-range potentials. COUL serves to calculate the Coulomb potential. POIDS is intended for calculations of the net parameters that is necessary for the quantization of the set of equations and calculation of the integration weights, which allow to change integrals in the right parts of the equations for sums. PLEG calculates Legendre polynomials. DEUT is intended for calculations of the net parameters and activation of the DEUTON subroutine which calculates the deuton wave function in the net knots. ELM transforms a set of differential equations into a set of difference equations specifying the net in the polar coordinate system which is constructed with subroutines POIDS and DEUT. This subroutine calculates also inhomogeneous terms in corresponding equations. HYPSY and HYPER calculate irregular and regular Coulomb wave functions [91]. In these programs the WHIT subroutine is used which calculates the Whitekker functions [91] which transform to the Hankel functions [91] at calculations of the tree-particle systems with neutral particles $n_i = 0$. TRIAN is intended for the solution of a set of linear equations by the Gauss method. RESOL calculates amplitudes of the elastic scattering and decay into three free particles.

The detailed description of the input data and their formates is reported in Ref. [60, 51], where the detailed information is presented about the performance peculiarities of the sofware package for specific computer systems. This package has been succesfully used in BESM-6, SAIBER-172, and ES-1045 computers.

However, notice that to achieve with a computer of the ES type the same calculation accuracy as with BESM-6 and SAIBER-172, it is necessary to change programs in order to operate with doubled accuracy (due to the difference in the length of the computer word). It introduces additional difficulties in performance with a given package. To avoid them in computers of the ES type, the special optimizing translator can be used which allows to change automatically any program for operation with a doubled accuracy [60].

2.4. Differential equations of the quantum theory of scattering in the model of boundary conditions.

The model of boundary conditions is now applied widely for a study both bound states and scattering states in atomic, molecular, and nuclear physics [15 - 28]. In this model the region of interaction between particles is separated into the external one, where the distance between particles $r \gg a$ is greater than a certain distance a. and the internal one $r < a$. In the external region the interaction is described with some potential, while the interaction in the internal region is considered introducing boundary conditions at $r = a$.

This model appears naturally in nuclear physics at consideration of strongly interacting particles, for example, quarks in adrons.

The introduction of the Dirichle boundary condition $f\,|_{u=\emptyset}$ (the hard spheres model) corresponds to the infinite strong repulsion inside all the internal region (hard core). In the case of the boundary conditions of the third kind at $\partial/\partial|x_\alpha| + \tau_\alpha(x_\alpha)f|_{x_\alpha=a_a+i_o}$ the corresponding potential has also an infinitely deep attraction part at $r = a$ [18 - 22].

Notice that when such potentials are used directly, the integral equations of the quantum theory of scattering in the system of an arbitrary number of bodies lose their meaning even when special procedures are employed. For example, in the case of three bodies after proceeding to the limit in the Faddeev equations to potentials corresponding to the boundary conditions, nonFredholm equations appear. Therefore in papers [15 - 28], simplified variants of the boundary conditions model are considered where the boundary conditions are used to describe the interaction in a finite number of the pairs orbital states (usually in s-states) which does not allow to consider the boundary conditions model in a general case. Notice, however, that the introduction of the boundary conditions poses a boundary-value problem of the second kind for the partial differential equation which can be solved by means of the potential theory [102]. In the case of two bodies, the surface where the boundary conditions are introduced is compact and smooth, hence in this case the potential theory equations allow to determine completely Green functions and wave functions.

In the case of three bodies, the boundary conditions are set on the infinite surface and give rise to non-Fredholmian equations of the potential theory which do not apply in the investigation of three-body systems.

In this model coordinates of three bodies are determined similarly to (2.1), (2.2), and a condition $|x_\alpha| > a_\alpha > 0$, $\alpha = 1, 2, 3$ forms the space Ω_c with a boundary Γ, and the system energy operator has the form

$$Hf(x) = \left[-\Delta + \sum_{\alpha} \vartheta_\alpha (x_\alpha) \right] f(x), \tag{2.127}$$

where $\vartheta_\alpha(x_\alpha)$ are fast decreasing paired potentials.

The system operator H_α consisting of two hard spheres of a pair α and a free particle is determined similarly, except that the region Ω and boundary Γ must be changed for Ω_α and Γ_α which is determined as follows: $|x_\alpha| > a_\alpha$, $\vartheta_\alpha(x_\alpha) \equiv 0$, if $\beta \not\equiv \alpha$

The resolvent kernels

$$R(X,\ X',\ Z) = (H - Z)^{-1},$$

$$R_\alpha(X,\ X',\ Z) = (H_\alpha - Z)^{-1},$$

$$R_0(X,\ X',\ Z) = (H_0 - Z)^{-1}$$

satisfy to the equations

$$R(X, X', Z) = R_0(X, X', Z) - \sum_{\alpha} \left[\int_{\Omega} dX'' R_0(X, X'', Z)\, \vartheta_\alpha(x''_\alpha), R(X'', X', Z) + \right.$$

$$+ \int_{\Gamma_\alpha \backslash \Gamma} d\sigma_S R_0(X, S, Z) \frac{\partial}{\partial n^\alpha_S} R(S, X', Z), \tag{2.128}$$

where $X, X' \in \Omega$, and n^α_s is a normal to the cylinder Γ_α directed to Ω_α.

It follows from these equations that the function R(X,X´,Z) can be restored explicitly by kernels

$\mu_\alpha(S, X'Z) = \partial/\partial n_S^\alpha, R(S, X', Z)$ и $\mu_\alpha(X, X', Z) = \upsilon_\alpha(x_\alpha)R(X, X', Z)$.

In the case of the absence of paired potentials $\vartheta_\alpha = 0$,«for any α in the right part remains only the surface integral which is a simple layer potential [102], and by differentiation over n_S^α Eqs. (2.128) are reduced to the integral equations of the potential theory for kernels μ_α, while the equations are found to be non-Fredholmian because the surface is infinite [44].

To rearrange these equations, we shall use the method suggested by L.D. Faddeev [40, 41, 44, 52]. For this purpose, we separate and reciprocate explicitly the diagonal part of the operator @ in terms of the two-body problem.

Then we obtain the set of equations

$$\mu_\alpha(S, X, Z) = V_\alpha R_\alpha + V_\alpha \rho_\alpha \mu_\alpha - V_\alpha R_\alpha \sum_{\beta \neq \alpha} \mu_\beta,$$

$$V_\alpha R_\alpha(S, S', Z) = \frac{\partial}{\partial n_S} R_\alpha(S, S', Z),$$

where V_α and R_α are the integral operators which kernels are a normal derivative of the density ρ_α determining the Green function R_α in the form of the double layer potential [102]:

$$R_\alpha(X, X', Z) = R_0(X, X', Z) - \int_{\Gamma_\alpha} d\sigma_S \rho_\alpha(X, S, Z) \frac{\partial}{\partial n_S} R_0(S, X', Z).$$

Since variables in H can be separated, R_α and ρ_α can be expressed explicitly in terms of the problem of two hard spheres.

The equations operator can be shown [41, 44, 52] to allow after a finite number of iterations the representation in the form of a sum of a compact operator and a limited one with a norm less than unity. The limited operator corresponds to the integration in the vicinity of singularities, where the surface Γ_α has fractures. Hence to the equations is applicable the Fredholm alternative providing their unique solvability. Notice that a set of singularities of the integral equations obtained consists of a discrete spectrum not only of the external, but also of some internal boundary-value problems. This condition is not a peculiarity of only given equations, since the absence of a complete equivalency between integral equations and corresponding boundary-value problems is a general consequence of the potential theory [44, 102], and due to the orthogonality of the free term in homogeneous conjugate equations, the solvability exists also in singularities.

The integral equations for kernels μ_α and M_α in the case $\vartheta_\alpha \neq 0$ are constructed similarly. As a result, the Green function R at Z = E ± i0 can be shown to have the form:

$$R(X, X', E \pm i0) = \sum_{\alpha_j} Q^{\pm}_{\alpha_j}(X, y'_\alpha, E)\, \overline{\psi}_{\alpha_j}(x'_\alpha) + Q^{\pm}_0(X, X', E),$$

similar to the Green function structure in the problem of three neutral particles. Here $\psi_{\alpha_j}(x_\alpha)$ are the wave functions of bound states of two bodies with energies ε_{α_j}.

At $X' \to \infty$ and $y'_\alpha \to \infty$ kernels $Q^{\pm}_0$ and $Q^{\pm}_{\alpha_j}$ have the form

$$Q^{\pm}_0(X, X', E) \sim \frac{1}{4\pi}\left(\frac{\sqrt{E}}{2\pi}\right)^{3/2} \frac{e^{\pm i\sqrt{E}|X'|}}{|X'|^{5/2}} U(X, P),$$

$$Q^{\pm}_{\alpha_j}(X, y'_\alpha, E) \sim \frac{1}{4\pi} \exp \pm \left\{ i\sqrt{E + \varepsilon_{\alpha_j}}\, |y'_\alpha| \right\} U^{\pm}_{\alpha_j}(X, p_\alpha),$$

where $U^{\pm}(X, P)$ is an amplitude which corresponds to the scattering process with three free particles in the initial state, and $U^{\pm}_{\alpha_j}(X, p_\alpha)$ is an amplitude of the process of the particle scattering on a bound pair

$$P = \mp \sqrt{E}X/|X'|,$$

$$p_\alpha = \mp \sqrt{E + \varepsilon_{\alpha_j}}\, y'_\alpha/|y'_\alpha|.$$

To derive differential equations, we define components R^α as follows:

$$R^\alpha(Z) = \delta_{\alpha_1} R^0(Z) - R^0_0(Z)\, \vartheta_\alpha R(Z),$$

where $\overset{o}{R}_o(Z)$ is a resolvent of the operator H in the absence of all paired potentials. Such a definition corresponds to the following division of wave functions:

$$U = \sum_{\alpha} U^{\alpha}, \qquad (2.129)$$

in which components U^{α}, $X \in \Omega$ satisfy to the set of equations:

$$(-\Delta + \vartheta_{\alpha}(x_{\alpha}) - E)\, U^{\alpha}(X) = \vartheta_{\alpha}(x_{\alpha}) \sum_{\beta \neq \alpha} U^{\beta}(X)$$

with boundary conditions

$$U^{\alpha}(X)\big|_{X \in \Gamma} = 0.$$

Alongside the division (2.129), consider the division defined with formulas

$$\widetilde{R}^{\alpha}(X, X', \Gamma) = \delta_{\alpha_1} R(X, X', Z) - \int d\delta_S R_0(X, S, Z) \times$$
$$\times \frac{\partial}{\partial n_S^{\alpha}} R(S, X', Z) - \int_{\Omega} dX'' R_0(X, X'', Z)\, \vartheta_{\alpha}(x''_{\alpha})\, R(X'', X', Z).$$

Components $\widetilde{U}^{\alpha}(X)$ of the wave functions corresponding to this division satisfy to the set of equations

$$(-\Delta + \vartheta_{\alpha}(x_{\alpha}) - E)\, \widetilde{U}^{\alpha}(X) = -\vartheta_{\alpha}(x_{\alpha}) \sum_{\beta \neq \alpha} \widetilde{U}^{\beta}(X), \ |x_{\alpha}| > a_{\alpha}, \qquad (2.130)$$

$$(-\Delta - E)\, \widetilde{U}^{\alpha}(X) = 0, \quad |x_{\alpha}| < a_{\alpha}, \qquad (2.131)$$

with boundary conditions $\sum_{\alpha} \widetilde{U}^{\alpha}(X) = 0, \quad X \in R^6 / \Omega.$ @, (2.132)

Eqs. (2.130), (2.131) with boundary conditions (2.132) are uniquely resolvable [44, 45, 52, 65].

It should be mentioned that instead of conditions (2.132) conditions of the type

$$\sum_{\alpha} U^{\alpha}(X)\,|_{X \in \Gamma} = 0, \quad \alpha = 1, 2, 3,$$

can be considered as well, and in this case Eqs. (2.131) are solvable uniquely for all scattering energies, except some countable set of E values which consists of the discrete spectrum of the internal Dirichle problems for the regions to which the set $R^6 | \Omega$ is devided by cylinders Γ_{α}. Similar statements are hold for conditions of the third kind [44].

Further, expanding the wave functions components into a series in terms of spherical harmonics (2.31) and separating spin-isospin angular variables, we obtain sets of two-dimensional integro-differential equations with boundary (in a general case two-sided) conditions at $|x| = a \pm 0$.

3. RESULTS OF CALCULATIONS OF VARIOUS PROCESSES IN ATOMIC MOLECULAR, AND NUCLEAR PHYSICS

Consider the application of the quantum theory of scattering in the system of several particles both in the integral and differential forms for calculations of various processes in atomic, molecular, and nuclear physics. The calculations results presented were obtained both on a basis of numerical methods developed and corresponding software packages for the numerical solution of sets of integral and integro-differential equations and on a basis of analytical solutions obtained in some limiting cases. A study were made of threshold peculiarities in scattering processes occuring in the systems of three particles at various collisional conditions.

3.1. Results of calculations of various processes in atomic physics

At present calculations of various processes of collisions of electrons with molecules are necessary for practical investigations of spectra of aurora and the night sky glow in polar ionosphere [6 - 12], active media of the electroionization, gas-discharge and chemical lasers [1 - 5], and various plasma-chemical units [7 - 12]. Therefore of great importance become the development and application of universal methods and corresponding software for calculations of cross-sections of these processes based on the quantum theory of scattering in the system of several particles [41 - 70]. In this case substantially different processes occurring in the systems of three particles are investigated, based on the same methods.

The results of calculations of the cross-sections of proceses going at collisions of electrons with two-atomic molecules in the ground and excited vibration-rotation states,

$$e + AB(v_1, J_1) \rightarrow \begin{cases} e + AB(v_1, J_1) - & \text{an elastic scattering,} \\ e + AB(v_2, J_2) - & \text{a vibration-rotation excitation} \\ A^-(eA) + B - & \text{a dissociative attachment (DA)} \\ A + B^-(eB) & \\ e + A + B - & \text{a molecular dissociation,} \end{cases} \quad (3.1)$$

performed on the basis of the quantum theory of scattering in systems of several particles both in integral and differential forms, are presented in Figs. 2 - 21. together with the experimental data and calculations of other authors [23 - 70, 105]. As concrete two-atomic molecules, molecules H_2, HD, D_2, F_2, Cl_2, Br_2, I_2, HF, DF, HCl, DCl, HBr, DBr, HI, DI, RbCl, RbBr, CsCl, CsBr, and KI were investigated.

The main approximation in these calculations is that the interaction of the impacting electron with electrons and nuclei of the target molecule is changed for the interaction of the electron with each atom as a whole, considering the atom to be a force center. Thus, the complicated many-particle problem of calculating cross-sections of the electron scattering on two-atomic molecules is reduced to the problem of collisions in the system of three bodies which is solved by means of Eqs. (1.2). This approximation appears to be reasonable when the impacting electron energy is less than the energy of the molecular excited state.

As the initial data in this problem, paired interaction potentials, masses, and energies of impacting particles are used. As paired potentials of interaction between electrons and atoms in a molecule, potentials of zero radius [20] and potentials of the type $V(r) = \lambda \exp(-\beta r)/r$ were used which parameters were determined on the basis of the empirical data: the electron bond energy in the negative field, the scattering wavelengths and an effective radius, while spin (in the case of homonucleous molecules) was accounted for as follows. As the scattering distance, a quantity

$$\frac{1}{a} = \frac{1}{a_1} = \frac{1}{a_2} = \frac{1}{4}\left(\frac{3}{a_t} + \frac{1}{a_s}\right),$$

was used [17, 20, 46 - 49, 55 - 59], where a_t and a_s are triplet and singlet scattering distances, respectively. Paired potentials of interaction between atoms in molecules were simulated with Morse potentials

$$V(r) = D\{1 - \exp[-\alpha(r - r_0)]\}^2, \tag{3.2}$$

with parameters determined from the spectroscopic data [106].

The results of calculations of cross-sections of the electron scattering on hydrogen molecules and on their isotopically substituted analogs are presented in Figs. 2 - 7.

A great ratio of proton and electron masses is a favorable condition which allows to obtain the analytical solutions for the set of Eqs.(1.2) and, hence, cross-sections of processes (3.1) which can be expressed through elementary and special functions (1.23 - 1.28). The results of calculations of processes (3.1) are presented in Figs. 2 - 9 which show that cross-sections of processes (3.1) are well described with formulae (1.23 - 1.28) giving both a quantitative description of processes (3.1) and a qualitative picture of the phonomenon which implies that in the framework of this approximation one can rather accurately take into account the effect of multiple scattering. In the same approximation, the isotopic effect shows up first predicted by Yu.N. Demkov [20]. Notice that the results of cross-sections calculations for processes (3.1) in the single scattering approximation differ considerably from the experimental data because the single scattering approximation is invalid and it is necessary to take into account multiple scattering.

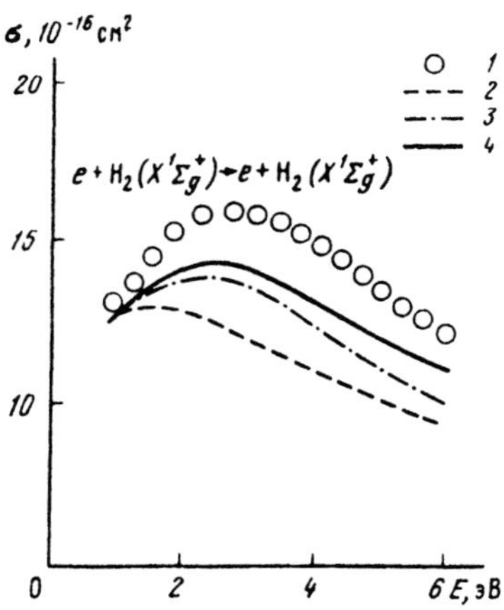

Fig. 2 Elastic scattering of electrons on hydrogen molecules

1- experimeantal data [97], 2 - calculation results in the ZRP approximaton [20], 3 - calculaton results in the strong coupling approximation [38,39], 4 - results of calculatons in this paper

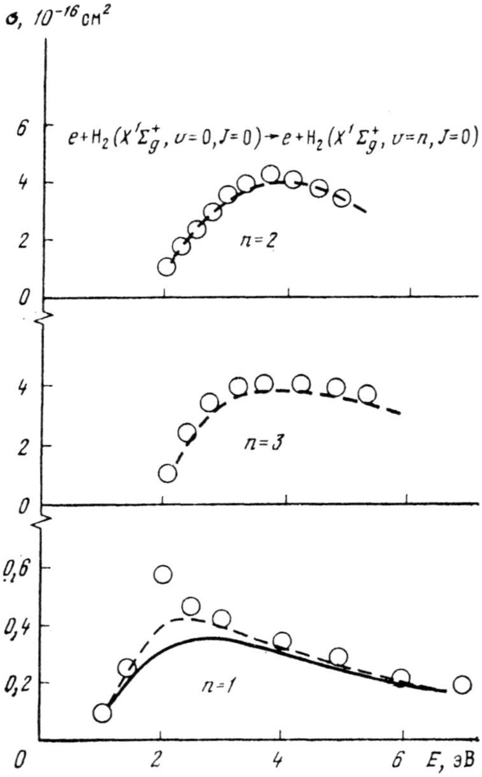

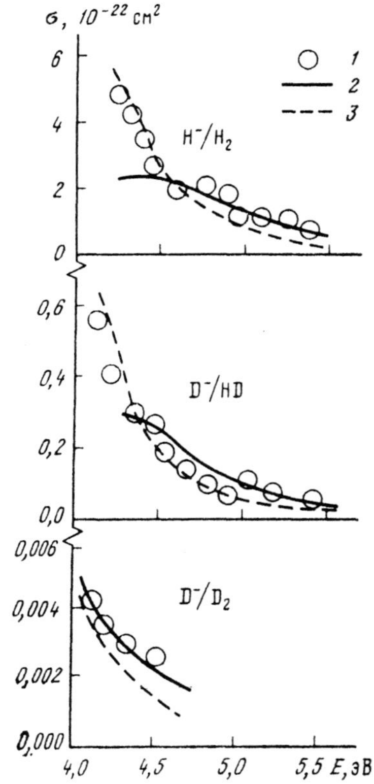

Fig. 3. Vibrational excitation of hydrogen molecules with electrons
1 - experimental data [97], 2 - calculation results in the ZRP approximation [20], 3 - results of calculations in this paper

Fig. 4. Dissociative attachment of electron to H_2, HD, and D_2 molecules
1 - experimental data [97], 2 and 3 - results of calculations based on analytical solutions of Eqs.(1.2), when the interaction between atoms in a molecule was simulated with Morse potentials and the harmonic oscillator potential, respectively

Figs. 4 and 5 show the calculation results for cross-sections of the electron dissociative attachment to hydrogen molecules and their isotopically substituted modifications performed by a numerical solution of the set of Eqs.(1.2) by means of the method described in the first part of this paper. Nonmonotone dependences of cross-sections of the dissociative attachment processes on the energy were obtained. This shows the existence of the threshold singularities in the reactions cross-sections which are the most prominent in systems consisting of two identical heavy particles and a single light one. Comparision of calculations performed with experimental data shows that simulating the electron interaction with each of the molecule atoms by means of potentials (1.11) and (1.12) in the framework of multiple scattering allows to obtain a satisfactory agreement with experimental data (coincidence of the cross-sections orders, including isotopic effects and threshold singularities).

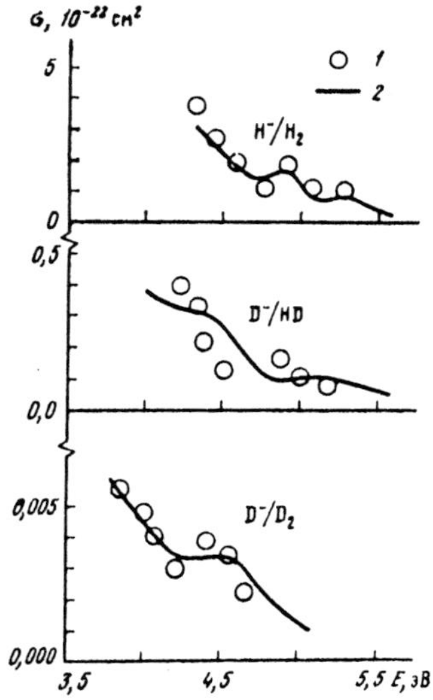

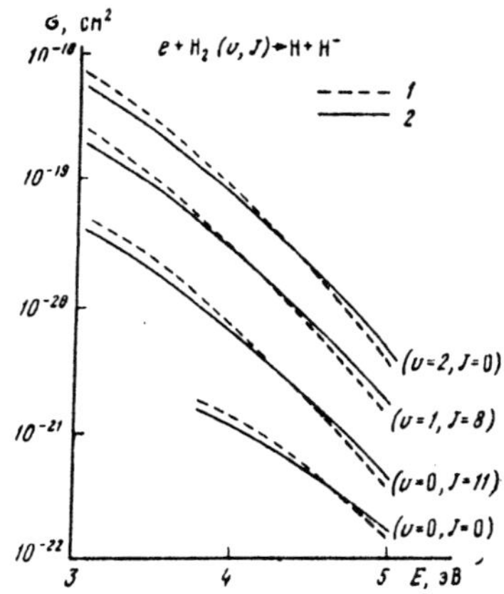

Fig. 5. Dissociative attachment of electrons to H_2, HD, and D_2 molecules
1 - experimental data [97], 2 - results of calculations based on the numerical solution of Eqs.(1.2)

Fig. 6. Dissociative attachment of electrons to hydrogen molecules in the excited vibration-rotation states
1 - results of calculations in the quasi-classical approximation [35], 2 - results of calculations in this paper

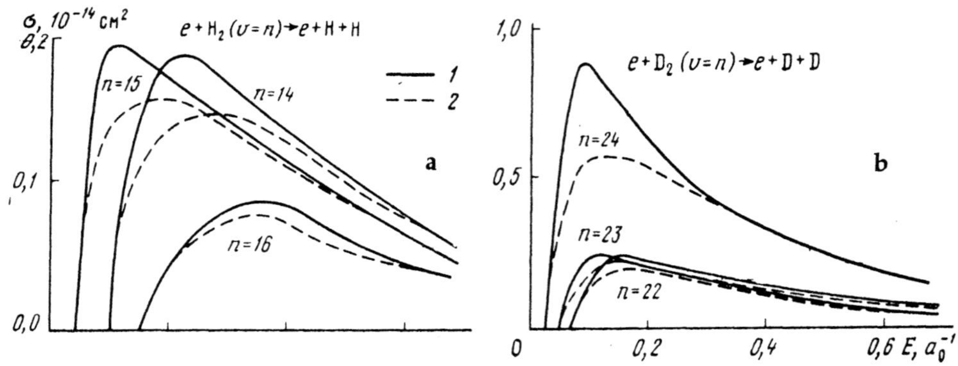

Fig. 7. Dissociation cross-sections for molecules of hydrogen (a) and deuterium (b) in the excited vibrational states
1 - results of calculations in the ZRP approximation [93], 2 - results of calculations in this paper

Notice that there are at present various theories for calculations of cross-sections of processes (3.1) going at the interaction of electrons with molecules [17-39, 77-85, 97] for which terms of molecular negative ions and a number of other empirical parameters are needed as the inital data. However, in many cases it is impossible to perform specific calculations according to these theories even for two-atomic molecules because these data are absent for the great majority of molecules. In addition, there is a great difference between the experimental results and results of these calculations. which is demonstrated clearly by calculations performed on the basis of the Firsov-Smirnov model [47] where the approximate energy value and widths of the quasi-stationary H_2^- term obtained in the ZRP approximation are used. For the energy lower than the dissociation threshold, too great cross-section values were obtained, while for the energy higher than this threshold, an agreement with the experiment and our calculations was considerably better.

Thus, namely, in this case the advantages are clearly manifested of the approach based on the quantum theory of scattering in the system of several particles, in which paired T-matrices are needed as initial data, which can be constructed on the basis of paired interaction potentials and WF corresponding to the initial molecular state.

Figs. 8 - 10 show results of cross-sections calculations for collisions of electrons with nitrogen molecules performed on the basis of quantum theory of scattering in the system of three particles, experimental data [97], quasi-classical calculations [35], and calculations made in the strong coupling approximation [107].

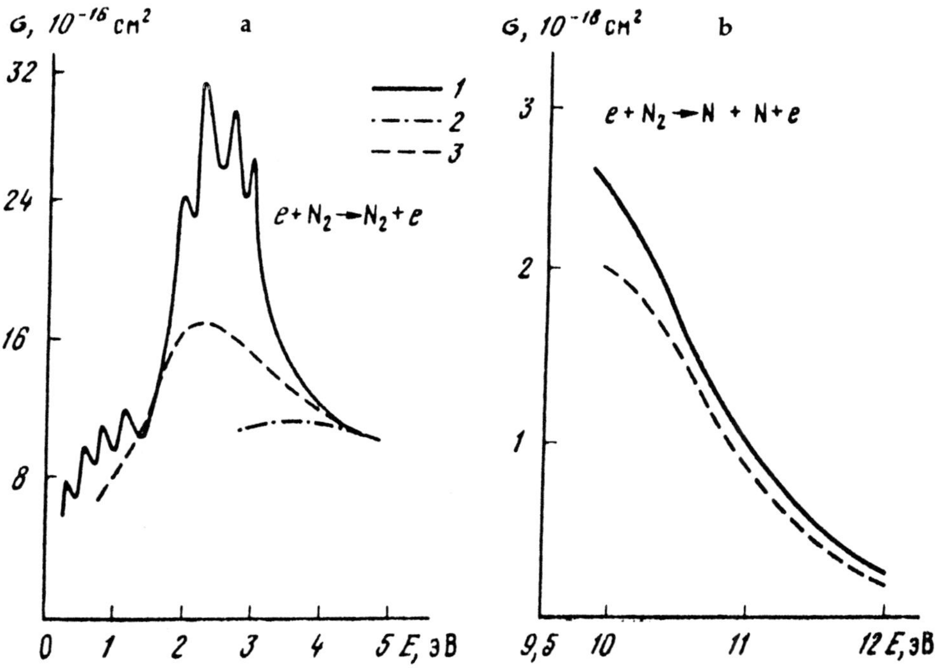

Fig. 8. Cross-sections (a) of the elastic scattering and (b) dissociation of nitrogen molecules by electrons
1 - experimental data [97], 2 - results of calculations in the strong coupling approximation based on Eqs. (1.2) [38, 39], 3 - resuts of calculations in this paper

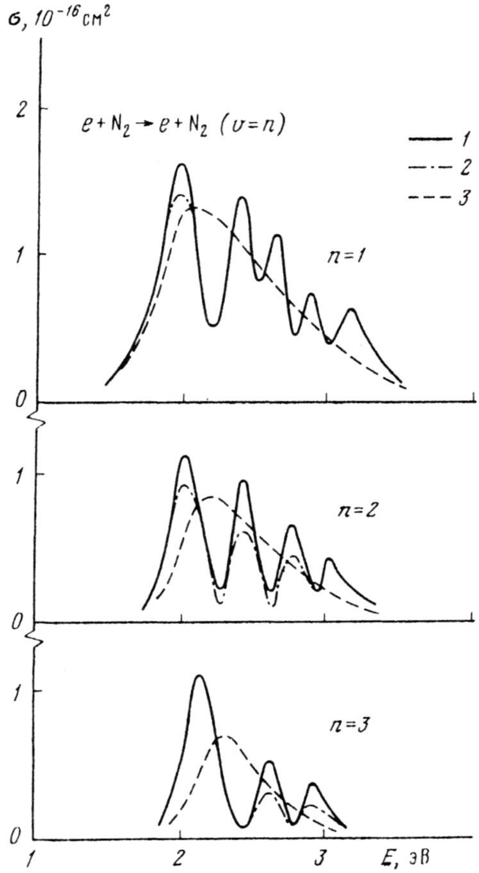

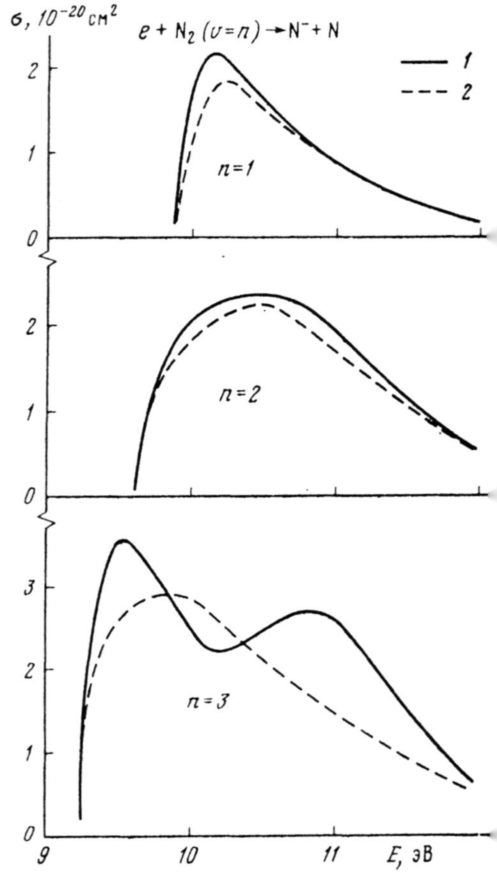

Fig. 9. Cross-sections of the vibrational excitation of nitrogen molecules by electrons
1 - experimental data [97], 2 - results of calculations in the quasi-classical approximation [35, 107], 3 - results of calculations in this paper

Fig. 10. Dissociative attachment of electrons to the vibrationally excited nitrogen molecules
1 - results of calculations in the quasi-classical approximation [107], 2 - results of calculations in this paper

The distinctive feature of the DA of electron to nitrogen molecule is the fact that in this reaction anegative nitrogen ion is produced in P-state

$$e + N_2 \rightarrow N(^4S) + N^-(^3P).$$

Therefore to describe such a reaction in the separable approximation, it is necessary to use potentials (1.11) with form-factors of the type

$$g(k) = [1 + \mu k Y_{10}(\hat{\rho})]/(k^2 + \beta^2). \qquad (3.3)$$

Formulae connecting parameters λ, μ, and β with the bond energy, scattering distance, and effective radius are presented in [20]. Unfortunally, at present only bond energies are known from the experiment [106], while the scattering distance was only theoretically evaluated [20]. As for the effective radius, there are for it neither experimental data nor theoretical estimations. For this reason the absolute value of the DA cross-section of electron to a nitrogen molecule has been used instead of the effective radius to determine parameter values [97]. The values of parameters λ, μ, and β determined by this method are 0.634, 0.240, and 0.871. The potential of interaction between atoms in a molecule was simulated with Morse potential. To construct the separable T-matrix with this potential, the Baytman method was used, and to solve the set of integral equations (1.2) in this approximation, the method described in (1.3) was applied. Comparision of the calculation results with experimental data [97] shows that simulating interaction of electrons with atoms in molecules by means of nonlocal separable potentials (1.11) and (1.12) allows to obtain a satisfactory agreement with experimental data in the average, which is natural because of the nonlocal potentials properties. Also notice that the choice of the separable potentials parameters using experimental data [106] allows to calculate all possible collision processes of electrons with nitrogen atoms which confirms the self-consistence of the model suggested.

Collisions of electrons with halogen and hydrohalogen molecules were extensively studied both experimentally and theoretically [17 - 33, 97, 103 - 109]. However, at present there are no comprehensive and reliable data for cross-sections of these reactions. For example, the absolute value of the DA cross-section for electron and iodine molecules measured by different authors between 1956 and 1978 was in the range of two orders. In addition, with increasing an accuracy and changing experimental methods, the unknown earlier structure was found in the process cross-section. Similar situation also exists in the case of other halogen molecules. Disadvantages of the experimental papers also are the absence of data about absolute cross-sections values, except single estimations.

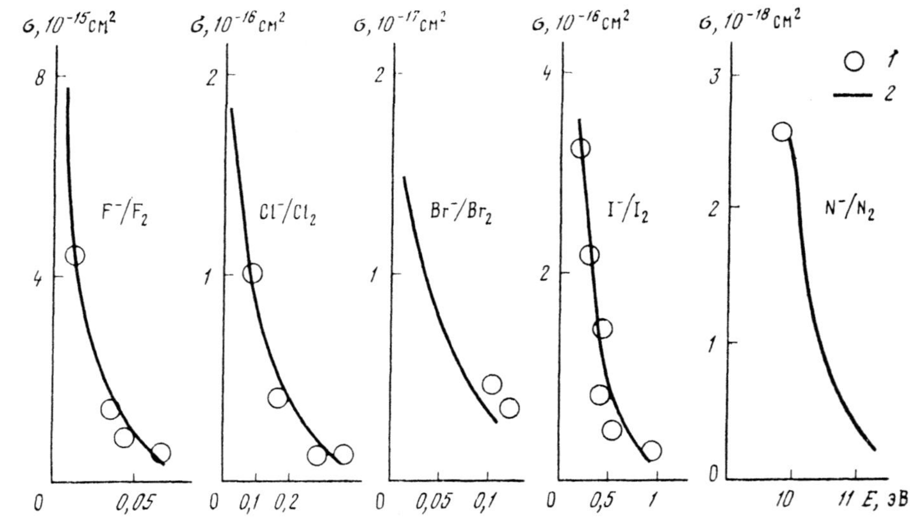

Fig. 11. Cross-sections of the dissociative attachment of electrons to halogen and nitrogen molecules
1 - experimental data [109, 110], 2 - results of calculations in this paper

A peculiarity of electron collisions with halogen molecules is that due to DA reaction halogen atoms are produced in P-states [103]. So in this case it is necessary to use form-factors of the type (3.3) as form-factors of separable potentials (1.11). Calculation of cross-sections for processes (3.1) occurring at electron collisions with halogen and hydrohalogen molecules on the basis of quantum theory of scattering in the system of three particles is complicated due to the absence of data about parameters of the potentials of interaction between electrons and halogen atoms, because for these atoms scattering distances and effective radii are unknown. However, these quantities can be determined using Tomas-Fermi or Titz statistical potential [114], and it is necessary to notice that halogen atoms come precisely into the range of the charge change, where the scattering distance is approaching infinity. Values of potential parameters determined by the above described method are presented in [46 - 49, 55 - 59]. The calculation results and experimental data for collisions of electrons with halogen molecules are given in Fig. 11.

The application of quantum theory of scattering in the system of three particles for calculations of the electron collisions with hydrohalogen molecules in the above described approximation shows the considerable discrepancy between the calculation results and experimental data for cross-sections of a realignment of the dissociative attachment (by a factor of 4 - 5). This means that calculations of processes of the electron DA to hydrohalogen molecules in this approximation is too rough because potentials of the type (1.11), (3.3) take effectively into account only the interaction in s- and p-states. In the case of hydrohalogen molecules, which have a dipole moment, the long-range interaction considerably influence on the cross-sections values [35]. To take into account the long-range interaction the model was used in which a hydrohalogen molecule was treated as a system consisting of proton (p), electron (e) and halogen atom (X) because of the large energy of the electron affinity of a halogen atom (4 eV)6 as opposed to a hydrogen atom (0.75 eV) [106]. Similar considerations can be also applied to modelling alkali halides molecules (RbCl, RbBr, CsCl, CsBr, KI) which can be treated as consisting of positively charged alkali ion (Y^+ = Rb^+, Cs^+, K^+), electron (e), and a halogen atom (X). Thus, to calculate cross-sections of the electron scattering on these molecules, one can use the quantum theory of scattering in the system of four particles

$$e + \mathrm{HX}\,(e + \mathrm{X} + p) \rightarrow \begin{cases} e + \mathrm{HX}, \\ \mathrm{X}^- \,(e + \mathrm{X}) + \mathrm{H}\,(e + p), \\ \mathrm{X} + \mathrm{H}^- \,(e + p + e), \\ e + p + e + \mathrm{X}, \\ e + \mathrm{X} + \mathrm{H}\,(e + p), \end{cases}$$

Notice that similar reactions occure also at the interaction of electrons with alkali halides ($Y^+ + X + e$).

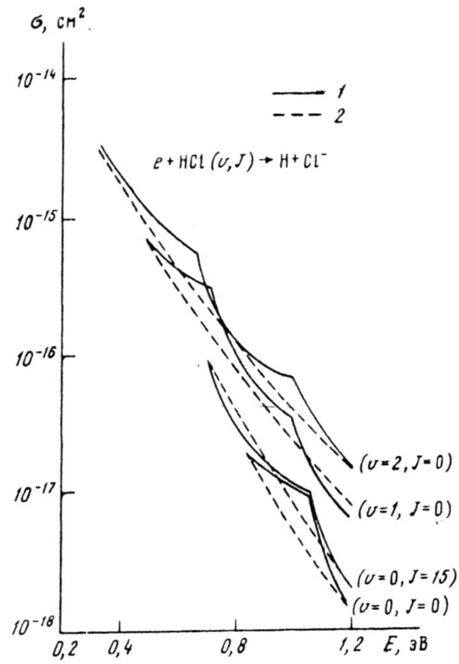

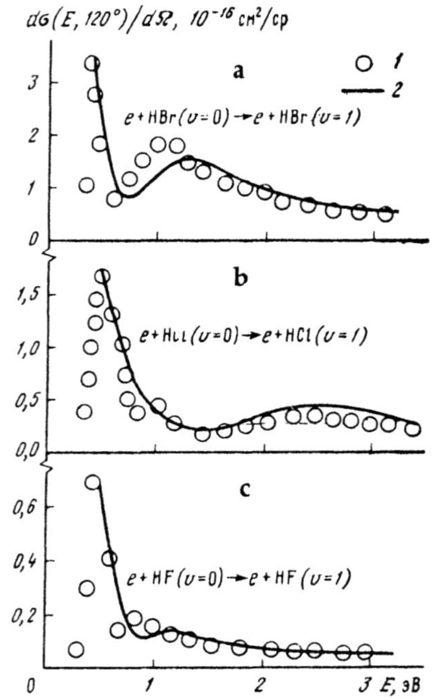

Fig. 12. Cross-sections of dissociative attachment of electrons to HCl molecules in the excited vibration-rotation states

1 - results of calculations in quasi-classical approximation [112, 35], 2 - results of calculations in this paper

Fig. 13. Differential cross-sections of the vibrational excitation of (a) HBr, (b) HCl, and (c) HF molecules by electrons

1 - experimental data [111], 2 - results of calculations in this paper

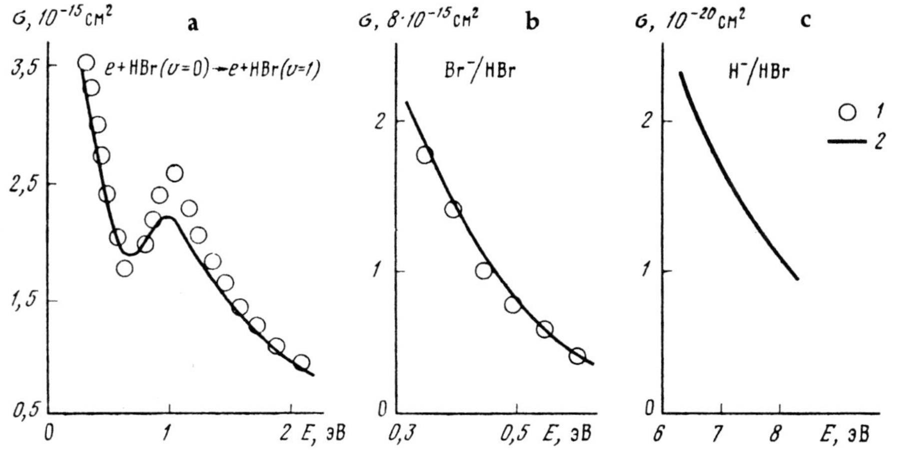

Fig. 14. Cross-sections of processes of (a) the vibrational excitation and (b, c) dissociative attachment of electrons to HBr molecules

1 - experimental data [111, 112, 113], 2 - results of calculations in this paper

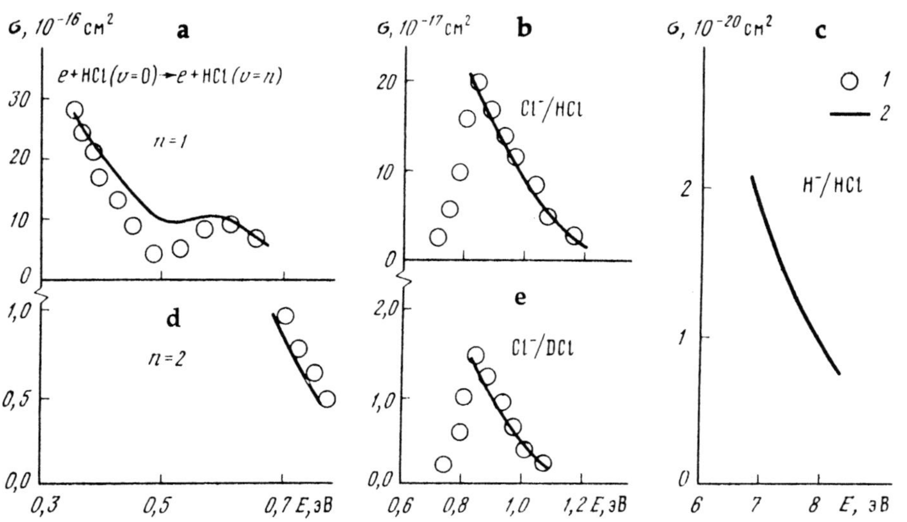

Fig. 15. Cross-sections of processes of (a, b) the vibrational excitation and dissociative attachment of electrons to (c, d) HCl and (e) DCl molecules

1 - experimental data [112, 113], 2 - results of calculations in this paper

To solve this problem at the energies lower than the dissociation energy of these molecules, one can use the integral equations (1.14) which allows analytical solutions in the approximation (1.15). At the impacting electron energy higher than the dissociation energy, it is necessary to reconstruct equations (1.15) or use equations (2.10) with appropriate boundary conditions (2.13 - 2.25). Results of these calculations are presented in Figs. 12 - 17 together with experimental data [97, 109 - 113].

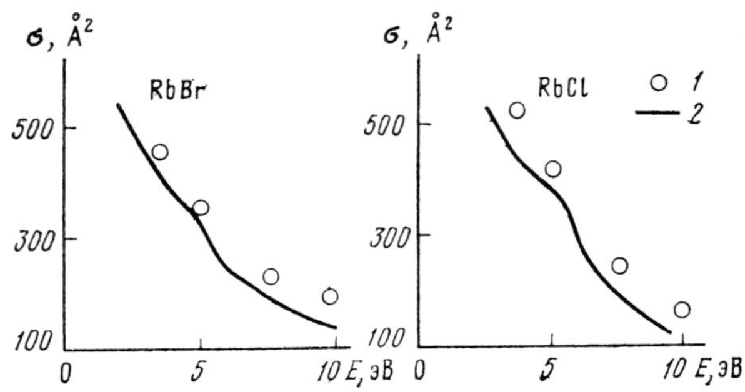

Fig. 16. Cross-sections of the elastic scattering of electrons on RbBr and RbCl molecules
1 - experimental data [113], 2 - results of calculations in this paper

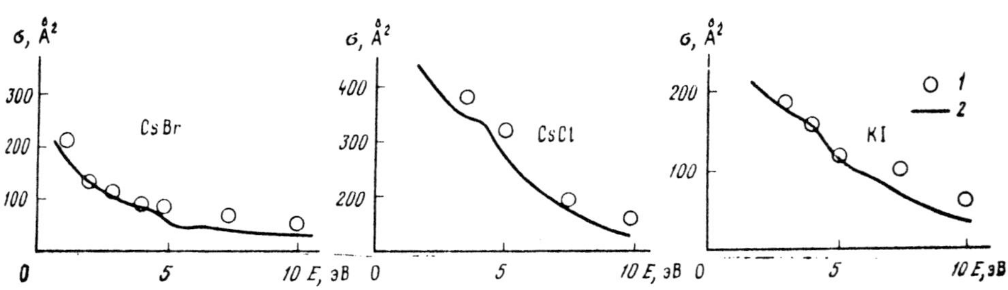

Fig. 17. Cross-sections of the elastic scattering of electrons on CsBr, CsCl, and KI molecules
Notations as in Fig. 16

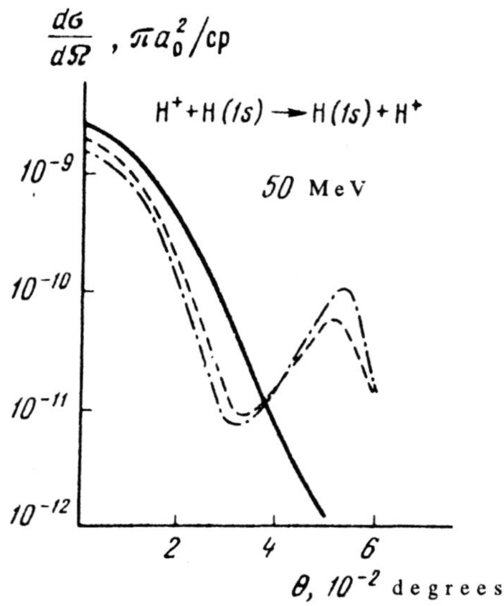

Fig. 18. Cross-section of proton recharge on hydrogen atoms 1 - experimental data [115 - 117], 2 - results of calculations in Born approximation [8 6 115], 3 - results of calculations in Glauber approximation [116], 4 - results of calculations in this paper

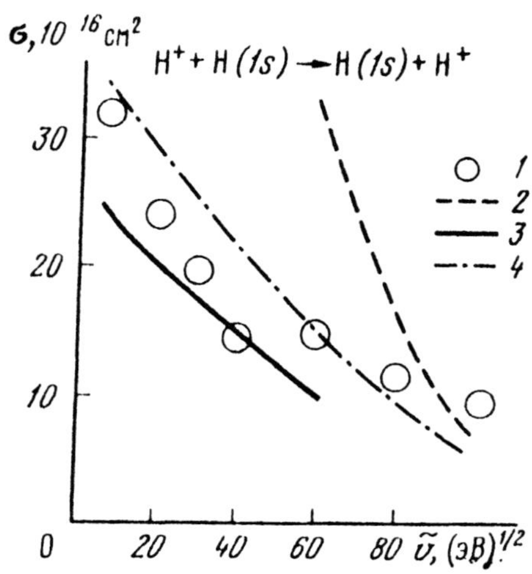

Fig. 19. Differential cross-section of proton recharge on a hydrogen atom

Notations as in Fig. 18

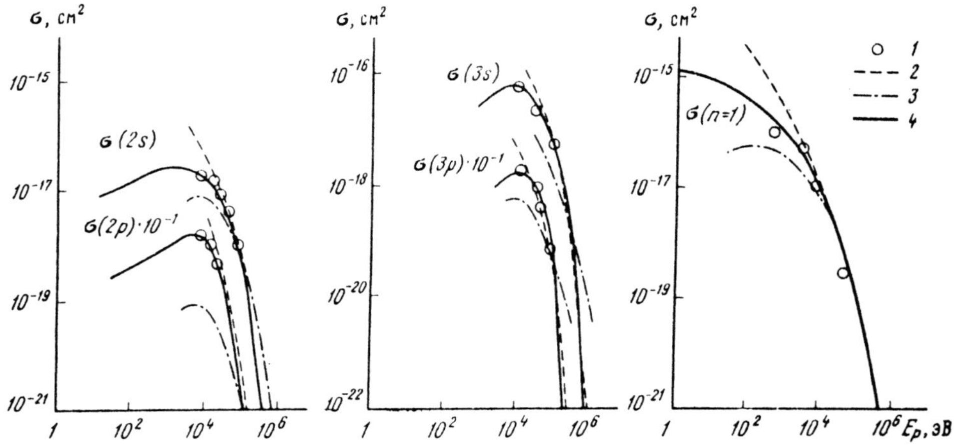

Fig. 20. Proton recharge on hydrogen atoms (excitation of 2s and 2p levels)
1 - experimental data [116], 2 - results of calculations in Born approximation [8, 115], 3 - results of calculations in quasi-classical approximation [116], 4 - results of calculations in this paper

Fig. 21. Proton recharge on hydrogen atoms (excitation of 3s-and 3p-levels)
Notations as in Fig. 20

Fig. 22. Proton recharge on hydrogen atoms with the excitation n = 1
Notations as in Fig. 20

The calculation results obtained for various processes of interaction of electrons with two-atomic molecules in the approximation of quantum theory of scatterinng in systems of three and four particles allow to make the following conclusions.

1. Chemical reactions induced by collisions of electrons with two-atomic molecules occure due to multiple scattering of electrons on the centers of force which are atoms in the molecule. Multiple scattering is the main factor determining

peculiarities in the processes cross-sections (3.1), such as isotopic effects and threshold oscillations both in the case of neutral and charged particles. These peculiarities in the scattering cross-sections (3.1), first predicted on the complitely different basis, were confirmed in the framework of the picture of multiple scattering.

2. The calculations presented do not pretend to reproduce the experimental data exactly, however in the framework of the approximation considered they allow to reproduce above mentioned peculiarities of the processes cross-sections (3.1) and terms of the quasi-stationary states of molecular negative ions.

Processes of theexcitation, ionization, recharge, and neutralization occuring at the interaction of protons with hydrogen atoms and ions and also between hydrogen atoms play an important role in many physical applications, including plasma diagnostics in tokamaks, a development of X-ray lasers etc. [6 - 11]. To describe these processes, one needs to know cross-sections and velocities of corresponding reactions in the wide range of energies and temperature. The systems of proton and a hydrogen atom (two protons and electron) and of proton and a negative hydrogen ion (two protons and two electrons) are nonmodel three- and four-particle systems, respectively. So the most natural and universal method of description of all possible processes in these systems is the quantun theory of scattering in systems of several particles based both on integral (1.2) and differential (2.10) equations. It should be noted here that namely for this specific system of proton and a hydrogen atom integral equations (1.2) and (1.14) allows analytical solutions in the approximation (1.15).

Results of calculations of the recharge and excitation processes occuring at collisions of protons with hydrogen atoms performed on the basis of analytical solutions of the set of equations (1.2) in the approximation (1.15), (1.27) are presented in Fig. 18 - 26 together with experimental data [115

- 117] and calculations in papers [8, 115, 116] made in Born, Glauber, and classical approximations. One can see from these figures that calculations in Born approximation are valid in the energy range $E_p \geq 100$ KeV, while at lower energies the only correct method of calculations of these processes is the quantum theory of scattering in the system of three bodies.

Fig. 27 shows results of calculations of the recharge cross-sections in reactions

$$\mathrm{H} + \begin{cases} \mathrm{C}^{k+} \\ \mathrm{Li}^{k+} \end{cases} \rightarrow \mathrm{H}^{+} + \begin{cases} \mathrm{C}^{(k-1)+} \\ \mathrm{Li}^{(k-1)+}, \end{cases} \quad k = 1, 2, 3$$

(at energies $E_p \geq 10$ KeV) performed in the modified Born approximation [118] and on the basis of the quantum theory of scattering in the system of three bodies. Notice that in the last case ionized atoms C^{k+} and Li^{k+} were treated as single centers. These calculations are consistent with that performed on the basis of various approximations [17 - 28, 117 - 120] and experimental data [118]. This results from the fact that the recharge problem has a single characteristic parameter [23, 24].

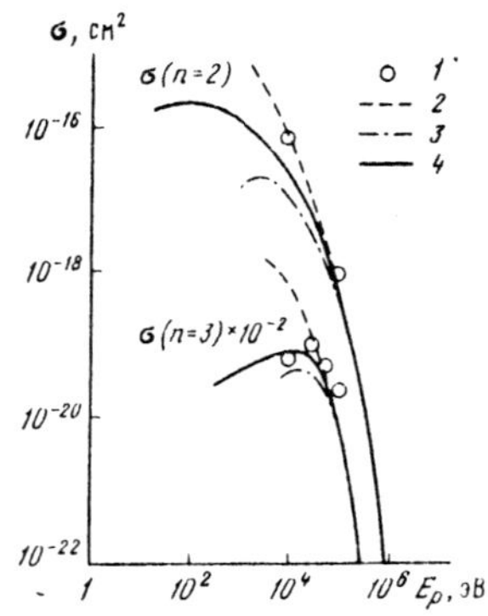

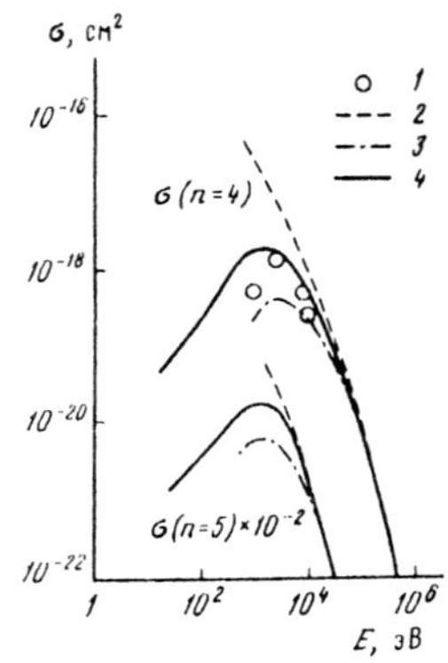

Fig. 23. Proton recharge on hydrogen atoms with the excitation n = 2, 3
Notations as in Fig. 20

Fig. 24. Proton recharge on hydrogen atoms with the excitation n = 4, 5
Notations as in Fig. 20

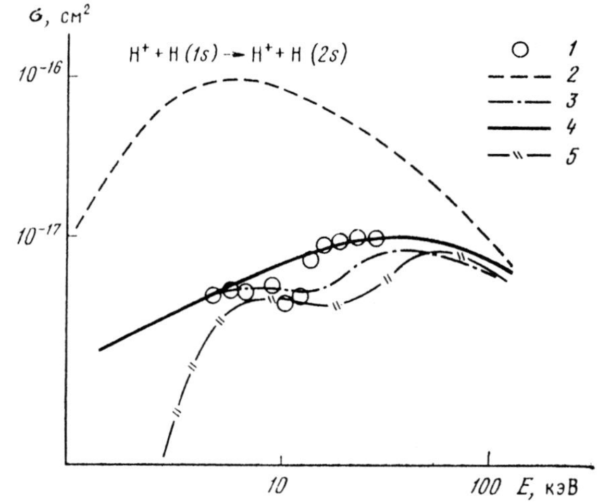

Fig. 25. Cross-sections of excitation of hydrogen atoms H (2s) by protons
1 - experimental data [116, 117], 2 - results of calculations in Born approximation [8, 115], 3 - results of calculations in the strong coupling approximation [117], 4 - results of calculations in this paper, 5 - results of calculations in Glauber approximation [8, 115, 117]

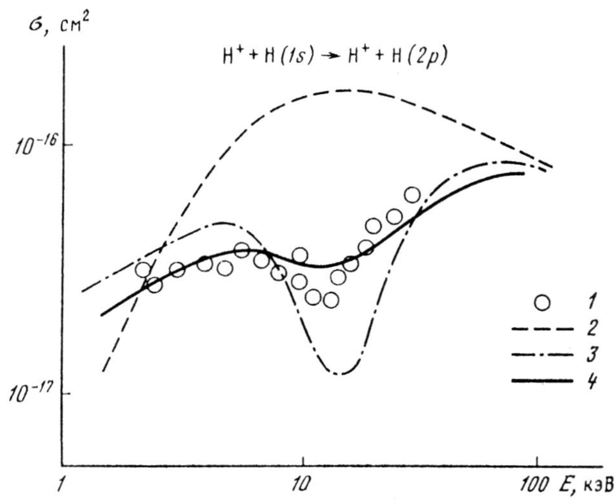

Fig. 26. Cross-sections of excitation of hydrogen atoms H (2p) by protons
Notations as in Fig. 25

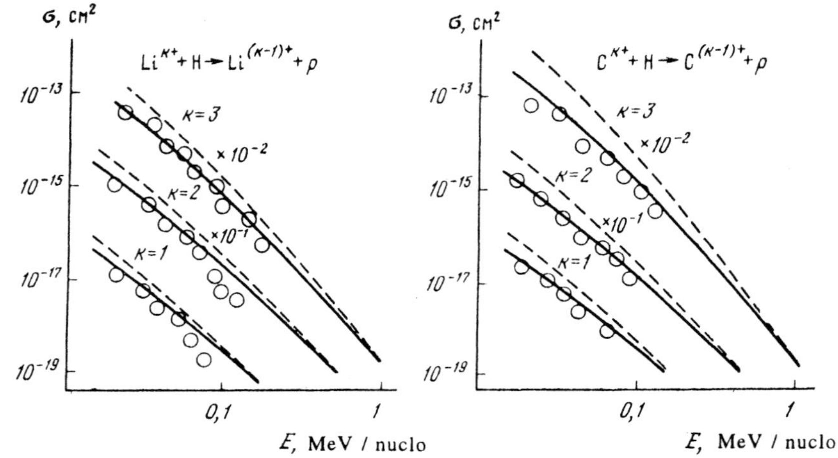

Fig. 27. Cross-sections of recharge of ionized litium and carbon atoms on hydrogen atoms
Circles are experimental data [118], dotted lines are results of calculations in the eikonal approximation [118], solid lines are results of calculations in this paper

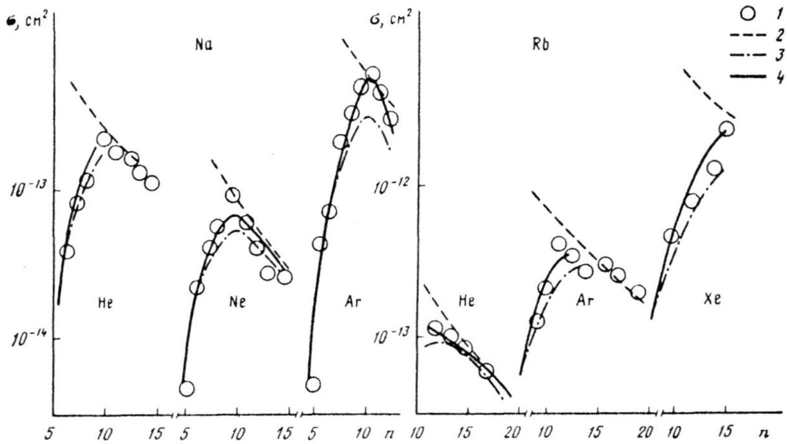

Fig. 28. Dependence on the main quantum number of the cross-section of collisions of highly excited Ridberg sodium and rubidium atoms with helium, neon, argon, and xenon atoms
1 - experimental data [119], 2 - results of calculations in the impulse approximation [119], 3 - results of calculations in the modified impulse approximation [119], 4 - results of calculations in this paper

The same approximation was used for calculations of collisions between He Ne, Ar, and Xe atoms and highly excited Ridberg Na and Rb atoms (Fi. 28), which in these calculations were treated as a positive ion and loosely bound electron on a high orbit. These processes are both of practical (Rydberg states were observed in the spectra of some astrophysical objects, such as solar protuberances, planetary nebulae etc.) and theoretical interest, which is connected with the fact that highly excited atoms, being quantum objects, have a number of classical properties [23]. For example, a hydrogen atom with n = 100 has a diameter on the order of 1 μ, ionization and excitation energies equal to 10^{-3} and 10^{-4} eV, respectively. So in some papers [17 - 21, 23] the classical approximation was used to describe these processes. The physical substantiation of these calculations consists in the fact that an atom with one electron in the highly excited orbit has a large size and behaves as a classical object relative to impacting particle, and, thus, in these conditions the main contribution to cross-sections give regions coinciding with the small vicinity of the classical trajectory.

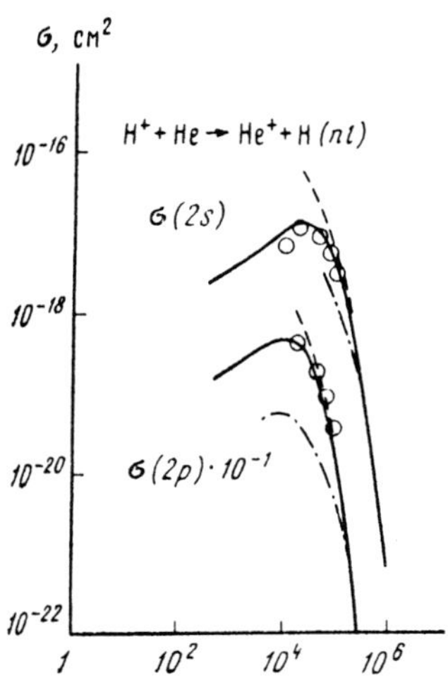

Fig. 29. Cross-sectons of the recharge processes of protons on helium atoms (excitation of 2s- and2p-levels of helium atoms) 1 - experimental data [120], 2 - results of calculatons in Born approximation [8, 120], 3 - results of calculations in the quasi-classical approximation [8, 120], 4 - results of calculations in this paper

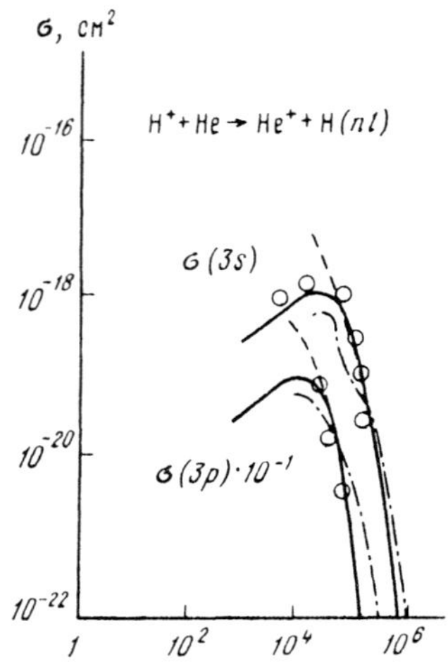

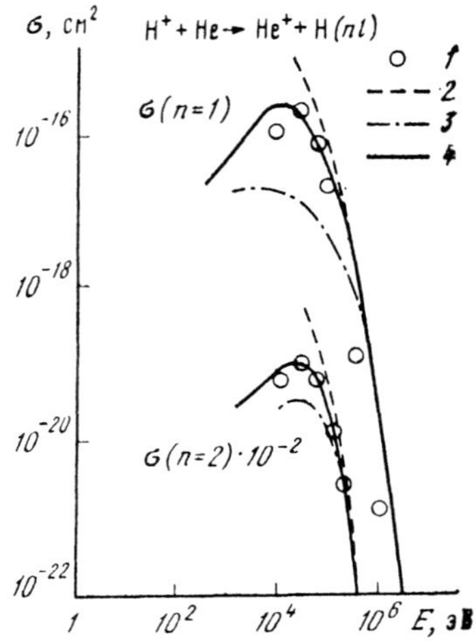

Fig. 30. Cross-sections of the recharge processes of protons on helium atoms (excitation of 3s-and 3p-levels)
Notations as in Fig. 29

Fig. 31. Cross-sections of the recharge processes of protons on helium atoms with the excitation of states with n = 1, 2
Notations as in Fig. 29

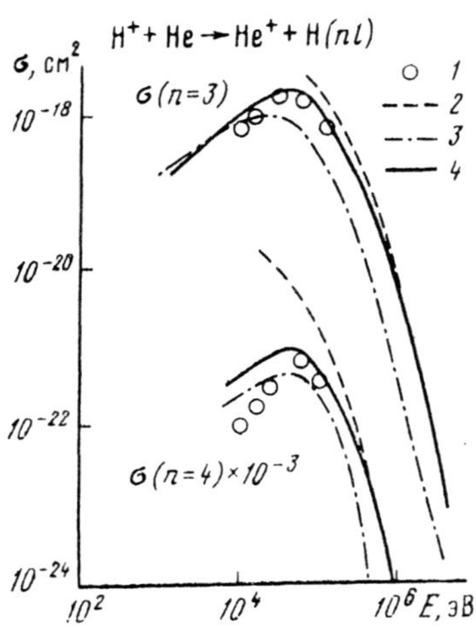

Fig. 32. Recharge of protons on helium atoms (excitation of states with n = 3, 4)
Notations as in Fig. 29

Calculations of the recharge cross-sections of protons on helium atoms were performed on the basis of quantum theory of scattering in the system of four particles (two protons and two electrons). In this case the analytical solutions of the set of equations (1.14) in the approximation (1.15) were used. The results of these calculations together with experimental data [120] and calculations from [8, 120] are presented in Figs. 29 - 32.

Calculations of cross-sections of reactions

$$Na + I \rightarrow Na^+ + I^-$$

were made on the basis of quantum theory of scattering in the system of particles Na^+, e, and I. In this case differential equations (2.10) with corresponding boundary conditions (2.13 - 2.22) were used. The results of these calculations are presented in Fig. 33 together with classical and quasi-classical calculations [29]. Based on these equations. calculations of cross-sections of the hydrogen atoms ionization by protons were done, which are presented in Fig. 34 together with experimental data [120] and calculations from [8, 20, 23, 120].

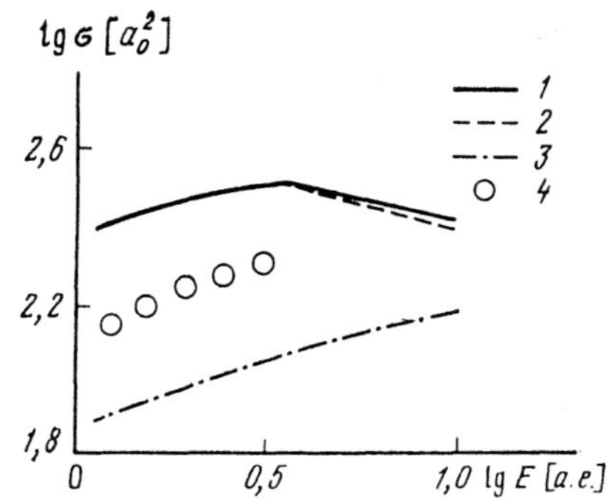

Fig. 33. Cross-sections of the reaction Na + I → Na + I^-
1 - results of classical calculations [29], 2 - results of calculations in the strong coupling approximation [28], 3 -results of calculations in this paper

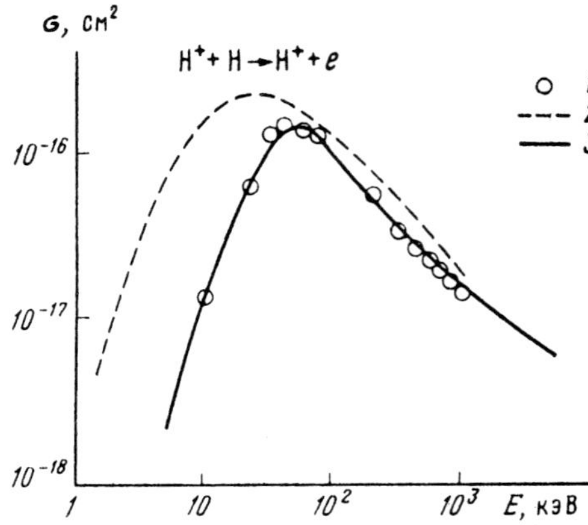

Fig. 34. Cross-sections of the hydrogen atom ionization by protons
1 - experimental data [120], 2 - classical calculations [20, 23], 3 - results of calculations in this paper

Calculations of the collision cross-sections of protons with negative hydrogen ions (the reaction of neutralization $H^+ + H^- \to H + H$) and of collision processes between hydrogen atoms (reactions reversed to the neutralization reaction: $H + H \to H^+ + H^-$) were made on the basis of quantum theory of scattering in the system of four charged particles. The results of these calculations together with experimental data [121 - 123] are presented in Figs. 35, 36. In the dissociative recombination (DR) process $e + AB^+ \to AB^*(n,\ l,\ m) \to A^* + B^*$,

at the collision of electron with a molecular ion an autoionization state is produced which decays into two atoms with one of them or both in the excited state. DR of electrons with H_2, HD, and D_2 molecules results from the excitation of electron in H_2^+ ion from the ground state ($1s\sigma_g$) to the excited one ($2p\sigma_u$) with a simultaneous capture of impacting electron into Ridberg state with quantum numbers

$$e + H_2^+ \to H_2(n \geqslant 3, l, m) \to H(n_1, l_1) + H(n_2, l_2).$$

Results of calculations made in the approximation of quantum theory of scattering are presented in Fig. 37 together with experimental data and calculations in quasi-classical approximation [124, 125].

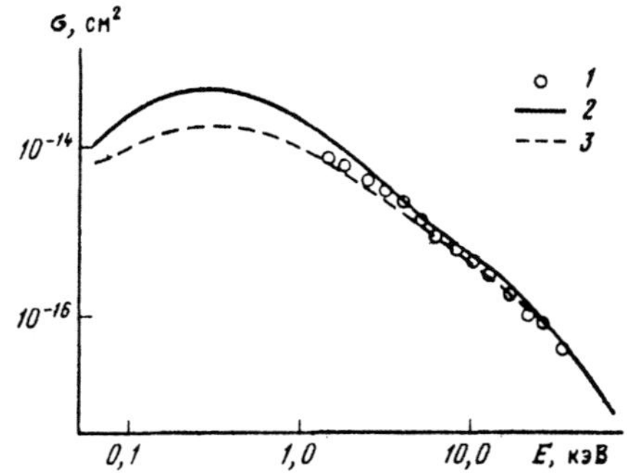

Fig. 35. Cross-sections of the neutralization reaction $H^+ + H^- \rightarrow H + H$
1 - experimental data [121], 2 - results of calculations in this paper, 3 - results of calculations in [122]

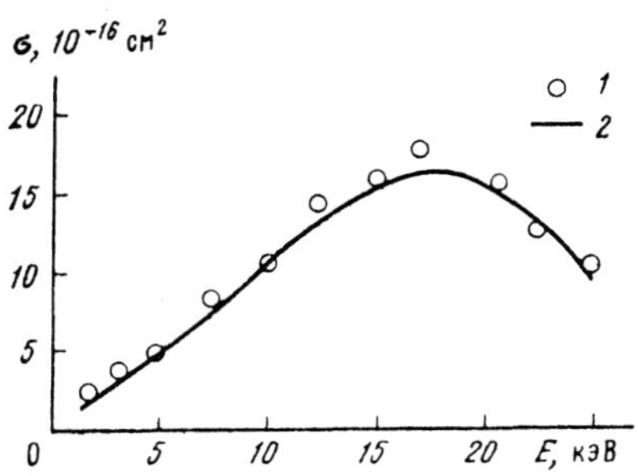

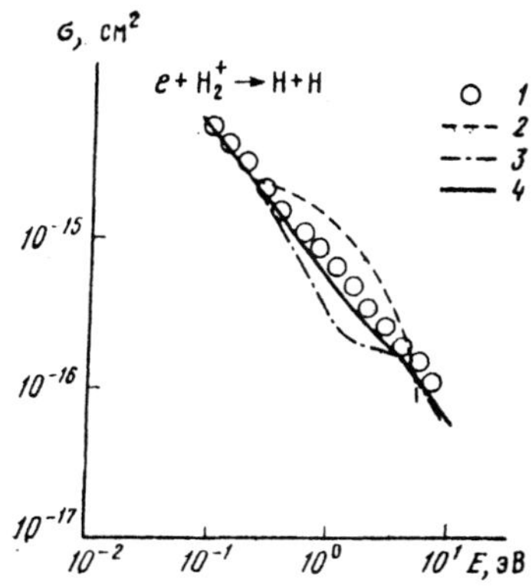

Fig. 36. Cross-section of the raction $H(1s) + H(1s) \rightarrow H^+ + H^-$
1 - experimental data [123], 2 - results of calculations in this paper

Fig. 37. Cross-sections of the dissociative recombination
1 - experimental data [124], 2 - results of calculations based on the quasi-classical approximation [124], 3 - results of calculations in [125], 4 - results of calculations in this paper

The calculations performed and experimental data show large cross-sections 10^{-17} - 10^{-15} cm^2 which is connected with the radiationless electronic transition and a dominant role of the coulomb interaction between colliding particles. The first factor is connected with the fact that decay times of autoionization states with $n \geq 3$ exceed nuclear scatter times (10^{-15} s), and, therefore, practically every act of producing autoionization state results in the dissociation of the molecular ion. This result is a general one and cofirmed with experimental and theoretical data on the cross-sections of DR for He_2^+, Ar_2^+, O_2^+, N_2^+, and NO^+ molecular ions and other.

3.2. Results of calculations of molecular reactions

In this section we consider the application of the quantum theory of scattering in the system of several particles for calculations of molecular reactions of the type

$$A + BC(v_1, J_1) \rightarrow \begin{cases} A + BC(v_1, J_1) & \text{- elastic scattering} \\ A + BC(v_2, J_2) & \text{- vibration-rotation excitation} \\ B + AC(v_3, J_3) & \text{- realignment reactions with} \\ C + AB(v_4, J_4) & \text{vibration-rotation excitation} \\ A + B + C & \text{- reactions of molecular dissociation (decay)} \end{cases}$$

on an example of calculations of collision cross-sections. Consider the examples of some real molecular systems.

1. Calculaton of the vibrational excitation of products in the reaction

$$O^3(p) + CS(X^1\Sigma^+) \rightarrow CO(X^1\Sigma^+, v = n) + S(^3P).$$

Notice the main features of this reaction: it goes without the intermediate complex formation, and the main part of the translational energy transfers into the vibrational energy of CO molecule. The approximate estimation of this state based on the model of collisions in the system of three hard spheres [19] gives the value of $E_v/E_t \sim 8/9$. Therefore, the main characteristics of this reaction are determined by the collisional dynamics which is consistently described by the approximation of the quantum three-body problem on the basis of equations (1.2). In this case the application of the classical. quasi-classical, and approximate quantum-mechanical calculation methods is made difficult by the above-mentioned disadvantages of these approximations. Calculation results fot this reaction together with experimental data [105] are presented in Fig. 38, a.

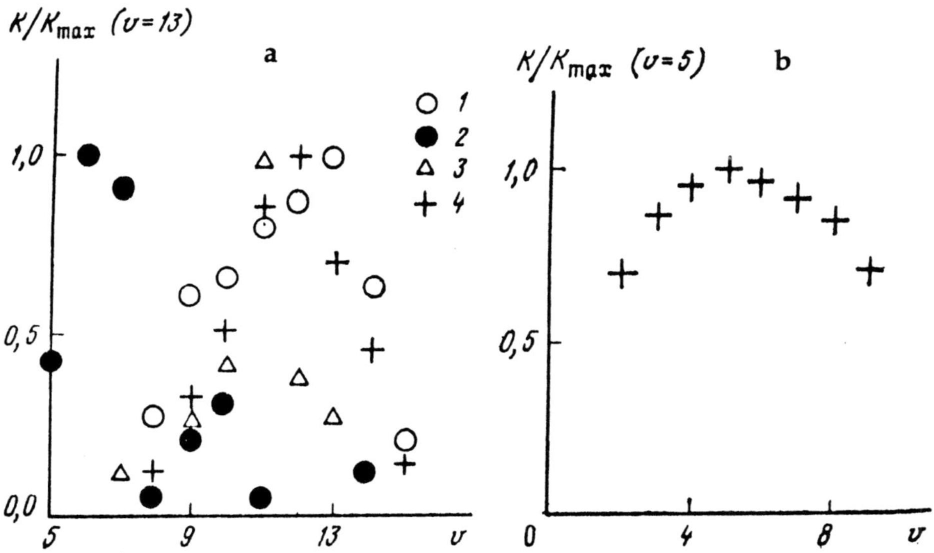

Fig. 38. Vibrational excitation of products in reactions

$$O(^3P) + CS(X^1\Sigma^+) \rightarrow CO(X^1\Sigma^+, v) \rightarrow S(^3P)\ (a)\ и\ H(^2S) + H(^2S) + H(^2S) \rightarrow H_2(X^1\Sigma_g^+, v) + H(^2S)\ (6)$$

1 - experimental data [105], 2 - results of classical trajectory calculations [18, 19], 3 - results of quasi-classical calculations based on the method suggested in [24], 4 - results of calculations in this paper

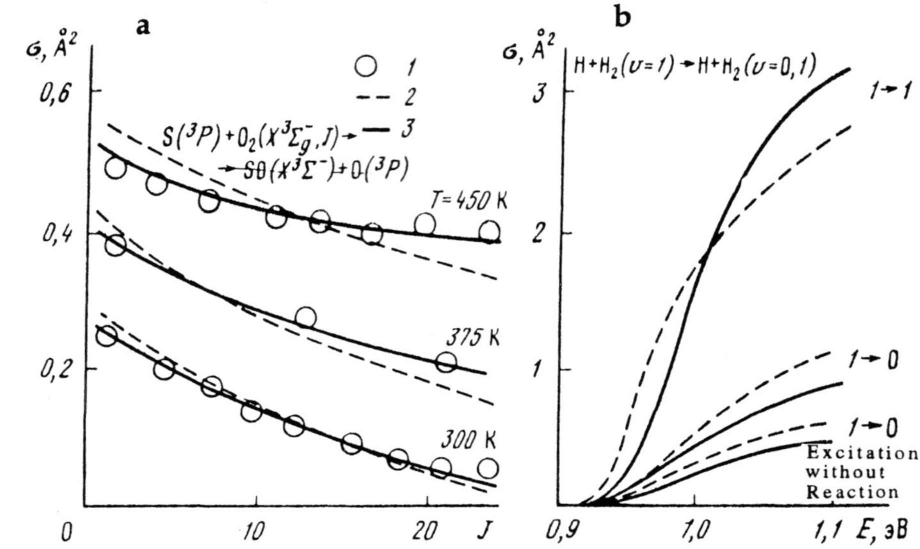

F g. 39. Vibrational and rotational excitation of products in reactions

$$S(^3P) + O_2(X^3\Sigma_g^-, J) \to SO(X^3\Sigma^-) + O(^3P)\ (a)\ и\ H + H_2(v = 1) \to H + H_2(v = 0, 1)\ (б)$$

1 - experimental data [30], 2 - results of calculations in the quasi-classical approximation [30], 3 - results of calculations in this paper

2. Fig. 38,b shows the calculation results for the vibrational excitation of products in the reaction

$$H + H + H \to H_2(v = n) + H$$

confirming the universality of the method described for the cross-sections calculations in chemical reactions involving three atoms being initially in a free state

$$A + B + C \to \begin{cases} AB(v_1, J_1) + C \\ AC(v_2, J_2) + B \\ BC(v_3, J_3) + A \\ A + B + C \end{cases}$$

- reactions with the coupled pairs formation in the ground and excited vibration-rotation states

scattering of free particles

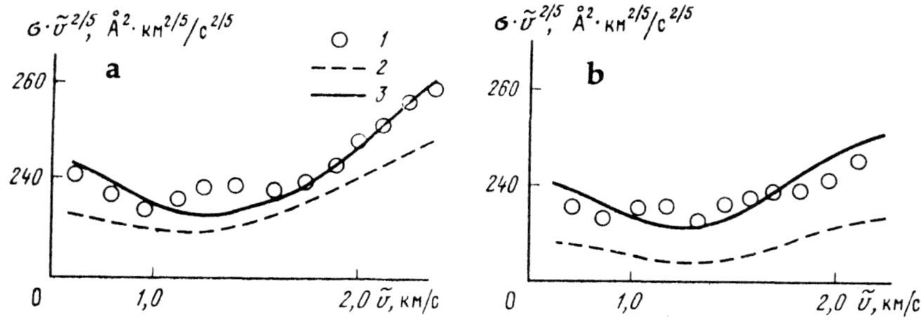

Fig. 40. Differential cross-sections of reactions

$O(^3P) + N_2\left(X^1\Sigma_g^+\right) \rightarrow NO(X^2\Pi) + N(^4S)$

(a) и $O(^3P) + O_2\left(X^3\Sigma_g^-\right) \rightarrow O_2\left(X^3\Sigma_g^-\right) + O(^3P)$ (б)

1 - experimental data [103], 2- classical calculations [103], 3 - results of calculations in this paper

3. The calculation results for the vibrational excitation of products in the reaction

$$S(^3P) + O_2(^3\Sigma_g, v = v_0, \ J = J_0) \rightarrow SO(^3\Sigma, v = v_n, \ J = J_n) + O(^3P)$$

together with the classical calculation results and experimental data [30] are presented in Fig. 39.

4. The calculation results for the cross-sectons of processes

$$H + H_2(v = 1) \rightarrow H_2(v = 0{,}1) + H,$$

$$O(^3P) + N_2\left(X^1\Sigma_g^+\right) \rightarrow NO(^2\Pi) + N(^4S),$$

$$O(^3P) + O_2\left(^3\Sigma_g^-\right) \rightarrow O_2\left(^3\Sigma_g^-\right) + O(^3P)$$

together with the quasi-classical calculations and calculations by the method of the strong coupling [26, 31, 103] are presented in Figs.39 - 40. Notice here that these calculations allow to evaluate the experimental data qualitatively and do not pretend to an accurate reproduction of these data. It is in this consists the main advantage of such calculations which allow to reproduce qualitatively the experimental data using rather crude approximations (the several bodies problem approximation, the application of paired potentials of the

Morse type etc.), which shows the calculations presented in Figs. 38 - 40.

The main approximation in these calculations is that the interaction of the impacting atom (consisting of electrons and nuclei) with electrons and nuclei of the target molecule is changed for the interaction with every atom considering the atom as a whole as a center of force.

Notice here that in the method proposed the inital data are paired potentials of the interaction between energies and masses of impacting particles rather than the potential energy surfaces (PES) which are needed to calculate cross-sections of chemical reactions by classical, semiclassical, and quasi-classical methods, while it is a sufficiently accurate calculation of PES which presents the main problem in the above-mentioned methods [18 - 19]. On the other hand, in the method considered PES can be obtained automatically when solving a set of equations, as it was realized in papers [32, 34, 55 - 59, 75, 86].

Notice that with increasing a number of particles the complexity of solving a set of equations increases considerably. In addition, one have to consider the inner structure of impacting particles which is made on the basis of reduction of the initial equations (2.1) to the set of dynamical equations with effective intercluster interaction, which is nonlocal in a general case.

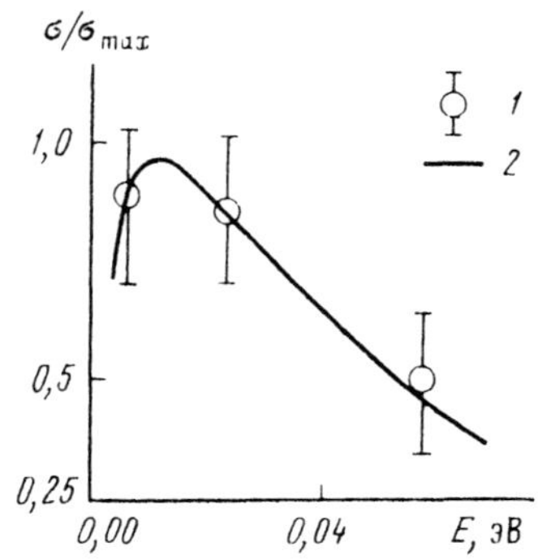

Fig. 41. Calculation results for the reaction $O + CF_3I \rightarrow CF_3 + IO$

1- experimantal data [126], 2 - results of calculations based on the cluster approximation of the quantum theory of scattering in the system of three bodies

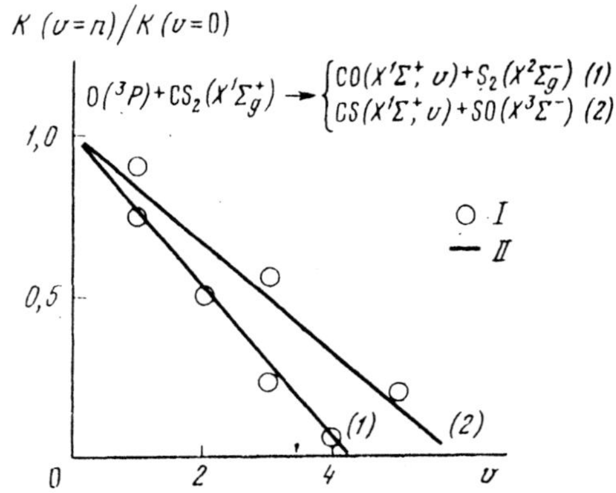

Fig. 42. Vibrational excitation in reactions

$$O + CS_2 \rightarrow \begin{cases} CO + S_2, & (1) \\ CS + SO, & (2) \end{cases}$$

I - experimental data [105], II - results of calculations in this paper

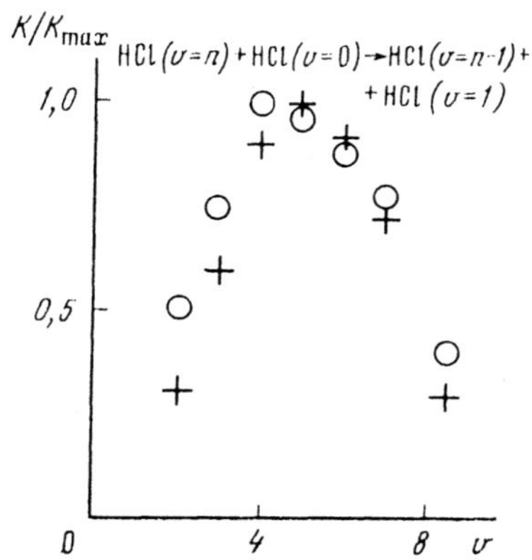

Fig. 43. Vibrational exchange in the collision reaction between HCl molecules
circles - experimental data [33], crosses - results of calculations in this paper

This method is effective when a number of particles in the impacting clusters is small and dynamics of the intracluster degrees of freedom can be treated nonrelativistically. So to study real physical systems, it is necessary to consider various approximations, for example, models of the interaction of composite particles [87- 90]. In this case the following problems arise:

(1) the construction of self-conjugate Hamiltonians for systems of several particles with inner structure which determine the inner and external dynamics of interaction;

(2) the construction of correspondin equations with a proof of their Fredholmity and unique solvability.

The above-mentioned problems were solved in papers [87, 88], while the problem of construction of self-conjugated Hamiltonians is solved on the basis of the theory of expansion of symmetrical self-conjugated operators; in this case the operators determining the inner and external dynamics can be of different types.

As an example of the application of cluster approximation in the framework of quantun theory of scattering in the system of three bodies, Fig. 41 shows the calculation results for the cross-section of the reaction

$$O + CF_3I \rightarrow IO + CF_3$$

together with experimental data [126].

The effective interaction between O, CF_3, and I clusters in cluster approximation was simulated with Morse potentials, which parameters were determined on the basis of the agreement between calculations and some single cross-section value obtained in the experiment. Calculations of the molecular reactions occuring at the collisions between two-atomic molecules

$$AB(v_1, J_1) + CD(v_2, J_2) \rightarrow \begin{cases} AB(v_1, J_1) + CD(v_2, J_2) & \text{- the elastic scattering,} \\ AB(v_3, J_3) + CD(v_4, J_4) & \text{- the vibration-rotation excitation,} \\ AC(v_5, J_5) + BD(v_6, J_6), & \text{- reactions of the realignment with} \\ CB(v_7, J_7) + AD(v_8, J_8) & \text{vibration-rotation excitation} \end{cases}$$

were performed based on the quantum theory of scattering in the system of four particles in the pole approximation. The results of calculations of the following specific chemical reactions

$$HCl\,(v = n) + HCl\,(v = 0) \rightarrow HCl\,(v = n - 1) + HCl\,(v = 1),$$

$$O + CS_2 \rightarrow \begin{cases} CO + S_2, \\ C + S_2O \end{cases}$$

$$O + OCS \rightarrow CO + SO,$$

together with experimental data [105] are presented in Figs. 42 and 43.

3.3. Results of calculations of various processes in nuclear physics

The necessity of developing correct mathematically rigorous methods of solution of the few-nucleon problems with the realistic nucleon-nucleon interactions [15] arose in connection with the studies of few-nucleon systems in the framework of the potential model, a consideration of many-paticle interactions, the mesonic catalysis and mesonic current, Δ-isobaric configurations etc. [15 - 34]. For this reason at present a great number of papers [15 - 45] are devoted to the improvement of the methods available and to the choice and development of new mathematically rigorous calculations for the system of several nucleons.

One of the most rigorous and mathematically justified methods of solution of problems of several nucleons in the region of nonrelative energies is the method of quantum theory of scattering in the system of several particles [36 - 54] which allows to calculate both bound states and scattering states in systems of several particles using the universal equations with proved unique solvability [41].

It has been shown in preceding sections that the method of quantum problem of scattering in the system of several particles based on equations (1.2) and (2.10) is the only method of calculations of various scattering processes. So at present this method is considered [18 - 45, 72] to be the most effective for the solution of scattering problems. However, we shall show below that calculations of bound states of systems of several particles performed on the basis of equations (1.2) and (2.10) give considerably better results compared to calculations using Schrodinger equation. This is explained by the fact that these equations are constructed for the components of the total WF and, thus, even by their type alone take into account all possible asymptotics of many-particle systems.

In this section in Figs. 44 - 48 and Table are presented based on equations (1.2) and (2.10) the results of calculations of bound and scattering states both for nonmodel systems, such as neutron-deuton, proton-deuton, positronium ion, and systems which at certain conditions can be treated as three-particle systems (for example, ^{12}C nuclei can be considered as three α-particles). In the same approximation one can consider the α-particles scattering on ^{8}Be nuclei which can be considered as two α-particles.

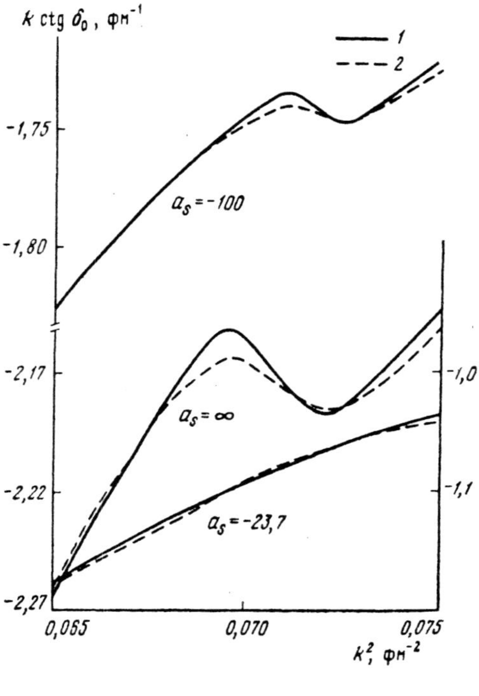

Fig. 44. dependance of k ctg δ on k at the elastic scattering of neutron on deuton for various values of the scattering distance a_s (fm)
1- results of calculatons from [73, 74], 2 - results of calculatons in this paper

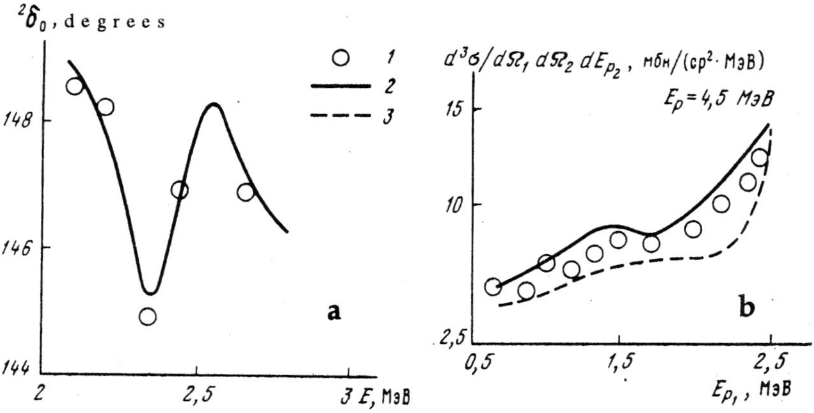

Fig. 45. Dependence of the elastic scattering phase on the impacting proton energy (a) and cross-section of the reaction p + d → p + p + n (b)
1 - experimental data [84], 2 - results of calculations in [84]

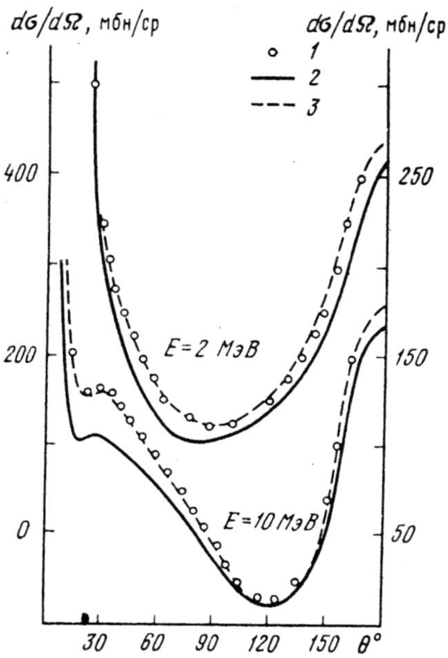

Fig. 46. Differential cross-section of the proton elastic scattering on deuton at E=2 MeV (the right scale) and E=10 MeV (the left scale)
1- experimental data [84], 2-results of calculations in [84, 86], 3-results of calculatons in this paper

Consider calculations of bound states in the system of three particles, for example, a positronium ion $(e^-e^+e^-)$. In this case we shall locate the coordinate system so that positron will be a particle 3 and electrons will be particles 1 and 2. Then the system Hamiltonian in the Jakobian coordinates x_α and y_α has the form

$$H = -\Delta_{x_\alpha} - \Delta_{y_\alpha} + \Sigma n_\alpha / |x_\alpha| = \Delta + V^C. \tag{3.4}$$

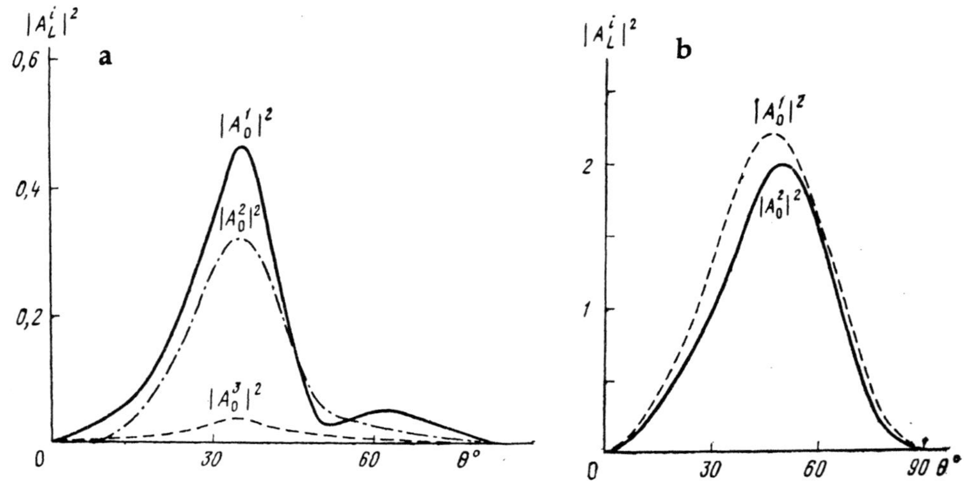

Fig. 47. Results of calculations of the squared amplitude modules $A_L^i(\theta)$

a - E = 10 MeV, L = 0, S = 1/2, b - E = 10 MeV, L = 0, S = 3/2

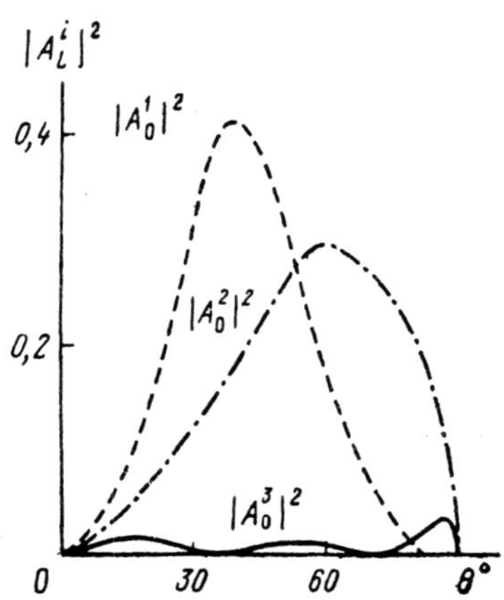

Fig.48. Results of calculations of the squared amplitude modules $A_L(\theta)$ in the approximation of the boundary conditions

model (E = 14.4 MeV, L =O

Thr ground state of the system is characterized by zero total spin of two electrons and the total orbital moment equal to zero (L = 0). Hence, the ground state wave function is symmetrical relative electrons

$$\Psi(r_1,\ r_2,\ r_3) = P_{12}\Psi(r_1,\ r_2,\ r_3). \qquad (3.5)$$

In (3.5) r_i are coordinates of particles, P_{12} is the transposition operator for particles 1 and 2. In terms of Jakobi coordinates the action of P_{12} is decribed by equalities

$$P_{12}f(x_\alpha,\ y_\alpha) = f(-x_\alpha,\ y_\alpha),$$
$$P_{12}f(x_3,\ y_3) = f(-x_3,\ y_3), \qquad (3.6)$$

which leads to the following relations between the wave function components

$$U_2 = P_{12}U_1, \qquad (3.7)$$
$$U_3 = P_{12}U_3. \qquad (3.8)$$

Due to relations (3.7) and (3.8) the set of equations (2.10) for a positronium ion takes the form

$$(-\Delta - |x_1|^{-1} - E)U_1 = (U_3 + P_{12}U_1)/|x_1|,$$
$$(-\Delta - |x_3|^{-1} - E)U_3 = (I + P_{12}U_1)/|x_3|.$$

For the angular analyses we shall use the bispherical basis

$$U_\alpha(x_\alpha, y_\alpha) = \sum_{\alpha} \frac{U^0_\alpha(|x_\alpha|, |y_\alpha|)}{|x_\alpha||y_\alpha|} Y_\alpha(|x_\alpha|, |y_\alpha|), \qquad (3.9)$$

where

$$Y_\alpha(|x_\alpha|, |y_\alpha|) = Y^0_{\lambda l}(|x|, |y|),$$

and L = 0 means that λ = 1, while (3.8) imposes a limitation on the quantum number l parity in the expansion (3.9):

$$\lambda = l = 2k,\ k = 0,\ 1,\ 2,\ \ldots,$$

which allows to present the expansion (3.9) in the form

$$U_1(x_1,y_1)=\sum_{l=0}^{\infty}\frac{U_1^l(|x_1|,|y_1|)}{|x_1||y_1|}Y_{ll}^{00}(|x_1|,|y_1|),$$

$$U_3(x_3,y_3)=\sum_{k=0}^{\infty}\frac{U_3^{2k}(|x_3|,|y_3|)}{|x_3||y_3|}Y_{kk}^{00}(|x_3|,|y_3|). \qquad (3.10)$$

After separating angular variables the set of equations transorms to the infinite set of coupled equations for radial components

$$(H_1^l-E)U_1^l=-V_1^c\sum_{l'}(-1)^l h_{ll'}\left[U_1^{l'}+(-1)^{l'}U_3^{l'}\right]/2,$$

$$(H_3^l-E)U_3^l=-V_3^c\sum_{l'}h_{ll'}U_1^{l'}, \qquad (3.11)$$

where V_α^c are coulomb potentials,

$$H_\alpha^l=-\Delta_x-\Delta_y+l(l+1)/(|x|^2+|y|^2)+V_\alpha^c,$$

$$h_{ll'}f(|x|,|y|)=\int_{-1}^{1}du\,h_{ll'}(|x|,|y|)f(|x|,|y|),$$

and nuclei $h_{ll'}$ are determined by the expression $h_{ll,ll'}^{00}$.

The set (3.11) must be cut off for numerical calculations. Because the ground state is described by quantum numbers λ = l = 0, (3.11) is reduced to two equations for U_1^0 and U_3^0, while the analyses of the configurational system $e^-e^+e^-$ shows that the component U_1^1 must be also accounted for because the main effects of the system $e^-e^+e^-$ realignment are determined by states with l = 0. At small distances between electron and positron, which corresponds to a consideration of the U_1 contribution in the U_1 channel, positron can not be far away from two other electrons (this state corresponds to a consideration of the U_1^1 contribution in the U_3 channel, and the next component which must be taken into account because of limitations on a parity is U_3^2) because otherwise there is no the bound state for the positronium ion. Thus, between channels U_1 and U_3 the component U_1^1 remains not taken into account which consideration transforms the set of equations (3.11) into the form

$$(H_1^0-E)U_1^0=-V_1^c\left(h_{00}U_1^0+h_{01}U_1^1+h_{10}U_3^0\right)/2,$$

$$(H_1^1-E)U_1^1=-\frac{V_1^c\left(h_{10}U_1^0-h_{11}U_1^1+h_{10}U_3^0\right)}{2}, \qquad (3.12)$$

$$(H_3^1-E)U_3^0=-V_1^c\left(h_{00}U_1^0-h_{01}U_1^1\right).$$

For unique solving the set (3.12), the boudary conditions are needed which determine the behaviour of components U_α^l at $x = y = 0$ and $|x| \to \infty$.

For this purpose we define the column vector $U = col\ \{U_1^0, U_1^1, U_0^0\}$ and denote

$\widehat{H} = \mathrm{diag}\,\{H_1^0, H_1^1, H_3^1\},\ V_l^c = \mathrm{diag}\,\{V_1^c, V_1^c, V_3^c\},\ H = \widehat{H} = V^c h,$ where

$$h = \begin{pmatrix} h_{00}/2 & -h_{01}/2 & h_{10}/2 \\ h_{10}/2 & -h_{11}/2 & -h_{10}/2 \\ h_{00} & h_{01} & 0 \end{pmatrix}.$$

Then the set (3.12) can be written in the form

$$(H - E) = 0. \qquad (3.13)$$

The boundary conditions in this case are

$$U(0,\ y) = U(x,\ 0) = 0, \qquad (3.14)$$

$$U(|x|,\ |y|)\ \text{при}\ |x| \to \infty. \qquad (3.15)$$

Using the finite-difference approximation, we get for equations (3.13) the relations $(H_{N_m} - E)\Psi = 0$, where H_{N_m} is a linear operator in the finite dimension space $\Psi \in R^{N_m}, N_m = 3N_\theta(N_\rho + 1)$. 139) is a vector corresponding to the function U values in the net knots, $R^{N_m}, \Psi = \{\Psi_1, \Psi_2, \ldots, \Psi_{N_m}\}$ Thus, the problem of determination of the system $(e^-e^+e^-)$ energy values is reduced to the finding zero of $\det(H_{N_m} - E)$ lying below the positronium bond energy $E(e^-, e^+) = 6.803$ eV. At $N_\theta = 40$, $N_\rho = 70$, and $\rho = 30$ fm the calculation accuracy get stabilized, and we obtain the value $E(e^-e^+e^-) = -0{,}264$ a. e.

Similarly were considered ^{3}H and ^{3}He systems, which main parameters are presented in [15, 21, 34, 100].

For calculation of these systems the equations (2.10) were used, where $\chi^L = 0$ and the asymptotic conditions have the following form: $\Phi_k^0(|x|, |y|) = 0$, $k = 1, 2.3$ at $\rho = \rho_{N_\rho}$ and

$\Phi_k^o(0, y) = \Phi_k^o(x, 0)$ which is achieved at $\rho_{N_\rho} \cong 50 - 60$ fm.

Using the finite-difference approximation for the problem (3.11), the equations (3.13) are reduced to the form with I = 0. From (3.13) we get $\left(H_{N_m} - E\right)\Psi = 0$,

where H_{N_m} is a linear operator in the space R^{N_m}, $\Psi = (\Psi_1, \Psi_2, \Psi_3, \ldots, \Psi_m)$ is a vector corresponding to the function Φ^o values in the net knots $\Psi \in R^{N_m}$, and the number

$$N_m = \begin{cases} 2N_\theta\,(N_\rho + 1) \text{ для } {}^3\text{H}, \\ 3N_\theta\,(N_\rho + 1) \text{ для } {}^3\text{He}, \end{cases}$$

ans it is taken into account that for the nnp (^{3}H) system in (3.13) the component Φ_3^o corresponding to the isospin T = 3/2 is absent.

Thus, the problem of determination of the systems energies is reduced to the finding zero of $\det(H_{N_m} - E)$ lying below the deuton bond energy $\varepsilon_\alpha = 2.2239$ MeV for the Yukava potential I - III [15]. $\Delta E = E(^3\text{H}) - E(^3\text{He})$ is the Coulomb nuclear bond energy equal to 0.75 MeV.

Results of calculations are presented in the Table. The calculation accuracy was controlled by the method of the defining the net knots and by the choice of the cut-off radius.

As an example of the application of the boundary condition model on the basis of differential formulation of the Faddeev equations, we present the calculation results for the neutron scattering on deuton when the total orbital moment is equal to zero L = 0 (see Fig. 48). In the numerical calculations the set of equations was used assuming that nucleons interact only in S-states via potentials of the type [15]

$$V(r) = -\lambda_A \frac{e^{-\mu_A r}}{r} + \lambda_\beta \frac{e^{-\mu_\beta r}}{r}, \qquad V(r) = V_0^{s,t} \exp\left[-(r-a)/R_{s,t}\right],$$

$$V(r) = -V_0^{s,t} \exp\left[-r/R_{s,t}\right].$$

Table
Results of calculations of the bond energy (MeV) for ^3H and ^3He systems for various potentials

the system energy / potential	$\Delta E(^3H)$	$\Delta E(^3He)$	the system energy / potential	$\Delta E(^3H)$	$\Delta E(^3He)$
MT I–III	8,43	0,75	GPT	8,3	0,62
RSC	7,11	0,633		8,48	0,764
SSC	7,50	0,653			

{Tab., p. 140, 1 - the system energy, 2 - potential}

3.4. Threshold singularities

A study of the threshold singularities of the amplitudes and cross-sections of nuclear, molecular, and chemical reactions allows to obtain sufficiently large information about properties of these reactions (the energy dependence of the elastic scattering cross-section near threshold allows to find both the cross-section value for inelastic scattering and the elastic scattering amplitude etc.). These singularities are investigated in detail in the case of scattering of two particles ineracting via both the short-range and long-range potentials [17 - 21]. The reaction threshold point is a singular one for the equations describing properties of the system considered, and at energies below the threshold these equations have a unique solution describing the elastic scattering, while above the threshold there are several solutions corresponding to possible reaction channels. The reason of the threshold anomalies may be either the appearance of the intermediate coupled (quasi-coupled) system or the influence of the closed channels on the scattering processes in the opened ones, or the appearance of the additional polarization interaction at the approaching the impacting particle toward the target. Hence, knowing the type of these singularities and their influence on the scattering processes. one can efficiently control various reactions choosing proper conditions. However, the application of methods [18 - 33] for calculations of threshold singularities in the systems of many bodies, which are efficiently treated as two-particle ones, causes serious complications [13 - 45]. For example, the scattering on the point dipole potential which is commonly used for the approximate description of the electron scattering on polar molecules [35]

$$V(r) = -D \cos \Theta / r^2, \tag{3.16}$$

in Born approximation leads to the following formula for the differential and diffusion scattering cross-sections

$$\sigma(\Theta) = 2D^2/3k^2(1 - \cos \Theta)^2, \ \Theta = (\widehat{k, k_0}), \ \sigma_d = 8\pi D^2/3k^2,$$

divergent at $k \to 0$. This divergence results not from Born approximation but from the long-range type of interaction and exists in all similar methods [18 - 29]. The investigations of the Schrodinger equation with the potential (3.16) give the following results:

there exists such $D = D_c = 0.6395$ a.u. that at $D > D_c$ the scattering particle falls on the center (similarly to the scattering on the potential α/r^2). This is explained by nonphysical behaviour of the potential at $r \to 0$;

at $D > D_c$ an infinite number of bound states appears which energies can be presented by the formula

$$E_n = -0{,}5 \exp [(-\pi(2n + 1) + \alpha\mu)], \quad n = 1, 2, \ldots,$$

where $\mu = \mu(D)$, α is a parameter characterizing the short-range interaction. This infinite number of states results not from the potential (3.16) behaviour at $r \to 0$ but from its long-range character. To consider the problem of scattering on the potential (3.16) at $D > D_c$, the potential is used which at $r \to r_0$ behaves like (3.16) and is a non-singular function at $r < r_0$. However, the main disadvantage ot this approach is that the calculation results depend on r_0 and the method of the potential approximation at $r < r_o$. So of much interest is the study of the threshold singularities in more natural approximation on the basis of quantum theory of scattering in the system of several bodies. In this approximation it is interesting to study the influence of long-range polarization potentials on these singularities and the role of intermediate virtual states [41 - 45, 54, 68, 73 - 90, 94, 95].

It has been shown in papers [17, 43, 73 - 85] that in the three-particle systems interacting via paired short-range potentials (for example, the zero radius potentials [20]) in the case of the presence of virtual states, the infinite number of levels appears in the decay threshold into three free particles. The reason of this effect is that if three particles have resonances in paired interactions, then these resonances generate the effective interaction $\alpha/R^2 (R^2 = r_1^2 + r_2^2 + r_3^2)$. This interaction does not depend on the paired forces properties and only determined by the fact that two particles have the resonance with the scattering distance a [73]. The number of these levels is determined by formula [73] $N \sim \pi^{-1}\ln(|a|/|r_o|)$, where r_o is the radius of action of two-particle forces. However, it is rather diffucult to observe such levels in real three-particle systems, and as estimations show [73 - 75], in the case of nuclear physics the existence of no more than one level can be expected. Calculations of the neutron-deuton, proton-deuton (see Figs. 44, 45) [54, 68], $^4He + (^4He)_2$ [77, 78] systems justify these assumptions. The most favourable case for the occurence of this effect is two heavy particles and a light one (for example, two neutral atoms and electron H_2^- with the neutral atoms taken to eliminate the influence of coulomb forces). Numerical calculations of scattering in the electron-hydrogen molecule system in the approximation of the three-body problem demonstrate oscillations in the threshold [59] (see Fig. 5).

We note here the paper [74], where based on the method of quantum theory of scattering in the three-body system, was first explained theoretically and confirmed by calculations the known correlation between the triton bond energy and the value of the doublet neutron-deuton scattering. This dependence appears because the bond energy of triton and deuton is small in the nuclear forces scale and corresponding sizes of triton and deuton exceed the nuclear forces radius.

Studies of this effect in the case of spin particles show that [73 - 85]:

the attraction between three particles has more complex structure than in the nonspin case which is connected both with the existence of different scatterinig distances for different resonances and the possibility of existence of several spin configurations for a specified total spin;

the value of this attraction is determined by masses and spins of particles, the total state spin, and also by the number of resonances and their spins;

the spectrum of levels looks like a superposition of several strongly interacting spectra for the nonspin case.

Notice that in the case of complex many-particle systems the virtual levels in paired subsystems do not result in such effect [81]. This does not mean that similar effect is absent because it can bb determined not by two-particle virtual levels but by many-particle ones. The first attempts to find experimentally such effects in many-particle systems in nuclear and atomic physics were made in [82, 83]. For example, in the case of the elastic scattering of electrons by CH_4 molecules, the peculiar Ramsauer-Townsen effect [82] has been discovered (see also experiments on the electron scattering by $CFCl_3$ and $2\text{-}C_4F_6$ molecules [83]) which allows to consider this assumption rather realistically.

In previous sections, the effective interaction potentials between clusters, for example, atoms, molecular complexes etc. considered as nonstructural objects were specified either empirically or on the basis of the perturbation theory or by the strong coupling method with the resulting effective potential of the form [44, 45]

$$V_{eff}(r) = q/r + \sum_{l=0}^{\infty} \mu_l(r)/|r|^{l+1} - \alpha/|r|^4 + \dots \qquad (3.17)$$

The terms of this expansion describe coulomb, multipole, and polarization interactions, respectively. Notice here that in calculations of cross-sections of various elementary processes in atomic, molecular, and nuclear physics it is important to know the rigorous proof of the expansion (3.17) in cases when the target is in the ground or one of the excited states.

The rigorous proof of the validity of the expansion (3.17) obtained on the basis of the method of quantum theory of scattering in the three-body system, when the target is in the ground state, is presented in papers [41 - 45, 94, 127 - 129].

In the case when the target is in the excited states with the total energy of the three-body system below the decay threshold, the expression for the effective matrix potential has the form

$$\widehat{V}_0(E) \sim \delta_{a_i a_j} V_C^{\alpha}(y_\alpha) + \sum_{l}^{\infty} \mu_{a_i a_j}(\widehat{y}_\alpha)/|y_\alpha|^{l+1} + \widehat{C}_0\chi/|y_\alpha|^4 + \dots, \qquad (3.18)$$

where χ is a vector which dimension is equal to the number of open channels with consideration for their multiplicity and degeneracy $N_\alpha(E)$, and at $N_\alpha(E) = 1$ (the state a_o is the ground one and only the elastic channel is open) the expression (3.18) coincides with (3.17). C_o is a numerical matrix of dimension $N_\alpha(E) \times N_\alpha(E)$. The operator $C_o\chi/|y_\alpha|^4$ in (3.18) has a meaning of the matrix polarization interaction which physical sense consists in that it determines the contribution of the closed channels of the target excitation into the total effective interaction.

In the general case of scattering of a particle α with the impulse p_α on the bound pair, the asymptotics of the three-body system wave function can be presented in the form [127 - 130]:

$$\Psi_{\alpha a}(X, p_\alpha) \sim \varphi_a^\alpha(x_\alpha)\chi_C^\alpha(y_\alpha, p_\alpha) + \sum_{\beta, b} f_{ba}^{\alpha\beta}(\widehat{y}_\beta, p_\alpha)\varphi_b^\beta(x_\beta)Q_{\beta b}(|y_\beta|, E) + + F_0(X, p_\alpha)Q(X, E), \qquad (3.19)$$

where χ_c^α is the wave function of the continuous spectrum of the two-particle Hamiltonian which describes pure coulomb interaction of the impacting particle with the total target charge $q_{\alpha\alpha} = \sum_\gamma q_\gamma / |S_{\alpha\gamma}|$,

$$Q_{\beta b}(|y_\beta|, E) = \exp\left\{i\sqrt{E - \varepsilon_b^\beta}\,|y_\beta| + + iq_{\beta b}\ln\left(2\sqrt{E - \varepsilon_b^\beta}\,|y_\beta|\right)/2\sqrt{E - \varepsilon_b^\beta}\right\}\Big/|y_\beta|,$$

$$Q_0(X, E) = \exp\left\{i\sqrt{E}X + i\sum_\gamma q_\gamma |X| \ln\left(2\sqrt{E}|X|/2\sqrt{E}|x_\gamma|\right)\right\}\Big/|X|^{5/2}.$$

The last term in (3.19) describes the decay process for the three-body system. Amplitudes $f_{ba}^{\beta\alpha}$ with $\beta = \alpha$ describe the inelastic channels of realignment, $f_{ba}^{\beta\alpha}$ describe the realignment processes, i.e. transitions of a pair α from the state φ_a^α to the state φ_b^α, whereas the elastic scattering is described by the amplitude

$$f_a(k, k') = A_s^\alpha(k, k') + A_{aa}^{\alpha\alpha}(k, k').$$

Here $A_{aa}^{\alpha\alpha}$ is a differentiable limited function, and A_s^α contains all singularities connected with the long-range part of paired potentials which coincide with singularities of the two-particle amplitude of scattering on the potentials of the form

$$V_i|y_i| = \sum_{i \neq j}\int n_i |\Psi_i(x_i)|^2 |c_{ij}x_i + s_{ij}y_i|^{-1} dx_i,$$

where Ψ_i are the wave functions of the bounded states of potentials (3.16). In the following, we shall omit indices α, $\alpha\alpha$, and aa for simplicity.

The physical sense of the amplitude A is that it corresponds to the particle scattering on the bonded pair with the total charge not equal to zero which produces the long-range potential

$$\widetilde{V}_i\,|\,y_i\,| \sim n_{ii}/|\,y_i\,| + \sum_{l=1}^{\infty} \mu_l\,(\widehat{y}_i)/|\,y_i\,|^{l+1}, \qquad (3.20)$$

where

$$\mu_l\,(\widehat{y}_i) = \sum_{j\neq i} \int |\,\Psi_i\,(x_i)\,|^2\, n_j\,|\,\widehat{x}_l\,|^l\, s_{ij}^{l+1} c_{ij}^{-1} P_l\,(\cos\theta)\, dx_i,\ \cos\theta = (\widehat{y_i, x_i}),$$

and $P_l(x)$ are Legendre polinomials. The absence of information about the explicit form of singularities in f does not allow to study theoretically and experimentally the threshold singularities in the scattering processes of three charged particles. Notice that in addition to thecommon angular singularities, the amplitude of scattering on potentials (3.20) acquires new singularities at $E \to 0$ which do not allow to determine the scattering distance in the processes with a participation of charged particles and cause oscillations in the cross-sections of the corresponding processes [129, 130].

It can be shown that the singular part of the scattering amplitude A_s on potentials of the type (3.20) can be presented in the form

$$A_s\,(k, k') = A_C\,(k, k') + \sum_{l=1}^{2} b_l\,(k, k')\,|\,k - k'\,|^{-2+l-2i\eta}, \qquad (3.21)$$

where

$$A_C\,(k, k') = -\,2\,(2\,|\,k\,|)^{2i\eta} \exp\,[2i \arg \Gamma\,(1 + i\eta)]/|\,k - k'\,|^{2+2i\eta},$$

is the scattering amplitude on the potential $n/|x|$, $\eta = n/2|k|$

$$b_1 = a_1, \qquad b_2 = a_2 + c + d,$$

$$a_l\,(k, k') = -\,(i/4)\,|\,k\,|^{2i\eta} e^{-\pi\eta}\,\Gamma\,(2 + l + 2i\eta)\,\widehat{a}_l\,(k, k'),$$

$$\widehat{a}_l\,(k, k') = \int d\widehat{x}\,\mu_l\,(\widehat{x})\,[1 - (px)]^{i\eta}\,((q, x) + i0)^{l-2-2i\eta},$$

$$C(k, k') = \frac{i}{4|k|}(2|k|)^{2i\eta}[2i \arg \Gamma(i\eta)]\left[\mu_1(\widehat{p}) + \mu_1(-\widehat{p})\right],$$

$$d(k, k') = -\frac{4^{-2i\eta}}{4\pi|k|} iB\left(\frac{1}{2} - 2i\eta, 2i\eta\right)\widehat{d}(k, k'),$$

$$\widehat{d}(k, k') = \int_0^{2\pi} d\theta''(\cos\theta'')^{-2i\eta} a_1(k + k'', k)\, a_1^0(k, k + k''),$$

$$q = k - k', \quad p = k + k', \quad k'' = |k - k'|,$$

$$\cos\theta'' = (\widehat{k'', q}), \quad a_1^0 = a_1|_{\eta=0},$$

$B(x, y)$ is Euler`s β-function.

The threshold behaviour of the scattering amplitude in the system of three charged particles at $E \to 0$ is determined by the representation [45, 128]

$$f(k, k') = f_C(k, k') + f_m(k, k') + \chi_s(k^2)\,\widetilde{f}(k, k'), \tag{3.22}$$

where f_C is the coulomb scattering amplitude, f_m corresponds to the scattering on the multipole part of potential (3.20) and in the senior order at $E \to 0$ is determined by the contribution of the dipole part

$$f_m(k, k') = (2i|k|)^{-1}\sum_{l=0}^{\infty}\left\{\sum_{\mathrm{Re}S_\alpha \in \Delta l} a(S_\alpha, \eta)\, e^{-i\pi S_\alpha} Q_\alpha(-k)\, Q_\alpha(k) - \right.$$

$$\left. - a(l, \eta)(2l + 1) P_l\left(\cos(\widehat{k, k'})\right)\right\}(1 + O(|k|)),$$

$$a(t, \eta) = \frac{\Gamma(t + 1 + i\eta)}{\Gamma(t + 1 - i\eta)}, \quad \Delta_l = [l - 1/2, \quad l + 1/2],$$

$Q_\alpha(x)$, $\lambda_\alpha = \widetilde{\lambda}_\alpha(\widetilde{\lambda}_\alpha + 1)$ are eigenfunctions and eigenvalues of the operator $\Delta_{\lambda^2} + \mu(x)$ [91]. The term $\chi_s(k^2)f(k, k')$ describes the contribution of the short-range part of potential

$$\chi_s(k^2) = \exp[i\varkappa(\eta) - \pi\eta(1 + \delta)]/|k|,$$

$$\varkappa(\eta) = 2\eta(\ln|n| - 1), \ \delta(n) = \begin{cases} 1, & n > 0, \\ 0, & n < 0, \end{cases}$$

and one can treat the coefficient f as the scattering distance in the system of three charged particles [41, 129]:

$$a = \lim_{k^2 \to 0} (f - f_C - f_m)/\chi_s(k^2).$$

Notice here that multipole terms are only present in the case when the bonded target state have no spherical symmetry which is often occures in the problems of atomic physics (scattering of electrons, atoms, and ions on molecules [17 - 28]). In the problems of nuclear physics the scattering occures mainly on nuclei in the ground states having spherical symmetry. In this case all statistical multipole moments $\mu_l = 0$. Then the amplitude $f(k, k')$ has only the coulomb singularity f_c in the forward scattering direction. In this case the effective potential $V_i(y_i)$ at $y \to \infty$ has the form [45, 127]

$$\widetilde{V}_i(y_i) \sim n_\alpha/|y| + V_p(y),$$
$$V_p(y) \sim -\widehat{a}(\widehat{y})/y^4(1 + O(|y|^{-1})),$$
$$\widehat{a}(\widehat{y}) = \sum_{\beta \neq \alpha} \frac{c^2_{\alpha\beta}}{S^4_{\beta\alpha}} \sum_{a'} \left| \int \Psi_a(x_\alpha)\, \Psi_{a'}(x_\alpha)\, |x_\alpha|\, P_1(\cos\theta)\, dx_\alpha \right|^2.$$

In the spherically symmetrical case $a(y) = \text{const}$, and energetical and angular singularities of the scattering amplitude are only determined by coulomb potential. At $p^2_\alpha \to 0$ the asymptotics of $f(k, k')$ has the form [127]

$$f(\widehat{y}, p') = f_C + b_p(\theta)\, e^{i\omega(r)}\, |p_\alpha|^3 (\sin\theta/2)^{-2i\eta} \times$$
$$\times (1 - O(1)) + \chi_s(p^2_\alpha)\, \widetilde{f}(p_\alpha, y)(1 + O(1)),$$
$$\cos\theta = (y, p_\alpha),\ \omega(\eta) = 2\eta(\ln n - 1),$$
$$\chi_s(p^2_\alpha) = \exp\{i\omega(\eta) - \pi\eta(1 - \varepsilon(\eta))\}/|p_\alpha|,$$
$$b_p(\theta) = -\frac{a\sin\theta}{4n^2_\alpha \cos[b(\theta/2)]}[(\pi - \varepsilon(\eta))(2 - \cos\theta) - 3\varepsilon(\eta)\sin\theta],$$
$$\varepsilon(\eta) = \begin{cases} 1, & \eta > 0, \\ 0, & \eta < 0. \end{cases}$$

The second term in f describes the contribution in the polarization potential amplitude V_p, the third one is the contribution of the fast decreasing part V_i, and it follows from the comparision of these two terms that at $n_\alpha > 0$ the contribution V_p dominates over V_s at $p^2 \to 0$ and leads to the break down of the analyticity of the function of effective radius, and, hence, the rigorous definition of the scattering distance became impossible. Coulomb interactions in the three-particle systems results in appearing a number of bonded states which energies are determined by the following expression [130]

$$E_n = C_0 \exp(-2\pi n\alpha), \quad n = 1, 2, \ldots, \tag{3.23}$$

where C_0 is a constant, and α depends only on the scattering distance in the short-range potential. This formula is of a universal type and contains the distance of scattering on the short-range potential which allows to determine it from the scattering experiments in the system of charged particles. This effect is similar to the Efimov effect [73], however, its physical origin is completely different and connected with the coulomb interaction in the three-particle system. Singularities (3.23) cause oscillations in the scattering processes characteristics in the threshold vicinity which are observed in calculations [59, 68, 74, 76].

The representation (3.21) is invalid, if potential (3.20) does not contain the coulomb part. In this case the scattering amplitude can be presented in the following form:

$$\widetilde{A}_s(k, k') = a_1^2(k, k')\,|k - k'|^{-1} + \left[a_1^0(k, k') + a_2^0(k, k')\right] \ln|k - k'|,$$

$$a_1^0 = -\frac{i}{4\pi\,|k}\int_0^{2\pi} d\theta'' a_1^2(k\,k' + '', k)\, a_1^2(k', k + k''),$$

$$a_2^0 = -\frac{1}{4\pi}\int d\widehat{x}\mu_2(\widehat{x}).$$

In the case of scattering on potentials $V(x) \sim n/|x| + a/|x|^{\alpha}$, $\alpha > 1$, it can be shown [127] that the scattering amplitude has the following asymptotics:

$$f(k, k') \underset{k^2 \to 0}{\sim} \begin{cases} fc + b_{\alpha} k^{2\alpha - 5} e^{i\varkappa(n)} (\sin\ \theta/2)^{-2i\eta} + \chi_s (k^2) \tilde{f}\, [1 + O(1)], & \alpha > 3/2, \\ \exp\{iW_{\alpha}(\theta, k^2)\} fc\, [1 + O(1)], & \alpha \leqslant 3/2, \end{cases}$$

$$\chi(k) = \exp[i\omega(\eta) - \pi\eta(1 + \mathrm{sgn}(n))]/k,$$

$$\omega(k) = 2k(\ln|k| - 1),$$

where functions W_{α} and b_{α} are expressed through the Legendre functions and have the form:

$$b_{\alpha}(\theta) = n\,(2\sin\theta/2)^{-1} D(\mathrm{sgn}(n)\,\mathrm{cosec}\,\theta/2),$$

$$W_{\alpha}(k, \theta) = \eta\,[\varphi_0(x_1(\theta, h)) - \varphi_0(x_0(\theta)] + 2k^{2\alpha - 3} D\left(\mathrm{sgn}(n)\sqrt{1 + \varkappa_1^2(\theta, k)}\right),$$

$$x_0(\theta) = \mathrm{ctd}\,(\theta/2);$$

$$x_1(\theta, k) = \mathrm{ctg}\,\frac{\theta}{2}\left\{1 + \frac{2}{|n|} k^{2\alpha - 2}\,\mathrm{cosec}\,\theta/2\right\} + \theta\left(k^{2\alpha - 1}\right),$$

$$\varphi_m(x) = -2\pi m x + 2x\,\mathrm{arctg}\,x^{-1} + \ln(1 + x^2),$$

$$D(t) = -\alpha\left(\frac{|n|}{2}\right)^{1-\alpha}\left(\frac{\pi}{\delta(|t|)}\right)^{1/2}\Gamma(\alpha - 1)(t^2 - 1)^{-2\alpha/4} P_{-1/2}^{3/2-\alpha}\left(t^{-1}\right).$$

CONCLUSION

In this paper are presented the most general and universal methods of solving a quantum problem of scattering in the system of several bodies interacting via arbitrary paired potentials which can be also specified numerically. These methods have provided the basis for realizing the software in the form of application packages designed for the successive serial calculations of main characteristics of various processes (the elastic scattering, excitation, excitation with the realignment and decay) occuring in the systems of several particles. This software satisfies to a number of special requirements, such as reliability, productivity, mobility, and, hence, allows to use effectively these application packages on various computers: BESM-6, SAIBER-172, SDS-6500, ES, VAX-11/780, and multiprocessor computers, for example, ES-1045 with a ES-2345 matrix printer.

Using the methods suggested and means realized on their basis, the calculations of main charǎcteristics of the following elementary processes have been performed.

In atomic physics: processes of the recharge, excitation, and dissociation by protons of hydrogen and heliun atoms in the ground and excited states; processes of the neutralization ($H^+ + H^- \to H\ \ H$), dissociation ($H + H \to H^+ + H^-$, $Na + I \to Na^+ + I^-$), recombination ($e + AB(n, l) \to A(n_1, l_1) + B(n_2, l_2)$), various processes (the elastic scattering, realignment, excitation and dissociation) of collisions of electrons with two-atomic molecules of hydrogen, hydrohalogens and their isotopic modifications, halogens, nitrogen, and alkali halides both in the ground and excited vibration-rotation states.

In molecular physics: reactions going on at collisions between atoms and two-atomic molecules

$$H + H_2(v = 1) \to H_2(v = 0,1) + H,$$
$$O + O_2 \to O_2 + O,$$
$$O + N_2 \to N + NO,$$
$$S - O_2(v_1, J_1) \to SO(v_2, J_2) + O,$$
$$O + CS(v_1) \to CO(v_2) + S,$$

(notice that for the last reaction, calculations were made on the basis of classical and quasi-classical methods) and three-atomic molecules

$$O + CS_2 \to \begin{cases} CO + S_2, \\ C + S_2O, \end{cases}$$
$$O + OCS \to CO + SO,$$

processes when there are three free particles in the initial state

$$(H + H + H \rightarrow H + H_2 (v = n)),$$

and also reactions going on at collisions between two-atomic molecules, in particular, the vibrational energy exchange in the collisional processes of HCl molecules was calculated. The results of calculations of $O + CF_3I \rightarrow IO + CF_3$ reaction were presented, made on the basis of the cluster approximation in the framework of quantum theory of scattering in the system of three particles for which were taken iodine and oxygen atoms, and CF_3 cluster.

In nuclear physics, scattering in neutron-deuton and proton-deuton systems is considered, scattering of α-particles on ^{8}Be nuclei, and also calculations of bound states of a positronium ion, tritium and helium-3 nuclei. These states are the most typical systems in nuclear physics, and in the region of nonrelativistic energies can be treated as nonmodel.

The above-listed elementary processes determine main characteristics of the active media for chemical lasers, various plasmo-chemical setups, processes of practical importance for the development of new means of plasma diagnostics, thermonuclear fusion, propagation of radiation in the ionosphere etc.

The method is presented of calculation of bound states and scattering states in the three-body systems for the boundary conditions model which was applied for calculations of processes of scattering of neutrons on deutons and bound states of tritium and helium-3 nuclei.

The application of the methods and means suggested in the framework of the cluster approximation allows to calculate various chemical and nuclear reactions which can not be calculated with standard quantum-mechanical methods. Thus, one can realize a purposeful search and construction of new compounds with pre-planned properties.

The results are presented of investigations of threshold singularities in systems of several particles and of calculations of these singularities in cross-sections of the elastic scattering in systems of three neutral (neutron-deuton) and charged (proton-deuton) particles, and also in the realignment reactions on the example of the reaction of dissociative attachment of electron to hydrogen molecules and their isotopically substituted modifications. Notice that a study of the threshold singularities in the cross-sections of various reactions is connected with the fact that a knowledge of the influence of these singularities on the character of the reactions course allows to realize the effective control mechanism of the reaction progress process which is especially important in the chain chemical and nuclear reactions.

The analytical solutions obtained for the quantum problem of scattering in the system of several particles are of theoretical interest in connection with the development of the total and noncontradictory theory of many-particle systems and allows to obtain simple relations reflecting explicitly characteristic peculiarities in cross-sections of various processes. This may be rather actual in investigations of the scattering processes when the experimental data are limited.

Thus, the presented results open new possibilities for the theoretical investigation and practical application of the wide scope of problems of great importance.

ABSTRACT

The quantum theory of many-body scattering based on the Faddeev-Yakubovskii equations in integral and differential forms is applied to the calculation of various processes of nuclear, atomic, and chemical physics. Elastic, inelastic, atom-exchange, and dissociative collisions are simultaneousl considered in this rigorous formalism. The results of the calculations are compared with available experimental data and other calculations (classical, adiabatic, impulse et al.). The method of numerical solution of the sets of integral and differential equations and software packages are presented.

REFERENCES

[1] A.S. Bakshin et al., Chemical Lasers, Nauka, Moscow, (1982)

[2] N.G. Basov et al., Trends in the development of chemical lasers on the electronic transitions, Preprint FIAN, No.36. Moscow, (1989). 40p.

[3] N.G. Basov et al., On chemical lasers of a new type on the electronic transitions with the chain excitation mechanism. Kvantovaya Elektronika, 14 (1987) 1772.

[4] N.G. Basov et al., On possible expanding spectral range of emission of chemical lasers on the electronic transitions. Ibid., 1787.

[5] N.G. Basov et al., Chemical lasers on the electronic transitions: analytical models. Theory of the Ba-Na exchange laser with the chain excitation mechanism, Preprint FIAN, No. 115, Moscow, (1988). 54p.

[6] I.I. Sobel`man, Introduction to the theory of atomic spectra, Nauka, Moscow, (1977). 362p.

[7] N.G. Basov, I.G. Lebo, and V.B. Rozanov, Physics of the laser thermonuclear fusion, Nauka, Moscow, (1988). 176p.

[8] L.A. Vainshtein and V.P. Shevel`ko, Structure and characteristics of ions in hot plasma, Nauka, Moscow, (1986). 207p.

[9] E.A. Oks and S.A. Pozdneev, On the mutual influence of effects of pressure and dynamical Stark effect in the plasma spectroscopy, Preprint FIAN, No. 278, Moscow. (1983). 12p.

[10] V.P. Gavrilenko, E.A. Oks, and S.A. Pozdneev, On new posibilities of the Langmuir turbulence diagnostics, Preprint FIAN, No. 1, Moscow, (1984). 17p.

[11] S.A. Pozdneev and V.A. Shcheglov, Methods of calculation of chemical reactions in physics of low-temperature plasma, Materials of III International workshop "Physics of low-temperature plasma, Minsk, (1988), p.103.

[12] N.D. Sokolov, Trends in the development and problems of modern quantum chemistryy, Uspekhi Khimii, 57, (1988) 177.

[13] S. Hogychi, Bio-mathematical approach to a new generation computer, Biomathematics and related computational problem., Klumer, L., (1988) p. 243.

[14] N.G. Rambidi and B.M. Zamarin, Molecular element base of perspective information logic devices, VINITI, Vol. 22 Nauka, Moscow, (1987). 170 p.

[15] D.E. Brown and A.D. Jackson, Nucleon-nucleon interactions, Atomizdat, Moscow, (1979). 272 p.

[16] Yu.A. Kuperin et al., Calculation of bound states in the system of three quarks, Yadernaya Fizika, 38, (1986) 702.

[17] G.F. Drukarev, Collisions of electrons with atoms and molecules, Nauka, Moscow, (1978). 255 p.

[18] R. Newton, The theory of scattering of waves and particles, Mir, Moscow, (1969). 607 p.

[19] E.E. Nikitin, The theory of molecular collisions, Khimiya, Moscow, (1969). 423 p.

[20] Yu. N. Demkov and V.N. Ostrovskii, Method of the zero-radius potentials in atomic physics, Leningrad State University, Leningrad, (1975). 240 p.

[21] A.I. Baz`, Ya.B. Zel`dovich, and A.I. Perelomov, Scattering, reactions, and decays in the nonrelativistic quantum mechanics, Nauka, Moscow, (1971). 544 p.

[22] V.P. Maslov and M.V. Fedoryuk, The quasi-classical approximation for the equations of quantum mechanics, Nauka, Moscow, (1976). 423 p.

[23] E.A. Solov`ev, Classical approximation for the negative ion ionization by the electron impact near a threshold, Zh. Eksp. Teor. Fiz., 72, (1977) 2072.

[24] E.A. Solov`ev, Nonadiabatic transitions in atomic collisions, Usp. Fiz. Nauk, 157, (1989) 437.

[25] V.P. Zhigunov and B.N. Zakhar`ev, The method of strong coupling of channels in the quantum theory of scattering, Atomizdat, Moscow, (1974). 268 p.

[26] N.F. Mott and H.S.W. Massey, The theory of atomic collisions, Vol. 1, 2, Clarendon Press, Oxford, (1987). 432 p.

[27] L.G. Christophorou, Electron-molecule interactions and their applications, Academic Press, N. Y., (1984) 681 p.

[28] T.D. Mark and G.H. Dunn, Electron impact ionization, Springer, N. Y., (1985) 395 p.

[29] S.K. Cray and W.H. Miller, Classical model for electronic degrees of freedom charge transfer in Na + J collisions, Chem. Phys Lett., 3, (1982) 341.

[30] J.H. Murrell and W. Craven, Classical dynamics of the reaction of $S(^3P)$ with $O_2(^3\Sigma_g)$, Mol. Phys., 49, (1983) 1077.

[31] G.C. Schatz, The quantum dynamics of H + H_2 (v = 1): A coupled states study of cross sections and rate constants, Chem. Phys., 120, (1981) 1007.

[32] D.A. Micha, Quantum theory of reactive molecular collisions, Adv. Chem. Phys., 30, (1975) 7.

[33] D.C. Clary, The theory of chemical reaction dynamics, Reidel, Dordrecht, (1986) 417 p.

[34] E. Hadjimichael and W.Orlert, Few-body problems, World Scientific, N. Y., (1986) 388 p.

[35] A.K. Kazanskii and I.M. Fabrikant, Scattering of slow neutrons on molecules, Usp. Fiz. Nauk, 143, (1975) 602.

[36] O.A. Yakubovskii, The Faddeev equations and method of strong coupling, Yadernaya Fizika, 25, (1977) 1311.

[37] I.I. Kholyavin and O.A. Yakubovskii, Integral equations of the scattering theory and method of strong coupling, Ibid, 26, (1977) 181.

[38] I.I. Kholyavin, Pole approximation for the Faddeev equations, Vestnik LGU. Fizika. Khimiya, 19, (1977) 83.

[39] I.I. Kholyavin, On explicit solution of integral equations of the scattering theory, Ibid, 7, (1983) 71.

[40] L.D. Faddeev, Mathematical problems of quantum theory of scattering fot the system of three particles. Nauka, Moscow, (1963) 122 p. (Trudi Mat. Inst. AN SSSR, v. 69).

[41] S.P. Merqur`ev and L.D. Faddeev, Quantum theory of scattering for the system of several particles, Nauka, Moscow, (1985) 398 p.

[42] A.M. Veselova and L.D. Faddeev, A singularity in coulomb three-particle scattering on the ionization threshold, Vestnik LGU. Fizika. Khimiya. 22, (1981) 42.

[43] A.M. Veselova, S.P. Merqur`ev, and L.D. Faddeev. Coulomb S-matrix and multiple scattering, In Diffraction interaction of adrons with nuclei, Naukova Dumka, Kiev, (1987) 107.

[44] A.A. Kvizinskii et al., Quantum problem of N bodies in the configurational space, Fizika elementar. chastiz i atom. yadra, 17, (1986) 267.

[45] A.A. Kvizinskii, V.V. Kostrikin, and S.P. Merqur`ev, Quantum theory of scattering for three-particle systems with fixed total orbital moment, Ibid, 21, (1990) 1301.

[46] S.A. Pozdneev, Application of the Faddeev equations for calculation of the cross-section of dissociative attachment of electron to a nitrogem molecule, Pis`ma Zh. Tekhn. Fiz. 6, (1980) 1206.

[47] S.A. Pozdneev, Adiabatic approximation in the problem of collision of a particle with coupled two-particle system at separable interaction, Zh. Eksp. Teor. Fiz., 77, (1979) 33.

[48] S.A. Pozdneev, Dissociative electron attachment to halogen molecules, Khim. visokich energii, 6, (1982) 442.

[49] S.A. Pozdneev, Application of quantum theory of scattering for calculation of cross-sections of dissociative attachment of electron to hydrohalogen molecules, Zh. Tekh. Fiz., 52, (1982) 1500.

[50] S.A. Pozdneev, Quantum-mechanical calculation of the reaction $H + H_2$ cross-section in the approximation of three-body problem, Khim. visokich energii, 2, (1979) 291.

[51] G.F. Drukarev and S.A. Pozdneev, Theory of dissociative electron attachment to two-atomic molecules, Vestnik LGU, Fizika. Khimiya. 16, (1982) 11.

[52] S.P. Merqur`ev and S.A. Pozdneev. Calculation of bond energies and scattering states of three bodies in the model of boundary conditions, Yadernaya Fizika, 29, (1979) 620.

[53] S.P. Merqur`ev and S.A. Pozdneev, Calculation of bond energies and scattering states of three bodies with coulomb interaction, Ibid, 30, (1979) 941.

[54] S.A. Pozdneev, The Efimov effect in the scattering of neutron on a deuton, Ibid, 35, (1981) 541.

[54a] S.P. Merqur`ev and S.A. Pozdneev, Calculations of p - d scattering, International symposium on the problem of few bodies in nuclear physics, Dubna, OIYI, (1979) 55.

[55] S.A. Pozdneev and V.A. Shcheglov, Application of quantum theory of scattering for calculations of simplest chemical reactions: scattering of electrons by two-atomic molecules, Khim. Fisika, 6, (1987) 1.

[56] S.A. Pozdneev and V.A. Shcheglov, Application of quantum theory of scattering for calculations of simplest chemical reactions: reactions with participation of three atoms, Ibid, 147.

[57] S.A. Pozdneev and V.A. Shcheglov, Application of quantum theory of scattering for calculations of simplest chemical reactions, Preprint FIAN, 18, (1986) 52 p.

[58] S.A. Pozdneev and V.A. Shcheglov, Application of quantum theory of scattering in the many-body system for calculations of simplest molecular reactions, In Kineticheskie i gasodinamicheskie prozessy v neravnovesnych sredach, MGU, Moscow, (1986) 31.

[59] N.A. Konoplev, S.A. Pozdneev, and V.A. Shcheglov, Application of cluster approximation for calculation of chemical reactions, Kratk. soobshch. Fiz, 6, (1987) 61.

[60] S.A. Pozdneev, Application package for solution of sets of integral and integro-differential equations of quantum problem of three bodies, In Application packages: functional filling, Nauka, Moscow, (1986) 48.

[61] S.A. Pozdneev and V.A. Shcheglov, Application package for calculations of cross-sections of elementary processes, In Kineticheskie i gasodinamicheskie prozessy v neravnovesnych sredach, MGU, Moscow, (1988) 183.

[62] S.A. Pozdneev, Application of matrix processor for solution of sets of integral and integro-differential equations, Upravlyayushchie sistemy i mashiny, 1, (1986) 95.

[63] S.A. Pozdneev, Software for the construction of surfaces specified with a function of two variables, Programmirovanie, 1, (1983) 93.

[64] S.A. Pozdneev, Complex of graphical softwares for the system of interactive computer graphics, Ibid, 6, (1984) 78.

[65] S.P. Merqur`ev and S.A. Pozdneev, New method for calculation of the scattering states, In XI Intern. conf. on the phys. of electr. and atomic collisions: (XI ICPEAC). Kyoto, (1979) 1024.

[66] G.F. Drukarev and S.A. Pozdneev, Calculation of dissociative attachment cross-sections to F_2, Cl_2, and J_2 molecules, Ibid, 346.

[67] S. Pozdneev, Application of quantum theory of three-body scattering to the calculation of theproblem of proton-deuteron collisions, J. Phys. G.: Nucl. Phys., 8, (1982) 1506.

[68] S. Pozdneev, The Efimov effect in neutron-deuteron scattering near deuteron breakup threshold, Phys. Lett. B, 125, (1983) 355.

[69] S. Pozdneev, Electron scattering on the nitrogen molecule, J. Phys. B: Atom Mol. Phys., 16, (1983) 867.

[70] S. Pozdneev and V. Shcheglov, Application of the quantum theory of scattering to the calculation of the simplest chemical reaction, Chem. Phys., 123, (1988) 27.

[71] S.T. Surzhikov, Computer-aided system for investigation of radiative and dynamical processes in the low-temperature plasma, Preprint IPM AN SSSR, 113, (1988) 23 p.

[72] V.B. Belyaev, Lectures on the theory of few-particle systems. Energoatomizdat, Moscow, (1988) 128 p.

[73] B. Efimov, Low-energy properties of three resonansly interacting particles, Preprint LIYAF, 436, (1978) 24 p.

[74] B. Efimov and E.G. Tkachenko, Phillips lines and the structure of few-nucleon systems. Yadernaya fizika, 47, (1988) 29.

[75] B. Efimov, Influence of the radius of forces on the long-range interaction in the system of three particles, Preprint LIYAF, 1486, (1989) 27 p.

[76] M.L. Zepalova, Effective long-range potential in the three-particle theory, Preprint ITF AN USSR, 87-21P, Kiev, (1987) 28 p.

[77] H.S. Huber and T.K. Lim, A study of the Efimov states and binding energies of the helium trimer through the Faddeev coordinate-momentum approach, J. Chem. Phys., 63, (1978) 1006.

[78] T.K. Lim and S.K. Duffy, Faddeev-UPE theory and the three ^{4}He atom system: Bound states and zero-energy atom-diatom scattering, Ibid, 68, (1978) 655.

[79] A.C. Fonseca, Efimov effect in analytically solvable model, Nucl. Phys. A, 320, (1979) 273.

[80] A.C. Fonseca and P.E. Shanley, Model three-body problem in the molecular mass limit, Ann. Phys., 117, (1979) 268.

[81] S.A. Vugal`ter and G.M. Zhislin, On the discrete spectrum of the Schrodinger operator for many-paricle systems with two-particle virtual levels, Dokl. Akad. Nauk SSSR, 267, (1982) 784.

[82] W. Sohn, K. Jung, and H. Ehrhart, Threshold structures in the cross sections of low-energy electron scattering of methane, J. Phys. B: Atom. Mol. Phys., 16, (1983) 891.

[83] N. Abusalby et al., Electron scattering by methane, J. Chem. Phys., 78, (1986) 1213.

[84] R.J. Slobodrian, The ^{4}He(α, p), ^{7}Li, ^{7}Li$^{*0.48}$ reaction near threshold, J. Phys. A: Atom. Nucl. Phys., 308, (1982) 15.

[85] H. Klar, Threshold fragmentation of atomic and molecular systems by charged particles impact, Ibid, A91, 75.

[86] V.E. Kuzmichev, Paryticle-two-particle interaction in configuration space, Nucl. Phys., A91, (1982) 173.

[87] Yu.A. Kuperin, The extension theory for three-particle Hamiltonians, Preprint LOMI, P-7-88, Leningrad,(1988) 18 p.

[88] Yu.A. Kuperin, Structure and resolvents of Hamiltonian of three-composite particle system, Preprint LOMI, P-8-88, (1988) 17 p.

[89] Yu.A. Kuperin, K.A. Makarov, and B.S. Pavlov, One-dimensional model of three-particle resonances, Teor. Mat. Fiz., 63, (1985) 78.

[90] Yu.A. Kuperin, K.A.Makarov, and B.S. Pavlov, Model of resonance scattering of composite particles, Ibid., 69, (1986) 100.

[91] Handbook of Mathematical Functions (Eds. M. Abramowitz and I. Stegun, National Bureau of Standarts, Applied Mathematical Series, 55 (1964); Nauka, Moscow, (1979) 832 p. (in Russian).

[92] L.V. Kantorovich and V.I. Krylov, Approximate methods of higher analysis, Fizmatgiz, Moscow, (1962) 708 p.

[93] R.I. Boyikova, Dissociation of vibrationally excited molecules by electrons, Vestnik LGU, Fizika, Khimiya, 10, (1977) 31.

[94] A.A. Kvizinskii, V.V. Kostyrkin, and S.P. Merqur`ev, Local points and properties of three-particle wave functions: Points of triple and paired collisions, Yader. Fizika, 49, (1989) 1273.

[95] Yu.A. Kuperin and K.A. Makarov, Adiabatic representation of the Faddeev equations, Vestnik LGU. Fizika. Khimiya, 11, 71.

[96] A.M. Veselova, Determination of scattering amplitudes in the problem of two and three charged particles, Teor. Mat. Fiz., 13, (1972) 368.

[97] G.J. Schulz, Resonances in electron impact on diatomic molecules, Rev. Mod. Phys., 45, (1973) 423.

[98] A.M. Veselova, Integral equations for three particles with coulomb long-range interaction, Teor. Mat. Fiz., 35, (1978) 180.

[99] G.A. Emel`yanenko and A.I. Machavariani, On standardization of softwares for numerical solution of the Faddeev equations by the example of the problem of elastic pion-deuteron scattering, Preprint OIYAI, R11-83-572, Dubna, (1983) 6 p.

[100] N.M. Larson and J.H. Hetherington, Solution of the Faddeev equation without contour rotation, Phys. Rev. C - Nucl. Phys., 9, (1974) 699.

[101] E.I. Dubovoi, Z.F. Kertkoev, and V.P. Yakovlev, On the accuracy of solution of the integral equation for three-nucleon problem, Izv. AN SSSR, ser. fiz., 46, (1982) 172.

[102] V.D. Kupradze, Boundary problems of the theory of vibrations and integral equations, Gostekhteoremizdat, Moscow - Leningrad, (1950) 280 p.

[103] B. Brunett, Study of the interactions of atomic and molecular oxygen with O_2 and N_2 by scattering data, J. Chem. Phys., 74, (1981) 6734.

[104] Y. Hahn, Thermal collisions of high Rydberg atoms with neutral particles. 1. General theory and modified impulse approximation, J. Phys. B: Atom. Mol. Phys., 14, (1981) 985; Higher-order corrections to the impulse approximation, Ibid., 15, (1982) 613.

[105] D.S.Y. Hsu et al., Dynamics of reactions of $O(^3P)$ atoms with CS, CS_2 and OSC, Chem. Phys., 44, (1979) 143.

[106] K.P.Huber and G.Gerzberg, Constants of diatomic molecules, Mir, Moscow, (1984) part 1, 408 p., part 2, 366 p.

[107] A. Huetz, F. Gresteau, and J. Mazeau, Initial vibrational state dependence of resonant excitation and "dissociative attachment" in electron-N_2 scattering, J. Chem. Phys., 72, (1980) 5297.

[108] D. Spence and P.D. Burrow, Resonant dissociation of N_2 by electron impact, J. Phys. B: Atom. Mol. Phys., 112, (1979) L179.

[109] W.C. Tam and S.F. Wong, Dissociative attachment of halogen molecules by 0-8 eV electrons, J. Chem. Phys., 68, (1978) 5626.

[110] M.V. Kurepa et al., Attachment rate coefficients of halogen molecules (F_2, Cl_2, Br_2, J_2) for mean electron energies 10^{-2} to 10^2 eV, Chem. Phys., 59, (1981).

[111] R. Abouaf et al., Resonances in HF:H^- formation around 10 eV, J. Phys. B. Atom. Mol. Phys., 18, (1985) 3017.

[112] D. Teilett-Billy and J.P. Gayacq, Dissociative attachment in e - HCl, DCl collisions, Ibid., 17, (1984) 4041.

[113] B. Jaduszliwer, Electron scattering by highly polar molecules, Phys. Rev., 30, (1984) 1269.

[114] N.B. Berezina and Yu.N. Demkov, Low-energy resonance scattering of electrons by Tomas-Fermi atoms, Zh. Eksp. Teor. Fiz., 68, (1975) 848.

[115] P.R. Simony and J.H. McGuire, Exact second Born calculations of 1s-1s electron capture in p + H, J. Phys. B: Atom. Mol. Phys., 14, (1981) L737.

[116] C.K. Tan and A.R. Lee, Electron capture into excited states of hydrogen, Ibid, 16, (1983) 1445.

[117] M.S. Shah and H.B. Gilbody, Ionization of H_2 by fast protons and multiply charged ions of He, Li, C, N, O, Ibid., 15, (1982) 3441.

[118] M.I. Karbovanetz, V.Yu. Lasur, and G.L. Yudin, Atomic collisions in plasma of the nuclear origin: Coulomb recharge, Preprint FEI, FEI-1509, Obninsk, (1984) 20 p.

[119] Y. Hahn, Thermal collisions of high Rydberg atoms with neutral particles. 1. General Theory and modified impulse approximation. 2. Higher order corrections to the impulse approximation, J. Phys. B: Atom. Mol. Phys., 14, (1981) 985; Ibid., 15, (1982) 613.

[120] M.B. Shah, P. McCallion, and H.B. Gilbody, Electron capture and ionization in collisions of slow H^+ and He^{2+} ions with helium, J. Phys. B: Atom. Mol. Phys., 22, (1989) 3037.

[121] W. Schon et al., Mutual neutralization in H^+ + H^- collisions, Ibid, 20, (1987) L759.

[122] R. Shingal and B.H. Brandsden, Mutual neutralization in H^+ + H^- collisions, Ibid, L533.

[123] J. Hill, J. Geddes, and H.H. Gilbody, H^- formation in electron capture by 4 - 25 keV metastable hydrogen atoms in the inert gases, Ibid, 12, (1979) L653.

[124] V.P. Zhdanov and M.I. Chibisov, Dissociative electron recombination on H_2^+ and D_2^+ molecular ions with formation of highly excited atoms, Zh. Eksp. Teor. Fiz., 74, (1978) 75.

[125] A.I. Mil`shtein, Total cross-section of the radiative electron-ion recombination, Preprint IYAI SO AN SSSR, 89-131, Novosibirsk, (1989) 11 p.

[126] H.K. Shin, Collisions in reaction O + CF_3J, J. Chem. Phys., 78, (1983) 795.

[127] A.A. Kvizinskii, Low-energy scattering by potentials contaning power corrections to coulomb interaction. Teor. Mat. Phys., 65, (1985) 226.

[128] A.A. Kvizinskii and S.P. Merqur`ev, Effective potential in the problem of scattering of three charged particles, Yader. Fizika, 48, (1988) 127.

[129] A.A. Kvizinskii, Scattering distance in the system of three charged particles, Pis`ma Zh. Eksp. Teor. Fiz., 36, (1982) 357.

[130] A.A. Kvizinskii and D.M. Latypov, Coordinate asymptotics of the Faddeev radial components for scattering in the system of charged particles, Yader. Fizika, 53, (1991) 1552.

Processes of Ionization of Rydberg Atoms at Thermal Collisions with Neutral Particles

V.S. Lebedev

1. INTRODUCTION

The study of ionization processes of highly excited atoms at thermal collisions with neutral particles is of great interest for the atomic spectroscopy, astrophysics, low-temperature plasma kinetics and gaseous discharge physics (see [1-3]). In recent years such processes are being studied intensively because of the development of new methods of selective excitation of the Rydberg atom states by tunable lasers radiation.

Previously the experimental investigations of the ionization processes were most often carried out for the Rydberg atoms of alkali metals and inert gases at collisions with two-atomic and polyatomic molecules. Such processes result from the mechanism of scattering of the perturbing molecule by the outer (quasi-free) electron of the Rydberg atom and are characterized by large values of rate constants. The basic results of the experimental and theoretical investigations of various ionization processes of highly excited atoms at collisions with molecules are reported in review papers [1, Ch. 8, 9].

At the thermal collisions between Rydberg atoms and atomic particles the traditional Fermi mechanism of scattering by the quasi-free electron turns out to be essential for ionization only at sufficiently high values of the principal quantum number n. In the range $n \leq 30$ to 40 usually investigated experimentally the ionization of Rydberg levels may occure effectively due to other physical mechanisms caused by the

*Throughout this chapter @ indicates that the Subsequent numbers or letters are subscripts

scattering by the atomic residue. This is illustrated, in particular, by the well studied resonance processes of the direct and associative ionization of highly excited atoms at collisions with the own gas atoms in the ground state. Such processes are caused by the dipole transitions between symmetric and antisymmetric terms of homonuclear quasi-molecular A_2^+ ion which is formed at the scattering of the perturbing A atom by the atomic residue A^+ of the Rydberg atom A(nl). The cro sections and ionization rate constants for this mechanism [4] were calculated in a number of papers [5-8] within the framework of the perturbation theory over the dipole interaction between the outer e^- and inner (recharging) electron of the quasi-molecule $A_2^+ + e^-$. A large body of experimental research in the resonance ionization processes mentioned above was carried out for Rydberg atoms of alkali metals (Li, Na, K, Rb, Cs) at thermal collisions with the own gas atoms in the ground state (see references to basic works in [3]).

Another mechanism of the direct and associative ionization caused by scattering of the perturbing atom B of the buffer gas by the atomic residue A^+ of theRydberg atom A(nl) has been suggested in [9]. In this mechanism the bound-free transitionof the outer electron e^- is accompanied by the nuclear transition within an electron term of the quasimolecular ion BA^+. Such transitions are caused by the interaction between the outer electron e^- of the heteronuclear $BA^+ + e^-$ system with the total dipole momentum of BA^+ ion at the electronic state involved. The above-mentioned mechanism proves to be the most effective for the light colliding atoms A(nl) and B in the case when the process of resonance energy transfer to the outer electron from the inner electron shell of quasi-molecule $BA^+ + e^-$ can not proceed. As it follows from [2, 9 - 11], such situation is realized, for example, in the case of thermal collisions of highly excited hydrogen atoms with light inert gas atoms.

In this paper the theory is developed of direct (1a) and associative (1b) ionization processes

$$A(nl) + B \rightarrow \begin{cases} A^+ + B + e^-, & (1a) \\ BA^+ + e^- & (1b) \end{cases}$$

of Rydberg atoms at thermal collisions with neutral particles. In Section 2 the traditional ionization mechanism (1a) is considered caused by the scattering of the perturbing atom or molecule by the quasi-free electron in the context of the impulse approximation. The general formulas are obtained connecting the total cross section σ_{nl} of this process with the differential cross section @of scattering of free electron e^- by the neutral particle B and with the distribution function $|g_{nl}(k)|^2$ of momenta of highly excited electron in the initial nl state. Contrary to the previous theory (see [1, Ch. 8]), these formulas are valid for the scattering amplitude @ of arbitrary form. They have been used to obtain simple analytic expressions for the ionization cross sections and rate constants in the approximation of scattering length ($f_{eB} = -L$). This expressions allow to estimate quickly the efficiency of the Fermi mechanism of ionization of highly excited atoms at collisions with atomic particles in the wide range of principal quantum numbers n and gas temperatures T.

In Section 3 processes of the direct and associative ionization of the Rydberg atom A(nl) are investigated caused by the scattering of the perturbing atom of the buffer (B) or own (A) gas by the atomic residue A^+. Various ionization mechanisms are considered corresponding to the nonadiabatic exchange of energy of the outer electron e^- of quasi-molecule $BA^+ + e^-$ with the electronic shell energy of BA^+ ion and also directly with the kinetic energy of relative motion of colliding atoms. The analysis of processes (1a) and (1b) is performed in the quasi-classical approximation in the framework of the stationary perturbation theory over the Coulomb interaction between the outer electron e^- and all inner electrons and nuclei of quasi-molecule $BA^+ + e^-$.

The analytic expressions for the total cross sections and ionization rate constants are obtained on the basis of the quasi-classical model [9 - 11] for processes (1a) and (1b) which proceed within an electron term of BA^+ ion (see Section 3.3). The specific calculations were performed for the case of thermal collisions between H(nl) and $He(1s^2)$ atoms. For the most effective ionization processes caused by the intersection of different electron terms the general formulas for the cross sections and rate constants were deduced using the Landau-Zinner model. These formulas are valid atthe arbitrary type and symmetry of the BA^+ ion terms converging to the same dissociation limit (see Section 3.2). In particular, they may be used (see Section 5) for the calculation of the rate constants of the dipole ionization of Rydberg levels of inert gas atoms $X[n_0p^5(^2P_{3/2})nl]$ by $X(^1S_0)$ atoms of own gas.

On the basis of the theory presented, the effective process was investigated of direct and associative ionization of Rydberg atoms $X[n_0p^5(^2P_{3/2})nl]$ of inert gas at collisions with atoms $Y(^1S_0)$ of more light buffer inert gas in the ground state. It was shown that in this case the ionization of the Rydberg electron e^- is followed by the transition $A_1|j_i = 3/2, \Omega_i = 3/2> \rightarrow X|j_f = 3/2, \Omega_f = 1/2>$ between two lower splitted terms of quasi-molecular ionYX^+. In this case the projection of the total momentum $j = L + S$ ($j = 3/2$) of the electron shell of heteronuclear ionYX^+ on the internuclear axis R changes: $\Omega_i = 3/2 \rightarrow \Omega_f = 1/2$. Due to selection rules (see the Appendix) the quadrupole and short-range parts of the Coulomb interaction between the outer electron e^- and electrons of the atomic residue $X^+(n_0p^5)$ play the main role in the ionization mechanism considered. Notice that the similar mechanism of collisional quenching of Rydberg atom levels in the buffer gas was suggested and studied in [12, 13]. In this paper specific numerical calculations of total rate constants for the direct and associative ionization processes were carried out for $Xe[5p^5(^2P_{3/2})nl]$ atoms at thermal collisions with He, Kr, and Xe atoms. In this case a detailed comparison of the efficiency of various ionization mechanisms was performed.

2. SCATTERING OF THE PERTURBING PARTICLE BY THE QUASI-FREE ELECTRON

2.1. Preliminary notes.

The ionization Fermi mechanism for the thermal collisions between Rydberg atoms and neutral atomic particles caused by the scattering by the quasi-free electron was originally studied by Flannery [14]. The semiclassical calculations [14] of the ionization rate constants are based on the method suggested by Pitaevskii [15] (see also [16]) for the describtion of the electron energy relaxation among the highly excited levels in the process of three-particle electron-ion recombination. In this case the approximation of the scattering length for the amplitude of elastic electron-atom scattering, i. e. $f_{eB}(k, \theta_{kk'}) = -L$ was used. The specific calculations were performed in [14] for the ionization process $H(n) + H \rightarrow H^+ + H + e^-$. The general semiclassical theory of the Rydberg atoms ionization by neutral particles was later formulated in apaper [17] (see also [1, Ch. 11]). However, at present the analytic expressions for the cross sections and rate constants of Rydberg atoms ionization by atomic particles are absent even for the simpliest case of the constant scattering amplitude $f_e = -L$. At the same time such expressions are of great interest for the estimation of efficiency of the scattering Fermi mechanism for the process of direct ionization of highly excited atoms[1] compared to other competitive mechanisms.

In the case of thermal collisions between A(nl) Rydberg atoms and B molecules the mechanism ofscattering by the

---

(Footnote 1: Simple analytical expressions for the cross sections and rate constants of inelastic transitions $nl \rightarrow n'$ and $n \rightarrow n'$ were obtained in the quasi-classical approximation based on the Fermi pseudopotential in [18]. Similar expressions were obtained concurrently in [19, 20] in the framework of the impulse approximation ($f_{eB} = -L$) using the binary approximation for atomic formfactors).

quasi-free electron plays the decisive role in the ionization process. This is due to the essential compensation of the energy $\Delta\varepsilon_{fi} = |\varepsilon_{nl}| + \varepsilon$ transferred to the outer electron e^- in $A(nl) + B(\upsilon \mathcal{J}) \to A^+ + B(\upsilon' \mathcal{J}') + e^-$ process from the inner vibration-rotation energy ($E_{\upsilon \mathcal{J}} > E_{\upsilon' \mathcal{J}'}$) of the perturbing molecular particle B = CD. In the case of molecules B = CD with the positive electron affinity (for example, F_2, SF_6, CCl_4 et al.) the efficiency of $A(nl) + CD \to A^+ + CD^-$ ionization process is determined by the great value of the cross section of the electron attachment to the perturbing particle.

The modern ionization theory (see [1, Ch. 8]) of Rydberg atoms by the polar and nonpolar molecules (such as NH_3, CO, CO_2, N_2 et al.) is based on the impulse approximation of Chew and Goldberger [21] (see also [22, 23]). In this theory, as a rule, Born approximation is used for the nonelastic scattering $e^- + B(\beta) \to e^- + B(\beta')$ of a free slow electron by a molecule. Here β and β' is the set of the quantum numbers (for example, vibration-rotation ones) which determine the inner state of the perturbing particle B with energies E_β and $E_{\beta'}$ (with $E_{\beta'} = E_\beta$ at the elastic scattering). The corresponding Born amplitude of the scattering $f_{eB}^{\beta\beta'}$ is either constant ($f_{eB}^{\beta\beta'}$ = const for the quadrupole molecules) or depends only on the value Q of momentum transferred to e^- electron at the collision with B molecule ($f_{eB}^{\beta\beta'} \sim Q^{-2}$ for the dipole molecules). Hence, the well-known original Matsuzava formula [24] for the Rydberg atom ionization cross section may be expressed in terms of atomic formfactors for the bound-free transition $nl \to \varepsilon$ of e^- electron in the Coulomb field of the A^+ atom residue. Note that the binary approximation for the formfactors was used in calculations of the ionization cross sections of highly excited atoms by the dipole and quadrupole molecules in [19] (see also [1, Ch. 8, 11] and references in them).

In many cases interesting in practice the Born approximation for the slow electron-molecular scattering and approximation of the scattering length for the electron-atomic scattering can not be applied (see, for example, [25 - 27]). Then the amplitude $f_{eB}^{\beta\beta'}(k, \theta_{kk'})$ of the elastic $\beta' = \beta$ or inelastic $\beta' \neq \beta$ scattering of e^- electron by the disturbing B particle depends both on the momentum k and scattering angle $\theta_{kk'}$. In these cases the present theory of ionization of highly excited levels (see [1, Ch. 8]) can not be used for the cross section calculations. Hence, the general formula[2] of the impulse approximation for the cross section of the direct ionization process is required, which can be used at the arbitrary form of the scattering amplitude $f_{eB}^{\beta\beta'}(k, \theta_{kk'})$.

2.2. General formulas of the impulse approximation for the ionization cross section.

Consider process of the inelastic collision of Rydberg atom A(nl) with a neutral particle B(β) followed by the bound-bound (excitation, deexcitation) or bound-free (ionization) transition of the highly excited electron

$$A(nl) + B(\beta) \rightarrow \begin{cases} A(n') + B(\beta'), & (2a) \\ A^+ + B(\beta') + e^-. & (2b) \end{cases}$$

According to the general scattering theory (see, for example, [22, 23]) the differential cross section of such process in the system of center of mass of colliding particles A(nl) and B(β) can be presented in the following form[3]:

$$d\sigma_{if}^{\beta\beta'}(\mathbf{q}, \mathbf{q}') = \frac{2\pi}{2l+1} \sum_m \left| \left\langle \left(\frac{\mu}{q}\right)^{1/2} \exp(i\mathbf{q}\mathbf{R}) \right| \langle \psi_i | \langle \beta | T | \beta' \rangle | \psi_f \rangle \right| \times \tag{3}$$
$$\times \exp(i\mathbf{q}'\mathbf{R})\rangle|^2 \delta(\varepsilon_f + E_{\beta'} + E' - \varepsilon_i - E_\beta - E)\, q'^2\, dq'\, dO_{\mathbf{q}'}/(2\pi)^3.$$

Here ψ_i, ψ_f are the eigenfunctions, ε_i, ε_f are energy eigenvalues for the Rydberg atom A^{**} in the initial $|i\rangle = |nlm\rangle$ and final $|f\rangle = |n_f l_f m_f\rangle$ (or $|f\rangle = |\varepsilon_f l_f m_f\rangle$) states of the

(Footnote 2: In the case of slow collisions $v \ll n$ the general formulas of the impulse approximation for cross sections of inelastic transitions $nl \rightarrow n'$ and $n \rightarrow n'$ between highly excited atomic levels were obtained in [28, 29]).

(Footnote 3: In the paper, the atomic system of units $e = m_e = \hbar = 1$ was used)

discrete or continuous spectrum; $q = \mu v$, $q' = \mu v'$ and $E = q^2/2\mu$, $E' = q'^2/2\mu$ are momenta and kinetic energies of the relative motion of colliding particles A^{**} and B ($\mu = M_A M_B/(M_A + M_B)$ is their reduced mass); E_β, $E_{\beta'}$ are values of the internal (for example, vibration-rotation) energy of the perturbing particle B before and after a collision. The scattering operator T in (3) is determined by the relation

$$T = V_{AB} + V_{AB}(\mathscr{E} - H + i\eta)^{-1}V_{AB}, \quad H = H_A + H_B + K_{AB} + V_{AB}, \qquad (4)$$

where K_{AB} is the kinetic energy operator for the relative motion of particles A^{**} and B; $H_A = K_e + H_{A^+} + U_{eA^+}$ and H_B are Hamiltoniansof the Rydberg atom A^{**} and perturbing particle B (atom or molecule), respectively, $V_{AB} = V_{eB} + V_{A^+B}$ is the operator of the potential energy of interaction between colliding particles.

In the framework of the quasi-free electron model, we shall only consider here the interaction of the perturbing particle B with the outer e^- electron of the Rydberg atom A^{**} and neglect its interaction with the atomic residue A^+, i.e. assume that $V_{AB} = V_{eB}$. Then corresponding matrix elements of the scattering operator $T = V_{eB} + V_{eB}(\mathscr{E} - H + i\eta)^{-1}V_{eB}$ may be presented in the impulse approximation [21] in the following form (see also [27]):

$$\begin{aligned} T_{if}(\mathbf{q}\beta \to \mathbf{q}'\beta') &= \langle \exp(i\mathbf{qR}) | \langle i | \langle \beta | T | \beta' \rangle | f \rangle | \exp(i\mathbf{q}'\mathbf{R}) \rangle = \\ &= \langle G_f(\mathbf{p}') | t_{eB}(\mathbf{k}\beta \to \mathbf{k}'\beta') | G_i(\mathbf{p}) \rangle_{\mathbf{p}}. \end{aligned} \qquad (5)$$

Here $G_i(p)$ and $G_f(p')$ are eigenfunctions of the Rydberg atom A^{**} in the impulse representation, $p = \mu_{eA} + v_{eA^+}$ and $p' = \mu_{eA} + v_{eA^+}$ are momenta of the relative movement of e^- electron and of the atom residue (core) A^+, $k = \mu_{eB} v_{eB}$, $k' = \mu_{eB} v'_{eB}$ are momenta of the relative movement of e^- electron and the perturbing particle B, and

$$t_{eB}(\mathbf{k}\beta \to \mathbf{k}'\beta') = \langle \exp(i\mathbf{kr}) | \langle \beta | t_{eB} | \beta' \rangle | \exp(i\mathbf{k}'\mathbf{r}) \rangle \qquad (6)$$

are matrix elements of two-particle operator t_{eB} of scattering of free electron e^- by B particle. The relation between the momenta and relative velocities of particles e^-, A^{**} and B in formula (5) is determined by the following expression:

$$\mathbf{k} = \frac{\mu_{eB}}{m_e}\mathbf{p} - \frac{\mu_{eB}}{\mu}\mathbf{q}, \qquad \mathbf{v}_{eB} = \frac{\mu_{eA^+}}{m_e}\mathbf{v}_{eA^+} - \mathbf{v}, \tag{7a}$$

$$\mathbf{k}' - \mathbf{k} = \frac{\mu_{eB}}{m_e}\left(\frac{M_{A^+}}{M_A} + \frac{m_e}{\mu}\right)\mathbf{Q}, \qquad \mathbf{p}' - \mathbf{p} = \frac{M_{A^+}}{M_A}\mathbf{Q}, \tag{7b}$$

where $Q = q - q'$ is the change in the momentum of the relative movement of colliding particles A^{**} and B, $M_A = M_{A^+} + m_e$ and $\mu_{eA} = m_e M_{A^+}/(m_e + M_{A^+})$, $\mu_{eB} = m_e M_e/(m_e + M_B)$ are the reduced masses (m_e = 1 a.u.).

Let us substitute expressions (5),(6) in the original formula (3) and integrate the result with respect to the directions $dO_{q'}$ of momenta of the relative movement of A^{**} and B particles. Then after summation over all finite f states of the Rydberg atom in the discrete spectrum and integration in the continuous spectrum, we get the following expression for the differential cross section $d\sigma_i^{\beta\beta'}(q, q')/dE'$ of the inelastic scattering into a unit interval of the kinetic energy $dE' = q'dq'/\mu$ of colliding particles

$$\frac{d\sigma_i^{\beta\beta'}(q, q')}{dE'} = \frac{\mu^2 q'}{(2\pi)^2 q(2l+1)} \sum_m \int dO_{\mathbf{q}'} \sum_f \langle G_i(\mathbf{p}) | t_{eB}^*(\mathbf{k}\beta \to \mathbf{k}'\beta') | G_f(\mathbf{p}')\rangle \times$$
$$\times \langle G_f(\mathbf{p}') | t_{eB}(\mathbf{k}\beta \to \mathbf{k}'\beta') | G_i(\mathbf{p})\rangle \, \delta(\varepsilon_f - \varepsilon_i + E_{\beta'} - E_\beta + E' - E). \tag{8}$$

Present next δ-function in the form of the Fourier integral

$$\delta(\varepsilon_f - \varepsilon_i + E_{\beta'} - E_\beta + E' - E) = \frac{1}{2\pi}\int_{-\infty}^{\infty} \exp[i(\varepsilon_f - \varepsilon_i + E_{\beta'} - E_\beta + E' - E)t]\, dt \tag{9}$$

and use the following known properties of Heisenberg evolution operators:

$$\exp(-iH_A t)|G_i(\mathbf{p})\rangle = \exp(-i\varepsilon_i t)|G_i(\mathbf{p})\rangle, \tag{10a}$$

$$\langle G_f(\mathbf{p}')|\exp(iH'_A t) = \langle G_f(\mathbf{p}')|\exp(i\varepsilon_f t). \tag{10b}$$

Hamiltonians of the Rydberg atom[4] in the impulse representaton have the form:

$$H_A \equiv H_A(\mathbf{p}, \hat{\mathbf{r}}) = \mathbf{p}^2/2\mu_{eA^+} + \hat{U}_{eA^+}, \tag{11a}$$

$$H'_A \equiv H_A(\mathbf{p}', \hat{\mathbf{r}}) = \mathbf{p}'^2/2\mu_{eA^+} + \hat{U}_{eA^+}, \tag{11b}$$

where p and $p' = p + (M_{A^+}/M_A)Q$ are momenta, $\hat{U}_{eA^+}(\hat{r})$ is the operator of the potential energy of interaction between Rydberg electron e^- and atomic residue A^+ and $\hat{r} = i\partial/\partial p$ is the operator of the electron radius vector.

Using Eqs.(9)-(11) and the property of completeness $\sum_f |G_f\rangle\langle G_f| = \hat{1}$ of the Rydberg atom wave functions, the expression (8) takes the form:

$$\frac{d\sigma_i^{\beta\beta'}(q, q')}{dE'} = \frac{\mu^2 q'}{(2\pi)^3 q(2l+1)} \sum_m \int dO_{\mathbf{q}'} \langle G_i(\mathbf{p})| \times$$
$$\times \int_{-\infty}^{\infty} dt \exp\left[i(E_{\beta'} - E_\beta + E' - E)t\right] t^*_{eB}(\mathbf{k}\beta \to \mathbf{k}'\beta') \times$$
$$\times \exp\left(iH'_A t\right) t_{eB}(\mathbf{k}\beta \to \mathbf{k}'\beta') \exp(-iH_A t) |G_i(\mathbf{p})\rangle. \tag{12}$$

Hence, the differential cross section $d\sigma_i^{\beta\beta'}(q, q')/dE'$ of the inelastic scattering of particles A^* and B in the unit energy interval of the continuous spectrum may be expressed in the form of the average value of some operator over the inintial state $|i\rangle = |nlm\rangle$ of the Rydberg atom.

In the framework of the impulse approximation in formula (12) the two-particle operator t_{eB} of the electron scattering by the perturbing particle may be considered to be commutative with the Rydberg atom Hamiltonian H_A (and, hence, with the operator $\exp(-iH_A t)$ of temporal evolution). In addition, according to the condition of applicability of the quasi-free electron model

(Footnote 4: Notice that Hamiltonian H_{A^+} can be eliminated from expressions (10), (11) since the inner energy E_{A^+} of the atomic residue A^+ does not change in the processes (2a) and (2b) considered in this section of the paper).

(see [28]) only the first member of the expansion for the operator may be used in (12)

$$\exp\left(iH'_A t\right)\exp\left(-iH_A t\right) = \exp\left[i\left(H'_A - H_A\right)t + \frac{1}{2}\left[H'_A, H_A\right]t^2 + \ldots\right] =$$
$$= \exp\left[it\left(\mathbf{p}'^2 - \mathbf{p}^2\right)/2\mu_{eA^+} + \ldots\right]. \quad (13)$$

Then after the integration over time dt in (12) the differential cross section of the inelastic process (2) takes the following form:

$$\frac{d\sigma_{nl}^{\beta\beta'}(q, q')}{dE'} = \frac{\mu^2 q'}{(2\pi)^2 q(2l+1)} \sum_m \int dO_{\mathbf{q}'} \int d\mathbf{p}\, |G_{nlm}(\mathbf{p})|^2 \times \quad (14)$$
$$\times |t_{eB}(\mathbf{k}\beta \to \mathbf{k}'\beta')|^2 \delta\left[(\mathbf{p}'^2 - \mathbf{p}^2)/2\mu_{eA^+} + E_{\beta'} - E_\beta + E' - E\right].$$

The kinetic energy $E' = q'^2/2\mu$ of the relative movement of colliding patricles A^{**} and B in (14) is determined from the law of conservation of energy for processes (2a) of excitation ($\varepsilon_{n'} > \varepsilon_{nl}$) or deexcitation ($\varepsilon_{n'} < \varepsilon_{nl}$) of Rydberg levels

$$E + E_\beta + \varepsilon_{nl} = E' + E_{\beta'} + \varepsilon_{n'},\ \varepsilon_{n'} = -1/2n'^2 < 0, \quad (15a)$$

or for the ionization process of highly excited atom

$$E + E_{\beta'} + \varepsilon_{nl} = E' + E_{\beta'} + \varepsilon,\ \varepsilon = K^2/2\mu_{eA^+}. \quad (15b)$$

where $\varepsilon_{nl} = -1/2n^2_{eff} < 0$ is the initial state energy of atom A^{**} $n_{eff} = n - \delta_l$, δ_l is the quantum defect); $\varepsilon_{n'}$ is the energy of the final hydrogen-like level n and ε is the energy of the emitted electron.

For the ionization process (2b) considered here the expression (14) may be presented as the differential cross section $d\sigma^{\beta\beta'}_{nl,\varepsilon}(E)/d\varepsilon$ of the ionization into the unit interval of the emitted electron energies

$$\frac{d\sigma_{nl,\varepsilon}^{\beta\beta'}(E)}{d\varepsilon} = \frac{\mu^2 q'}{\mu_{eB}^2 q(2l+1)} \sum_m \int dO_{\mathbf{q}'} \int d\mathbf{p}\, |G_{nlm}(\mathbf{p})|^2 \times \quad (16a)$$
$$\times |f_{eB}^{\beta\beta'}(\mathbf{k}, \mathbf{k}')|^2 \delta[(\mathbf{p}'^2 - \mathbf{p}^2)/2\mu_{eA^+} + \varepsilon_{nl} - \varepsilon].$$

Here @ is the amplitude of the elastic $\beta' = \beta$ or inelastic $\beta \neq \beta$ scattering of e^- electron by the perturbing particle B. The total ionization cross section $\sigma_{nl}^{\beta\beta'}(E)$ is determined from formula (16a) after its integration over all possible values of the emitted electron energy $\varepsilon = K^2/2\mu_{eA^+}$

$$\sigma_{nl}^{\beta\beta'}(E) = \int_0^{\varepsilon_{\max}} \frac{d\sigma_{nl,\varepsilon}^{\beta\beta'}(E)}{d\varepsilon} d\varepsilon, \qquad \varepsilon_{\max} = E - (|\varepsilon_{nl}| - \Delta E_{\beta\beta'}), \tag{16b}$$

where @ is the change of internal energy of the perturbing particle B. Possible values of the initial kinetic energy $E = q^2/2\mu$ of colliding particles A^* and B are for the ionization process (2b) lie in the range $E_{min} \leq E < \omega$ ($E_{min} = \max(0, |\varepsilon_{nl}| - \Delta E_{\beta\beta'})$).

Note that the result for the cross section of inelastic transition $nl \to n'$ from Rydberg nl level to all hydrogen-like $n'l'm'$ sublevels of another energy level n' follows from (16a) after the substitution $d\sigma_{nl,\varepsilon}^{\beta\beta'}(E)/d\varepsilon \to \sigma_{nl,n'}^{\beta\beta'}(E)$ of the discrete spectrum energ $\varepsilon_{n'} = -1/2n'^2 < 0$ for the value $\varepsilon = K^2/2\mu_{lA^+} > 0$ (see also (15a)). Such substitution corresponds to the approximation of quasi-continuous spectr of highly excited energy levels of an atom.

Slow collisions $n \ll \upsilon^{-1}$

Consider the range of principal quantum numbers $n \ll \upsilon^{-1}$ which is of primary interest for the case of thermal collisions between Rydberg atoms and neutral particles. In this case the following approximate relations can be used for the relative velocities and momenta of particles e^-, A^+ and e^-, B (see expressions (7a), (7b))

$$\mathbf{v}_{eB} \approx \mathbf{v}_{eA^+}, \quad \mathbf{p} \approx \mathbf{k}, \quad \mathbf{p}' - \mathbf{p} \approx \mathbf{q} - \mathbf{q}' = \mathbf{Q} \tag{17}$$

in the general formula (16) for the ionization cross section. Let us represent the initial wave function of the Rydberg atom as the product of the radial and angular parts $G_{nlm}(\mathbf{p}) = g_{nl}(\mathbf{p}) Y_{lm}(\theta_p, \varphi_p)$. Then substituting in (16a) and (16b) the integration over momenta transfered for the integration over solid angles @ of scattered particles A^{**} and B we get

$$\sigma_{nl}^{\beta\beta'}(E) = \frac{2\pi}{(2l+1)v^2} \sum_m \int_0^{\varepsilon_{max}} d\varepsilon \int_{Q_1(\varepsilon)}^{Q_2(\varepsilon)} Q\,dQ \int_{p_{\beta\beta'}}^{\infty} p^2\,dp\,|g_{nl}(p)|^2 \times \tag{18}$$

$$\times \int_0^{2\pi} d\varphi_p \int_{-1}^{1} d(\cos\theta_p)\,|Y_{lm}(\theta_p,\varphi_p)|^2\,|f_{eB}^{\beta\beta'}(p,Q)|^2 \times$$

$$\times\,\delta(pQ\cos\theta_{pQ} + Q^2/2 - |\varepsilon_{nl}| - \varepsilon).$$

Here $v = q/\mu = (2E/\mu)^{1/2}$ is the initial velocity of the relative movement of colliding particles A^{**} and B, $p_{\beta\beta'} = 1$ and $p_{\beta\beta'} = [2(E_{\beta'} - E_{\beta})]^{1/2}$ at $E_\beta \gg E_{\beta'}$ and $E_\beta < E_{\beta'}$, respectively. The lower and upper limits of integration over the transferred momentum are determined by the following relation:

$$Q_{1,2}(\varepsilon) = |q' \mp q| = (2\mu)^{1/2}\,|E^{1/2} - [E - (\varepsilon + |\varepsilon_{nl}| - \Delta E_{\beta\beta'})]^{1/2}|. \tag{19}$$

Note also that the scattering amplitude $f_{eB}^{\beta\beta'}$ in (18) is treated as the function of the electron momentum $k \approx p$ and transferred momentum Q.

Let us direct z axis of the coordinate system along Q vector (so that $\cos\theta_{pQ} = \cos\theta_p$) and perform the summation over magnetic quantum numbers m in (18) using the relation

$$\sum_m |Y_{lm}(\theta_p,\varphi_p)|^2 = (2l+1)/4\pi. \tag{20}$$

Then the integration over $d(\cos\theta_p)$ in (18) leads to the following formula for the cross section of the Rydberg atom ionization by the neutral particle:

$$\sigma_{nl}^{\beta\beta'}(E) = \frac{\pi}{v^2} \int_0^{\varepsilon_{max}} d\varepsilon \int_{Q_1(\varepsilon)}^{Q_2(\varepsilon)} dQ \int_{p_{min}}^{\infty} p\,dp\,|g_{nl}(p)|^2\,|f_{eB}^{\beta\beta'}(p,Q)|^2, \tag{21a}$$

$$p_{min} = \max(p_{\beta\beta'},\ |p_0(Q,\varepsilon)|),\ p_0(Q,\ \varepsilon) = (\varepsilon + |\varepsilon_{nl}| - Q^2/2)/Q. \tag{21b}$$

Changing the order of integration over $d\varepsilon$ and dQ, we obtain

$$\sigma_{nl}^{\beta\beta'}(E) = \frac{\pi\Theta(\Delta E_{\beta\beta'} - |\varepsilon_{nl}|)}{v^2} \int_{Q_1(0)}^{Q_2(0)} dQ \int_{\varepsilon_1(Q)}^{\varepsilon_2(Q)} d\varepsilon \int_{p_{\min}(Q,\varepsilon)}^{\infty} p\,dp\,|g_{nl}(p)|^2 \,|f_{eB}^{\beta\beta'}(p,Q)|^2 +$$

$$+\frac{\pi}{v^2}\int_{Q_1(0)}^{Q_2(0)} dQ \int_{0}^{\varepsilon_2(Q)} d\varepsilon \int_{p_{\min}(Q,\varepsilon)}^{\infty} p\,dp\,|g_{nl}(p)|^2 \,|f_{eB}^{\beta\beta'}(p,Q)|^2. \qquad (22)$$

Here $\Theta(x)$ is theta-function (i. e. $\Theta(x) = 1$ and $\Theta(x) = 0$ at $x \geq 0$ and $x < 0$, respectively) and values $\varepsilon_{1,2}(Q)$ are determined by the relation

$$\varepsilon_{1,2}(Q) = v(\mp Q - Q_0 - Q^2/2\mu v), \quad Q_0 = (|\varepsilon_{nl}| - \Delta E_{\beta\beta'})/v. \qquad (23)$$

The internal energy of the perturbing particle B does not change in the ionization process (2b) at the collisions between Rydberg atoms and neutarl particles, i. e. $E_{\beta\beta'} = 0$, $p_{\beta\beta'} = 0$ and $Q_0 = |\varepsilon_{nl}|/v$. In this case the change of kinetic energy $E_{if} = E - E'$ of colliding particles A^{**} and B is determined by the magnitude of the Rydberg electron energy change $\Delta\varepsilon_{fi} = |\varepsilon_{nl}| + \varepsilon$, i.e. $\Delta E_{if} = \Delta\varepsilon_{fi}$ (see Eq. (15b). Thus, the ionization threshold $(E_{\min} = |\varepsilon_{nl}| = 1/2n_{eff}^2 > 0)$ exists at any values of the principal quantum number n.

The ionization process (2b) at the collisions of Rydberg atoms with molecular particles is found to be more effective when the defect of the kinetic energy $\Delta E_{if} = \Delta\varepsilon_{fi} - \Delta E_{\beta\beta'}$ (see (15b)) is far less than the energy $\Delta\varepsilon_{fi} = |\varepsilon_{nl}| + \varepsilon$, tranferred to the outer electron. In this case the ionization of the Rydberg electron is followed by the deexcitation of the inner vibration-rotation energy of the perturbing molecular particle $\Delta E_{\beta\beta'} = E_\beta - E_{\beta'} > 0$, so that $p_{\beta\beta'} = 0$ (see (18)). Hence, the ionization threshold $(E_{\min} = |\varepsilon_{nl}| - \Delta E_{\beta\beta'} > 0)$ exists only in the range of quantum numbers $n_{eff} < n_{eff}^{(0)} = (2\Delta E_{\beta\beta'})^{-1/2}$. For great values of $n_{eff} \geq n_{eff}^{(0)}$ (when) $|\varepsilon_{nl}| - \Delta E_{\beta\beta'} < 0$) we have $E_{\min} = 0$.

Hence, it follows from (21a) and (21b) that the lower limit of integration over the electron momenta $p_{\min} = |p_0(Q, \varepsilon)$ in both cases mentioned above. Note further that in the considered range $n \ll v^{-1}$ we can approximately assume $\approx Q/2$ for all

Q and $\mathcal{E}$ values (see Eqs. (21a), (21b) and (22)). Then, integrating over the emitted electron energy $d\mathcal{E}$ in(22), we obtain the following result for the total ionization cross section:

$$\sigma_{nl}^{\beta\beta'}(E) = \frac{2\pi\Theta(\Delta E_{\beta\beta'} - |\varepsilon_{nl}|)}{v} \int_0^{Q_1(0)} Q\,dQ \int_{Q/2}^{\infty} p\,dp\,|g_{nl}(p)|^2 |f_{eB}^{\beta\beta'}(p, Q)|^2 +$$

$$+ \frac{\pi}{v} \int_{Q_1(0)}^{Q_2(0)} dQ\,(Q - Q_0 - Q^2/2\mu v) \int_{Q/2}^{\infty} p\,dp\,|g_{nl}(p)|^2 |f_{eB}^{\beta\beta'}(p, Q)|^2. \tag{24}$$

After substituting the variables p, $Q \to p$, $\theta_{pp'}$ ($Q = 2p\sin(\theta_{pp'}/2)$ and $\theta_{pp'}$ is the angle of the free electron scattering by the disturbing particle) the expression (24) may be presented in the following equivalent form:

$$\sigma_{nl}^{\beta\beta'}(E) = \frac{\Theta(\Delta E_{\beta\beta'} - |\varepsilon_{nl}|)}{v} \left[\int_0^{p_1} p^3 \sigma_{eB}^{\beta\beta'}(p)\,|g_{nl}(p)|^2\,dp + \right. \tag{25}$$

$$\left. + 2\pi \int_{p_1}^{\infty} p^3\,dp\,|g_{nl}(p)|^2 \int_{1-2p_1^2/p^2}^{1} d(\cos\theta_{pp'})\,|f_{eB}^{\beta\beta'}(p, \theta_{pp'})|^2 \right] +$$

$$+ \frac{\pi}{v} \left[\int_{p_1}^{p_2} p^2\,dp\,|g_{nl}(p)|^2 \int_{-1}^{1-2p_1^2/p^2} d(\cos\theta_{pp'})\,|f_{eB}^{\beta\beta'}(p, \theta_{pp'})|^2 F(p, \theta_{pp'}) + \right.$$

$$\left. + \int_{p_2}^{\infty} p^2\,dp\,|g_{nl}(p)|^2 \int_{1-2p_2^2/p^2}^{1-2p_1^2/p^2} d(\cos\theta_{pp'})\,|f_{eB}^{\beta\beta'}(p, \theta_{pp'})|^2 F(p, \theta_{pp'}) \right].$$

The values of the upper and lower integration limits $p_{1,2}$ and and $F(p, \theta_{pp'})$ function are determined by the following expressions:

$$p_{1,2} = Q_{1,2}(0)/2 = (\mu/2)^{1/2} \left| E^{1/2} \mp [E - (|\varepsilon_{nl}| - \Delta E_{\beta\beta'})]^{1/2} \right|, \tag{26a}$$

$$F(p, \theta_{pp'}) = p - (Q_0/2^{1/2})(1 - \cos\theta_{pp'})^{-1/2} - (p^2/2^{3/2}\mu^2 v^2)(1 - \cos\theta_{pp'})^{1/2}, \tag{26b}$$

$\sigma_{eB}^{\beta\beta'}(p)$ is the total cross section of the elastic $\beta' = \beta$ and inelastic $\beta' \neq \beta$ scattering of slow e^- electron by the perturbing particle B.

One should note that the first term in (25) is zero in the case of ionization of Rydberg atoms by atomic particles (when $\Delta E_{\beta\beta'} = 0$)) and also by molecular particles (if $|\varepsilon_{nl}| = 1/2n_{eff}^2 \gg \Delta E_{\beta\beta'}$). In this case the general formula (25) for the ionization cross section takes a particular simple form in a wide range of values of the colliding particles kinetic energy $E = \mu v^2/2$ (except the narrow region near the ionization threshold $E \gtrsim E_{min} = |\varepsilon_{nl}| - \Delta E_{\beta\beta'} > 0$)

$$\sigma_{nl}^{\beta\beta'}(E) \approx \frac{\pi}{v}\int_{p_1}^{\infty} p^2\,dp\,|g_{nl}(p)|^2 \int_{-1}^{1-2p_1^2/p^2} d(\cos\theta_{pp'})\,|f_{eB}^{\beta\beta'}(p,\theta_{pp'})|^2 \times \tag{27}$$

$$\times\left[p - \frac{2^{1/2}p_1}{(1-\cos\theta_{pp'})^{1/2}}\right],$$

where $p_1 \approx ||\varepsilon_{nl}| - \Delta E_{\beta\beta'}|/2v$. To derive this expression we have assumed $p_2 \to \infty$, since $p_2 \approx \mu v \gg p_1$.

In the opposite case $|\varepsilon_{nl}| = 1/2n_{eff}^2 < \Delta E_{\beta\beta'}$ (where $\Delta E_{\beta\beta'} > 0$) the similar formula for the ionization cross section can be presented in the region $\gg \Delta E_{\beta\beta'} - |\varepsilon_{nl}|$ in the following form:

$$\sigma_{nl}^{\beta\beta'}(E) \approx \frac{1}{v}\int_{0}^{\infty} p^3\sigma_{eB}^{\beta\beta'}(p)\,|g_{nl}(p)|^2\,dp - \frac{\pi}{v}\int_{p_1}^{\infty} p^2\,dp\,|g_{nl}(p)|^2 \times \tag{28}$$

$$\times \int_{-1}^{1-2p_1^2/p^2} d(\cos\theta_{pp'})\,|f_{eB}^{\beta\beta'}(p,\theta_{pp'})|^2\left[p - \frac{2^{1/2}p_1}{(1-\cos\theta_{pp'})^{1/2}}\right].$$

Hence, it follows that at $n_{eff} = n_{eff}^{(0)} = (2\Delta E_{\beta\beta'})^{-1/2}$ (when $|\varepsilon_{nl}| = \Delta E_{\beta\beta'}$,) the ionization cross section of the Rydberg atom by the molecular particle is described by the simple expression

$$\sigma_{nl}^{\beta\beta'}(E) \approx \frac{1}{2v}\int_{0}^{\infty} p^3\sigma_{eB}^{\beta\beta'}(p)\,|g_{nl}(p)|^2\,dp = \frac{\langle p\sigma_{eB}^{\beta\beta'}(p)\rangle_{nl}}{2v}. \tag{29}$$

Formulas (24) - (25) obtained in this paper are valid if the following condition is met:

$$\left[\left(2^{5/2}/a^{1/2}\right)v + 2\,\Delta E_{\beta\beta'}\right]^{-1/2} \ll n \ll v^{-1}. \qquad (30)$$

The lower limit of acceptable values of the principal quantum number in (30) is determined by the condition of applicability of the quasi-free electron model to the process of the direct ionization (2b):

$$r_{eA^+} \sim \frac{8v^2}{\left(1/2n_{eff}^2 - \Delta E_{\beta\beta'}\right)^2} \gg a = \max\left\{\left|f_{eB}\right|,\ \left|f_{A^+B}\right|\right\}. \qquad (31)$$

Here $r_{eA^+} \sim 2/p_1^2$ — is the characteristic dimension of the region of interaction between the Rydberg electron e^- and the atomic residue A^+, which makes the main contribution to the ionization cross section; f_{eB} and f_{A^+B} are the amplitudes of scattering of the perturbing particle B by e^- electron and the atomic residue A^+, respectively.

Fast collisions $n \gg v^{-1}$

Consider now the asymptotic region of high values of principal quantum number $n \gg v^{-1}$ (i.e. $n \gg 10^3 - 10^4$ at the thermal energies). The relative momentum and velocity of the electron and perturbing particle may be approximately presented for fast collisions (see (7a) and (7b)) in the form

$$\mathbf{v}_{eB} \approx -\mathbf{v}, \quad \mathbf{k} \approx -(\mu_{eB}/\mu)\mathbf{q}, \quad \mathbf{Q} = \mathbf{q} - \mathbf{q}' \approx \mathbf{k}' - \mathbf{k}. \qquad (32)$$

Let us substitute the integration over transferred momenta dQ in the general formula (16) for the integration over solid angles dO_q using (32). Furthermore, we change the order of integration in (16) and assume the upper limit of integration over the emitted electron energy to be equal to infinity. This is valid for the case of fast collisions $v = (2E/\mu)^{1/2} \gg 1/n$ due to the expression $E \gg |\varepsilon_{nl}| = 1/2n_{eff}^2$. As a result, it follows from the general formula (16) the known asymptotic expression for the cross section of ionization of Rydberg atom by the perturbing neutral particle

$$\sigma_{nl}^{\beta\beta'}(v) \underset{n\to\infty}{\to} \int |f_{eB}^{\beta\beta'}(k,\theta_{\mathbf{kk}'})|^2 d\Omega_{\mathbf{kk}'} = \sigma_{eB}^{\beta\beta'}(k \approx \mu_{eB}v). \tag{33}$$

Hence, at $n \gg v^{-1}$ the limiting value of the ionization cross section $\sigma_{nl}^{\beta\beta}(v)$ is determined by the total cross section $\sigma_{eB}^{\beta\beta'}$ ($v_{eB} \approx v$) Of the elastic $\beta' = \beta$ or inelastic $\beta \neq \beta$ electron scattering by the perturbing particle. The expression (33) corresponds completely to the preceding results [30, 31] obtained by the semiclassical method or using the optical theorem (see also Chs. 8 and 11 in [1]).

Approximation of the scattering lengh

For the ionization process (2b) of Rydberg atoms by the neutral atomic particles ($\Delta E_{\beta\beta'} = 0$) the general formula of the impulse approximation (25) is valid in the range of quantum numbers (see the condition (30))

$$(a^{1/4}/2^{5/4})v^{-1/2} \ll n \ll v^{-1}. \tag{34}$$

Simple analytic expressions for the ionization cross section σ_{nl} may be derived from (25) in the scattering length approximation $f_{eB}(p, \theta_{pp'}) = -L$. In this case, after calculation of the integral in (25) over the scattering angles $d(\cos\theta_{pp'})$ of the quasi-free electron, we get

$$\sigma_{nl}(E) = \frac{2\pi L^2}{v}\left\{\int_{p_1}^{p_2} p\,dp\,|g_{nl}(p)|^2\left[p^2 - p_1^2 - \frac{2}{3q}\left(p^3 - p_1^3\right) - Q_0(p - p_1)\right] + \right.$$

$$\left. + \left[p_2^2 - p_1^2 - \frac{2}{3q}\left(p_2^3 - p_1^3\right) - Q_0(p_2 - p_1)\right]\int_{p_2}^{\infty} p\,dp\,|g_{nl}(p)|^2\right\}, \tag{35}$$

where $p_{1,2} = (\mu/2)^{1/2}[E^{1/2} \mp (E - |\varepsilon_{nl}|)^{1/2}]$ and $Q_0 = |\varepsilon_{nl}|/v$. In view of the condition $n \ll v^{-1}$ we can use in (35) the asymptotic expression

$$|g_{nl}(p)|^2 \approx 4/\pi n_{eff}^3 p^6, \quad np \geqslant np_1 \approx {}^1/_4 nv \gg 1 \tag{36}$$

for the quasi-classical distribution function of momenta of highly excited electron

$$|g_{nl}(p)|^2 = \frac{4n_{\rm eff}}{\pi p^2\left(1+n_{\rm eff}^2p^2\right)^2}, \quad l \ll n, \tag{37}$$

averaged over the period of fast oscillating component.

Hence, after calculation of the integral over momenta dp (35) using the relation (36) we come to the following result for the ionization cross section of highly excited atom A(nl) from a given selectively excited nl level with $l \ll n$:

$$\sigma_{nl}(E) = (8/3\pi)\,\sigma_{\rm eB}^{\rm el} n_{\rm eff}\,(2E/\mu)^{1/2}\,(1-|\varepsilon_{nl}|/E)^{3/2}. \tag{38}$$

Here $\sigma_{eB}^{el} = 4\pi L^2$ is the total cross section of the elastic scattering of the ultraslow electron e^- by the perturbing atom B.

The expression for the ionization rate constant $K_{nl}(T) = \langle v\sigma_{nl}\rangle_T$ follows from (38) after the integration over the Maxwell distribution $F_m(E)dE = (2/\pi^{1/2})(E/T)^{1/2}\exp(-E/T)d(E/T)$ of kinetic energy $E = \mu v^2/2$ of colliding atoms A(nl) and B in the limits $|\varepsilon_{nl}| \le E < \infty$ and has the following form:

$$K_{nl}(T) = \frac{8\sigma_{\rm eB}^{el}T}{\pi\mu}\,n_{\rm eff}\exp\left(-\frac{1}{2n_{\rm eff}^2T}\right), \tag{39}$$

where T is the gas temperature.

When degenerate nlm-sublevels of theRydberg atom A(n) are equally populated within a given level n and the distribution function of the quasi-free electron momenta is described by the known Fok formula

$$|g_n(p)|^2 = \sum_{l=0}^{n-1}\frac{(2l+1)}{n^2}|g_{nl}(p)|^2 = 2^5n^3(1+n^2p^2)^{-4}\pi^{-1},$$

the similar consideration after calculation of the corresponding integral over dp gives the following result for the cross section

$$\sigma_n(E) = \frac{2^{10}\sigma_{\rm eB}^{el}}{15\pi}n^3\left(\frac{2E}{\mu}\right)^{3/2}\left(1-\frac{|\varepsilon_n|}{E}\right)^{1/2}\left(1-\frac{7}{6}\frac{|\varepsilon_n|}{E}+\frac{1}{6}\frac{|\varepsilon_n|^2}{E^2}\right) \tag{40}$$

and rate constants

$$K_n(T) = \frac{2^{10}\sigma_{\rm eB}^{\rm el}n^3T^2}{\pi\mu^2}\left(1+\frac{1}{6n^2T}\right)\exp\left(-\frac{1}{2n^2T}\right) \tag{41}$$

of ionization of highly excited atom by the perturbing atomic particle.

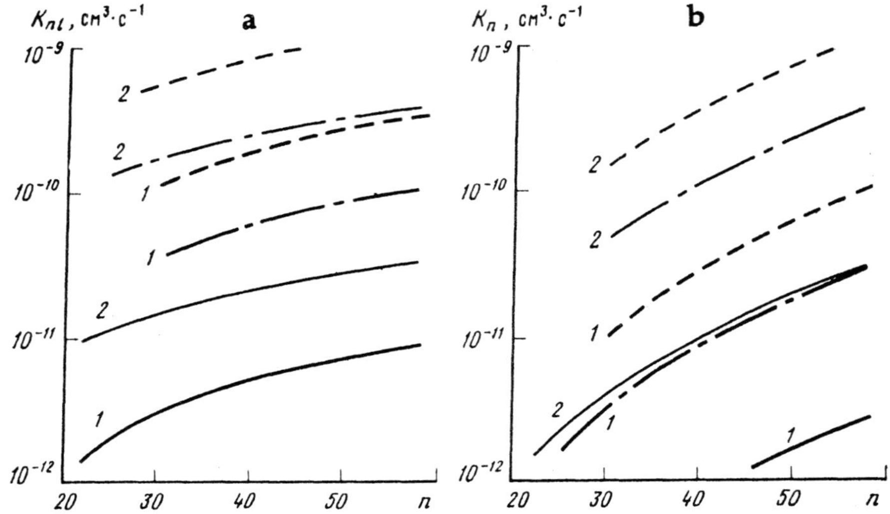

Fig.1. Rate constants $K_{nl}(T)$ and $K_n(T)$ (cm^3 sec^{-1}) of ionization of Rydberg atoms of hydrogen H (nl) at $l \ll n$ (a) and H(n) (b) at collisions with atoms of He (solid curves), Kr (dashed-dotted curves) and Xe (dashed curves) calculated by formulas (39) and (41) within the framework of the quasi-free electron model (L_{He} = 1.19 a.u., L_{Kr} = - 3.6 a.u. and L_{Xe} = 6.5 a.u.). Curves 1 and 2 correspond to the temperatures T = 300 K and T = 1000K)

Hence, as one can see from formulas (38) - (41), in the region of principal quantum numbers $n \ll (2T)^{-1/2}$ (when the electron bond energy $|\varepsilon_{nl}| \gg T$) the ionization rate constants calculated using the quasi-free electron model decrease sharply $K_{nl}(T), K_n(T) \propto nT \exp(-1/2n^2T)$ with decreasing the principal quantum number n. In the range of quantum numbers $(2T)^{-1/2} \ll n \ll$ the ionization rate constants obey the degree law $K_{nl} \propto n_{eff}$ $(l \ll n)$ and $K_n \propto n^3$ as n changes. This behaviour of the ionization rate constants is illustrated in Fig. 1a, b which presents calculations results for values $K_{nl}(T)$ and $K_n(T)$ in in the case of thermal collisions between highly excited hydrogen atoms and He, Kr, Xe inert gas atoms in the ground state.

In the range of high values of $n \gg v^{-1}$ formulas (38) - (41) deduced above become invalid. In this case the increase of the ionization cross sections (or the rate constants) with increasing the principal quantum number n becomes slow compared to their behaviour in the region of $n \ll v^{-1}$. According to (33) the limiting values of cross sections σ_{nl} or σ_n are determined by the value of the total cross section @ of the elastic electron scattering, i.e.

$$\sigma_{nl} \rightarrow \sigma_{eB}^{el} = 4\pi L^2 \qquad K_{nl} = \langle v \rangle_T 4\pi L^2, \tag{42}$$

where @ is the average thermal velocity of colliding atoms A(nl) and B.

3. SCATTERING OF THE DISTURBING PARTICLE BY THE ATOMIC RESIDUE

3.1. Statement of the problem. Original formulas for the ionization cross sections.

Consider now ionization processes of the Rydberg atom A(nl) caused by the perturbing atom B flying by the ion core A^+. Such processes occure if internuclear distances R in the quasi-molecular ion BA^+ fall far short of the radius $\langle r_e \rangle \sim 2n^2$ of highly excited electron e^- orbit. Hence, we present the total Hamiltonian H of the whole system $BA^+ + e^-$ in the form:

$$H = T_{\mathbf{R}} + H_{BA^+}(\mathbf{r}_\varkappa, \mathbf{R}) + H_e(\mathbf{r}_e) + V, \tag{43}$$

$$H_e | \psi_{nlm}(\mathbf{r}_e)\rangle = \varepsilon_{nl} | \psi_{nlm}(\mathbf{r}_e)\rangle, \quad H_e | \psi_{\mathbf{K}}^{\pm}(\mathbf{r}_e)\rangle = \varepsilon | \psi_{\mathbf{K}}^{\pm}(\mathbf{r}_e)\rangle, \tag{44}$$

$$H_{BA^+} | \varphi_{i,f}(\mathbf{r}_\varkappa, \mathbf{R})\rangle = U_{i,f}(R) | \varphi_{i,f}(\mathbf{r}_\varkappa, \mathbf{R})\rangle, \tag{45}$$

$$V = \sum_{\varkappa=1}^{N} |\mathbf{r}_e - \mathbf{r}_\varkappa|^{-1} - Z_B |\mathbf{r}_e - \mathbf{R}_B|^{-1} - Z_{A^+} |\mathbf{r}_e - \mathbf{R}_{A^+}|^{-1} + r_e^{-1}. \tag{46}$$

Here $T_R = -\Delta_R/2\mu$ is the kinetic energy operator of the relative movement of nuclei B and A^+ ($\mathbf{R} = \mathbf{R_B} - \mathbf{R_A}$ is the radius vector connecting particles B and A^+); $H_e = -\Delta_{r_e}/2 - 1/r_e$ is the Hamiltonian of the outer electron e^- in the Coulomb field of the ion core A^+, and

$\psi_{nlm}(r_e)$ and $\psi^{\pm}_K(r_e)$ are its wave eigenfunctions with certain values of the principal n, orbital l and magnetic m quantum numbers and with energy $\varepsilon_{nl} = -1/2n^2_{eff}$ in the discrete spectrum ($n_{eff} = n - \delta_l$, δ_l is the quantum defect) and the wave vector $K = (K, \theta_k, \varphi_k)$ and energy $\varepsilon = K^2/2$in the continuous spectrum; H_{BA^+} is the Hamiltonian of the ion BA^+ electron shell ($r_{\varkappa}$ are coordinates of the inner electrons) and $\varphi_{i,f}(\mathbf{r}_{\varkappa}, \mathbf{R})$ are its eigen functions corresponding to the adiabatic terms $U_i(R)$ and $U_f(R)$ between which considered transitions $|i\rangle \to |f\rangle$ occure (for transitions within a term $U_f(R) = U_i(R)$ we have $\varphi_f = \varphi_i$); V is the operator of Coulomb interaction of the outer electron (r_e) with all inner electrons $r(\varkappa)$ and nuclei (Z_B and Z_A) of the

$BA^+ + e^-$ system (except the term $(-r_e^{-1})$ included in the

Hamiltonian $H_e + H_{BA^+}$ of the zero approximation).

Consider first process (1a) of the direct Rydberg atom A(nl) ionization. Let us normilize the initial wave function $(\mu/q)^{1/2}\Psi^+_{iq}(\mathbf{R})$ of the relative motion of perturbing atom and atomic residue A^+ with a momentum $\mathbf{q} = (q, \theta_q, \varphi_q)$ and kinetic energy $E = q^2/2\mu$ (at $R \to \infty$) to the unit flow of impacting particles and the final nuclear wave function $(2\pi)^{-3/2}\Psi^-_{fq'}(\mathbf{R})$ and the wave function $(2\pi)^{-3/2}\psi^-_{\mathbf{K}}(r_e)$ of the emitted electron e^- to the corresponding δ-functions of momenta. In this case in the framework of the stationary perturberation theory (see [32]) for the Coulomb interaction of the outer electron e^- with inner electrons and nuclei of BA^+ ion we can write for the differential cross section of the process (1a)

$$d\sigma^{i,f}_{nl,\mathbf{K}}(\mathbf{q},\mathbf{q}') = \frac{2\pi g_e g_f}{g_{nl} g_i} \sum_m \left| \left\langle \left(\frac{\mu}{q}\right)^{1/2} \Psi^+_{i\mathbf{q}} \right| \langle \psi_{nlm} | \langle \varphi_i | V | \varphi_f \rangle | \psi^-_{\mathbf{K}} \rangle | \Psi^-_{f\mathbf{q}'} \rangle \right|^2 \times$$

$$\times \delta(E + \varepsilon_{nl} - \varepsilon - E') \frac{d\mathbf{K}\, d\mathbf{q}'}{(2\pi)^6}. \qquad (47)$$

Here $g_{nl} = 2(2l+1)$; $g_e = 2$ and g_i, g_f are the statistical weights of the outer electron e^- and quasi-molecular ion BA^+ in the initial $|i\rangle$ and final $|f\rangle$ states, respectively.

Let us expand wave functions $\Psi^+_{i\mathbf{q}}(\mathbf{R})$ and $\Psi^-_{f\mathbf{q}'}(\mathbf{R})$ of the nuclear relative motion, respectively, in $U_i(R)$ and $U_f(R)$ terms of BA^+

ion and also expand the Coulomb wave function $\psi_{\mathbf{K}}^{-}(\mathbf{r}_e)$ of the outer electron e^- over spherical harmonics (see [32, 33])

$$|\Psi^{\pm}_{i(f),\mathbf{q}}(\mathbf{R})\rangle = \frac{(2\pi)^{3/2}}{q}\sum_{JJ_z}\varkappa^{\pm}_{JJ_z}(\theta_{\mathbf{q}},\varphi_{\mathbf{q}})\,|\chi^{i(f)}_{qJ}(R)\rangle\,|Y_{JJ_z}(\theta_{\mathbf{R}},\varphi_{\mathbf{R}})\rangle, \qquad (48a)$$

$$|\psi_{\mathbf{K}}^{-}(\mathbf{r}_e)\rangle = \frac{(2\pi)^{3/2}}{K}\sum_{l'm'}\beta^{-}_{l'm'}(\theta_{\mathbf{K}},\varphi_{\mathbf{K}})\,|R_{Kl'}(r_e)\rangle\,|Y_{l'm'}(\theta_{\mathbf{r}_e},\varphi_{\mathbf{r}_e})\rangle. \qquad (48b)$$

Here $\chi^{i}_{qJ}(R)$ and $\chi^{f}_{qJ}(R)$ are nuclear radial wave functions with certain values of momentum q (normilized to $\delta(q - q')$) and of the orbital momentum J of the relative motion (J_z is its projection on the stationary axis R); $R_{Kl}(r_e)$ is the radial wave function of electron e^- in the Coulomb field (normalized to $\delta(K - K')$ and @ and @ are the expansion coefficients satisfying the orthogonality relation

$$\left\langle \varkappa^{\pm}_{JJ_z}\middle|\varkappa^{\pm}_{J'J'_z}\right\rangle = \int\left[\varkappa^{\pm}_{JJ_z}(\theta_{\mathbf{q}},\varphi_{\mathbf{q}})\right]^{*}\varkappa^{\pm}_{JJ_z}(\theta_{\mathbf{q}},\varphi_{\mathbf{q}})\,dO_{\mathbf{q}} = \delta_{JJ'}\delta_{J_zJ'_z}, \qquad (49a)$$

$$\langle\beta^{-}_{lm}|\beta^{-}_{l'm'}\rangle = \int\left[\beta^{-}_{lm}(\theta_{\mathbf{K}},\varphi_{\mathbf{K}})\right]^{*}\beta^{-}_{l'm'}(\theta_{\mathbf{K}},\varphi_{\mathbf{K}})\,dO_{\mathbf{K}} = \delta_{ll'}\delta_{mm'}. \qquad (49b)$$

Then, integrating (47) over nuclear momenta @ and solid angles dQ_k of the emitted electron and also averaging it over all initial directions dQ_q of colliding particles, considering the relation (49) we have

$$\frac{d\sigma^{i,f}_{nl,K}}{dK} = \frac{4\pi^3\mu^2 g_f}{q^3 q' g_i(2l+1)}\times$$
$$\times\sum_{JJ_z,J'J'_z}\sum_{m,l'm'}|\langle\chi^{i}_{qJ}|\langle JJ_z|\langle nlm|\langle\varphi_i|V|\varphi_f\rangle|Kl'm'\rangle|J'J'_z\rangle|\chi^{f}_{q'J'}\rangle|^2, \qquad (50)$$

where @. $q' = [2\mu(E + \varepsilon_{nl} - \varepsilon)]^{1/2}$.

Notice here that at thermal velocities the Cariolis interaction (see [32, 34]) of momentum J of the nuclear relative motion with momentum j of the quasi-molecular ion BA^+ electron shell and momentum l of the outer electron e^- is negligible compared to the Coulomb interaction (46). For this reason one can neglect when deducing (50) the weak effect of the internuclear axis R rotation about the static axis z. In this approximation the nuclear orbital momentum and its projection may be considered to be conserved $J' = J$ and $J'_z = J_z$

which allows to perform readilythe summation over J_z, J_z' and J' in the formula (50). As a result, by going in (50) to radial wave functions $\chi^{i}_{EJ}(R) = (\mu/q)^{1/2}\chi^{i}_{qJ}(R)$ and $\chi^{f}_{E'J'}(R) = (\mu/q')^{1/2}\chi^{f}_{q'J'}(R)$ of nuclei and of the outer electron $R_{\varepsilon l'}(r_e) = K^{-1/2}R_{Kl'}(r_e) = c$ and normalizing energies E - E′ or ε - ε′ to δ-functions, we obtain for the differential cross section $d\sigma^{i,f}_{nl,\varepsilon}(E)$ of direct ionization of the Rydberg atom A(nl) by atom B in the range of energies ε« ε + dε of the continuos spectrum the expression

$$\frac{d\sigma^{i,f}_{nl,\varepsilon}}{d\varepsilon} = \frac{4\pi^3 g_f}{g_i(2l+1)q^2}\sum_J(2J+1)\sum_{m,m'l'}\left|\langle\chi^{i}_{EJ}|\langle nlm|\langle\varphi_i|V|\varphi_f\rangle|\varepsilon l'm'\rangle|\chi^{f}_{E'J}\rangle\right|^2. \tag{51}$$

Consider now process (1b) of the associative ionization of A(nl) Rydberg atom at the collision with the perturbing atom B. In this case the differential ionization cross section with the formation of molecular ion $BA^+(vJ)$ on the vibration-rotation level of energy E_{vJ} is defined in the framework of the perturbation theory by the expression

$$d\sigma^{i,f}_{nl,\mathbf{K}}(\mathbf{q},vJ) = \frac{2\pi g_e g_f}{g_{nl}g_i}\sum_m\sum_{J_z}\left|\left\langle\left(\frac{\mu}{q}\right)^{1/2}\Psi^{+}_{i\mathbf{q}}\right|\langle nlm|\langle\varphi_i|V|\varphi_f\rangle|\psi^{-}_{\mathbf{K}}\rangle|\Psi^{f}_{vJJ_z}\rangle\right|^2\times$$
$$\times\,\delta(E+\varepsilon_{nl}-\varepsilon-E_{vJ})\frac{d\mathbf{K}}{(2\pi)^3}. \tag{52}$$

Here $\Psi^{f}_{vJJ_z}(\mathbf{R}) = \chi^{f}_{vJ}(R)\cdot Y_{JJ_z}(\theta_\mathbf{R},\varphi_\mathbf{R})$ is the wave function of the vibration-rotation motion of nuclei at the final term $U_f(R)$ of $BA^+(vJ)$ ion. By averaging (52) over directions dO_q of momentum-of-nuclei in the initial state, integrating it for all possible values of the emitted electron momentum $d\mathbf{K} = K^2dKdO_\mathbf{K}$ and using (48) and (49), we get for the crosssection of the associative ionization (1b) (similar to the deduction of (51)):

$$\sigma^{i,f}_{nl}(E,vJ) = \frac{4\pi^3 g_f(2J+1)}{g_i(2l+1)q^2}\sum_{m,m'l'}\left|\langle\chi^{i}_{EJ}|\langle nlm|\langle\varphi_i|V|\varphi_f\rangle|\varepsilon l'm'\rangle|\chi^{f}_{vJ}\rangle\right|^2, \tag{53}$$

where $\varepsilon = E - E_{vJ} + \varepsilon_{nl}$.

It is seen from (51) and (53) that to determine the cross sections of the ionization processes (1a) and (1b) one has to calculate the radial integrals $\langle\chi^{i}(R)|V^{f,\varepsilon l'm'}_{i,nlm}(R)|\chi^{f}(R)\rangle$ of the electron matrix element of Coulomb interaction (46)

$$V_{i,nlm}^{f,\varepsilon l'm'}(R) = \langle \psi_{nlm}(\mathbf{r}_e) | \langle \varphi_i(\mathbf{r}_\varkappa, \mathbf{R}) | V | \varphi_f(\mathbf{r}_\varkappa, \mathbf{R}) \rangle | \psi_{\varepsilon l'm'}(\mathbf{r}_e) \rangle \qquad (54)$$

with respect to wave functions $\chi_{EJ}^{i}(R)$ and $\chi_{E'J'}^{f}(R)$ (or $\chi_{EJ}^{i}(R)$ and $\chi_{vJ}^{f}(R)$) of the relative motion of nuclei in the continuous or discrete spectrum. To solve this problem we shall use the quasi-classical aproximation

$$\chi(R) = CR^{-1}v^{-1/2}(R)\cos\left[\int_b^R q(R)\,dR - \pi/4\right],$$
$$q(R) = \mu v(R) = \{2\mu[E - U(R) - (J + 1/2)^2/2\mu R^2]\}^{1/2}. \qquad (55)$$

Here $q_i(R)$ and $q_f(R)$ are momenta, $v_i(R)$ and $v_f(R)$ are relative velocities of nuclei; b_i and b_f are the left classical turning points at the movement A^+ and B particles in the initial $U_i(R)$ and final $U_f(R)$ terms ofquasi-molecular (or molecular) BA^+ ion, respectively; C is the normalization constant equal to $C_E = (2/\pi)^{1/2}$ in the continuous spectrum ($E > 0$) and to $C_{vJ} = 2/T_{vJ}^{1/2}$ in the discrete spectrum ($E = E_{vJ} < 0$), where T_{vJ}^{f} is the period of the vibration-rotation nuclear motion on E_{vJ} energy level in $U_f(R)$ term.

Integrating the result (51) over all possible values (@) $(0 \leqslant \varepsilon \leqslant \varepsilon_{max}^{d.i})$ of the energy $d\varepsilon$ of the emitted electron e^-, substituting integration over dJ (or over the aiming parameters $\rho = J/q$ of colliding particles) for summation over J and using (54), we get the following quasi classical expression for the direct ionization cross section $\sigma_{nl}^{d.i}(E) = \int_0^{\varepsilon_{max}} \left(d\sigma_{nl,\varepsilon}^{i,f}(E)/d\varepsilon\right) d\varepsilon$

$$\sigma_{nl}^{d.i}(E) = \frac{4\pi^3 g_f}{g_i(2l+1)q^2} \int_0^{\varepsilon_{max}} d\varepsilon \int 2J\,dJ \sum_{m,m'l'} |\langle \chi_{EJ}^{i}(R) | V_{i,nlm}^{f,\varepsilon l'm'}(R) | \chi_{E'J}^{f}(R) \rangle|^2, \qquad (56)$$

where @. $E - E' = |\varepsilon_{nl}| + \varepsilon$.

Similarly, for the total cross section @ of the Rydberg atom associative ionization we obtain:

$$\sigma_{nl}^{a.i}(E) = \frac{4\pi^3 g_f}{g_i(2l+1)q^2} \int dv \int 2J\,dJ \sum_{m,m'l'} |\langle \chi_{EJ}^{i}(R) | V_{i,nlm}^{f,\varepsilon l'm'}(R) | \chi_{vJ}^{f}(R) \rangle|^2, \qquad (57)$$

where the integration over dv and dJ is substituted for the summation over vibrational v and rotational J quantum numbers. It corresponds to the approximation of the quasi-continuous spectrum of energy levels E_{vJ} of the molecular ion BA^+.

3.2. Ionization with changing of the electronic state of quasi-molecular ion

In this case ionization processes (1a) and (1b) of Rydberg atom A(nl) at the collision with atom B occure at the cross-point of different electronic terms $U_{i,nl}(R) = U_i(R) + \varepsilon_{nl}$ ($\varepsilon_{nl} = -1/2n_{eff}^2 < 0$) и $U_{f,\varepsilon}(R) = U_f(R) + \varepsilon$ ($\varepsilon = K^2/2 > 0$) of the quasi-molecule $BA^+ + e^-$. Hence, we obtain within the framework of the known quasi-classical Landau theory (see [32, par. 90]) the following expression for nuclear transition matrix elements:

$$|\langle\chi_{EJ}^{i}|V_{i,nlm}^{f,\varepsilon l'm'}|\chi_{E'(v),J}^{f}\rangle|^2 =$$

$$= \frac{C_E^2 C_{E'(v)}^2 \pi |V_{i,nlm}^{f,\varepsilon l'm'}(R_\omega)|^2 \cos^2 S_0(J)}{2\mu q \Delta F_{if}(R_\omega)\left[1 - U_i(R_\omega)/E - (J+1/2)^2/q^2R_\omega^2\right]^{1/2}},$$

$$S_0(J) = \int_{b_i}^{R_\omega} q_i(R)\,dR - \int_{b_f}^{R_\omega} q_f(R)\,dR + \pi/4. \quad (58)$$

Here $\Delta F_{if}(R_\omega) = |dU_f/dR - dU_i/dR|_{R=R_\omega}$ is the difference of slopes of $BA^+ + e^-$ quasi-molecule terms in their cross-point R_ω,[5] which is determined from the condition $U_{i,nl}(R) = U_{f,\varepsilon}(R)$ when the frequency ω of the bound-free transition nl → ε of the outers electron e^- becomes equal to the splitting of terms of BA^+ ion, i.e.

$$\omega = |\varepsilon_{nl}| + \varepsilon = U_i(R_\omega) - U_f(R_\omega) \equiv \Delta U_{if}(R_\omega), \quad (59)$$

(Footnote 5: The form of electronic terms $U_{i,nl}(R)$ and $U_{f,\varepsilon}(R)$ of quasi- molecule $BA^+ + e^-$ (which are the eigenvalues of Hamiltonian $H_{BA^+} + H_e$, see (43)) coincides in the zero approximation in interaction with that of corresponding terms $U_i(R)$ and $U_f(R)$ of the isolated BA^+ ion. Corrections to the Rydberg terms caused by the long-range non-Coulomb part of the interaction potential V of the outer electron e^- with BA^+ ion were calculated in [35]).

where $\omega = E - E'$ and $\omega = E + |E_{vJ}|$ for processes of the direct (1a) and associative (1b) ionization, respectively.

Let us insert (58) to (56) and perform the integration over the orbital momentum dJ in the limits $0 \leq J \leq J_{max}(R_\omega)$ substituting the fast-oscillating quantity $\cos^2 S_0(J)$ by its average value equal to 1/2. As a result, we get for the direct ionization cross section of the Rydberg A(nl) atom

$$\sigma_{nl}^{d.i}(E) = \frac{8\pi^2 g_f}{g_i(2l+1)\mu v^2} \int_0^{\varepsilon_{max}} \sum_{m,m'l'} |V_{i,nlm}^{f,\varepsilon l'm'}(R_\omega)|^2 \frac{J_{max}(R_\omega) R_\omega}{\Delta F_{if}(R_\omega)} d\varepsilon, \quad (60)$$

where $J_{max}(R_\omega) = \mu v R_\omega [1 - U_i(R_\omega)/E]^{1/2}$; a $v = q/\mu = (2E/\mu)^{1/2}$ is the relative velocity $v_i(R)$ of A(nl) and B atoms at $R \to \infty$. However, the final result for $\sigma_{nl}^{d.i.}(E)$ is more conveniently presented as the integral over transition frequencies $d\omega = d|\Delta U_{if}(R_\omega)| = \Delta F_{if}(R_\omega) dR_\omega$ (see (59)) or over internuclear distances

$$\sigma_{nl}^{d.i}(E) = \frac{8\pi^2 g_f}{g_i(2l+1)v} \int_{R_{min}^{d.i}}^{R_{max}^{d.i}} \sum_{m,m'l'} |V_{i,nlm}^{f,\varepsilon l'm'}(R_\omega)|^2 \left[1 - \frac{U_i(R_\omega)}{E}\right]^{1/2} R_\omega^2 \, dR_\omega. \quad (61)$$

Here $E \gg |\varepsilon_{nl}|$, while upper $R_{max}^{d.i} = R_{nl}$ and lower $R_{min}^{d.i} = \max(R_E, b_E^i)$ on limits of integration correspond to minimum $\omega_{min}^{d.i}$ and maximum $\omega_{max}^{d.i}$ values of the BA^+ ion term spllitings in the process (1a)[6] BA^+:

$$\omega_{min}^{d.i} = \Delta U_{if}(R_{max}^{d.i}) = \Delta U_{if}(R_{nl}) = |\varepsilon_{nl}|, \quad (62)$$

$$\omega_{max}^{d.i} = |\varepsilon_{nl}| + \varepsilon_{max}^{d.i} = \Delta U_{if}(R_{min}^{d.i}) = \begin{cases} \Delta U_{if}(b_E^i), & R < b_E^i, \quad (63a) \\ \Delta U_{if}(R_E) = E, & R_E > b_E^i. \quad (63b) \end{cases}$$

Relations (62) and (63) are obtained on the basis of thelaw of conservation of energy and take into account the fact that

---

(Footnote 6: For heteronuclear YX^+ and homonuclear X_2^+ ions of inert gases considered in the paper the splitting value of lower electronic terms $\Delta U_{if}(R)$ decreases monotonically with increasing of the internuclear distance R).

colliding particles A^+ and B do not penetrate into the classically forbiden region $R < b_E^i$ behind the left turning point in the initial therm $\left(U_i\left(b_E^i\right) = E\right)$.

In a similar manner, using (57) and (58) we obtain the following expression for the cross section @ of the Rydberg atom associative ionization process (1b):

$$\sigma_{nl}^{a,i}(E) = \frac{8\pi^3\mu g_f}{g_i(2l+1)q^3}\int\int\sum_{m,m'l'}\left|V_{i,nlm}^{f,\varepsilon l'm'}(R_\omega)\right|^2\times$$
$$\times\left[1 - \frac{U_i(R_\omega)}{E} - \frac{J^2}{q^2R_\omega^2}\right]^{-1/2}\frac{2J\,dJ\,dv}{T_{vJ}^f\Delta F_{if}(R_\omega)}, \qquad (64)$$

where $\omega = E + |E_{vJ}|$ and R_ω is determined by theexpression (59). Furthemore, having substituted, considering the Bohr-Zommerfeld relation, the integration over vibrational quantum numbers $dv = \left(T_{vJ}^f/2\pi\right)d\left(|E_{vJ}|\right)$ by the integration over the quasi-continuous spectrum of the energy levels $d(|E_{vJ}|)$ of molecular ion BA^+ in the final state |f> and then having integrated over dJ similar to the deduction of (61), we obtain the result:

$$\sigma_{nl}^{a.i}(E) = \frac{8\pi^2 g_f}{g_i(2l+1)v}\int_0^{|E'|_{max}} d|E'|\frac{R_\omega^2}{\Delta F_{if}(R_\omega)}\times$$
$$\times\left[1 - \frac{U_i(R_\omega)}{E}\right]^{1/2}\sum_{m,m'l'}\left|V_{i,nlm}^{f,\varepsilon l'm'}(R_\omega)\right|^2. \qquad (65)$$

The energy $E' = E_{vJ} < 0$ in the discrete spectrum is measuredin (65) from the dissociation boundary $(U_i(\infty) = U_f(\infty) = 0)$ of BA^+ ion for considered therms $U_i(R)$ and $U_f(R)$, while the kinetic energy $\varepsilon = K^2/2$ of the emitted electron is determined from the law of conservation of energy $\varepsilon = E + |E'| - |\varepsilon_{nl}|$ for the process (1b). Accordingly at fixed values of kinetic energy $E = \mu v^2/2$ of colliding atoms A(nl) and B and of Rydberg electron bond energy $|\varepsilon_{nl}| = 1/2(n-\delta_l)^2$ we have

$$\omega_{max}^{a.i} = |\varepsilon_{nl}| + \varepsilon_{max}^{a.i} = \Delta U_{if}\left(R_{min}^{a.i}\right) = \Delta U_{if}\left(b_E^i\right), \qquad (66)$$

$$\omega_{min}^{a.i} = |\varepsilon_{nl}| + \varepsilon_{min}^{a.i} = \Delta U_{if}\left(R_{max}^{a.i}\right) = \begin{cases}\Delta U_{if}(R_{nl}) = |\varepsilon_{nl}|,\ E \leqslant |\varepsilon_{nl}|, & (67a)\\ \Delta U_{if}(R_E) = E,\ E > |\varepsilon_{nl}|. & (67b)\end{cases}$$

Hence, going to the integration over internuclear distances dR_ω in (61), we get for the associative ionization cross section the result which differs from (61) only by concrete values of the upper $R_{max}^{a.i.}$ and lower $R_{min}^{a.i}$ integration limits.

Table 1 presents values R_{min} and R_{max} for processes (1a) and (1b) at different relationships between bonding energy value $|\varepsilon_{nl}| = 1/2n_{eff}^2$ of the Rydberg electron e^-, kinetic energy value $E = \mu v^2/2$ of atoms A(nl) and B and characteristic values $\Delta U_{if}(R_0^i)$ and $\Delta U_{if}(R_0^f)$ of the splitting energy of electronic terms $U_i(R)$ and $U_f(R)$ converging to the same dissociative limit $U_i(\infty) = U_f(\infty)$ (points R_0^i and R_0^f are determined by conditions $U_i(R_0^i) = 0$ and $U_f(R_0^f) = 0$, see Figs.3 and 7 below). It follows from this table and (61) that the total rate constant $K_{nl}(T) = \langle v\sigma_{nl}\rangle_T$ of the process (1) (where $\sigma_{nl} = \sigma_{nl}^{d.i} + \sigma_{nl}^{a.i}$) has for the Maxwell distribution over velocities the form:

$$K_{nl}(T) = \frac{16\pi^{3/2} g_f}{g_i(2l+1)} \int_{E_{min}}^{\infty} d\left(\frac{E}{T}\right) \exp\left(-\frac{E}{T}\right) \int_{b_E^i}^{R_{nl}} R^2\,dR \times$$
$$\times \left[\frac{E}{T} - \frac{U_i(R_{nl})}{T}\right]^{1/2} \sum_{m,m'l'} |V_{i,nlm}^{f,el'm'}(R)|^2, \qquad (68)$$

where $E_{min} = \max(0, U_i(R_{nl}))$. Changing the order of integration in (68) and evaluating the integral over dE, we obtain the result:

$$K_{nl}(T) = \frac{8\pi^2 g_f}{g_i(2l+1)} \int_0^{R_{nl}} \sum_{m,m'l'} |V_{i,nlm}^{f,el'm'}(R)|^2 \exp\left[-\frac{U_i(R)}{T}\right] \Phi(R) R^2\,dR. \qquad (69a)$$

Here a point R_{nl} is determined from the condition $\Delta U_{if}(R_{nl}) = 1/2n_{eff}^2$ (see (62) or (66a)) and the form of function $\Phi(R)$ turns out to be different in the repulsive region $R < R_0^i$ of the initial term $U_i(R)$ and at $R \geq R_0^i$, when a point R is located in the attraction region $U_i(R) \leq 0$ of this term, i.e.

$$\Phi(R) = \begin{cases} 1, & R < R_0^i, \\ \Gamma(3/2, |U_i(R)|/T)/\Gamma(3/2), & R \geqslant R_0^i, \end{cases} \qquad (69b)$$

where $\Gamma(3/2, Z) = \int_Z^{\infty} t^{1/2} e^{-t}\,dt$ is the incomplete gamma-function (see [36]) and $\Gamma(3/2) = \pi^{1/2}/2$.

Table 1. R_{min} and R_{max} values for processes of direct and associative ionization.

bond energy, $\lvert\varepsilon_{nl}\rvert = 1/2n_{eff}^2$	kinetic energy, $E = \mu v^2/2$	direct ionization $R_{min}^{d.i}$	$R_{max}^{d.i}$	associative ionization $R_{min}^{a.i}$	$R_{max}^{a.i}$
$\lvert\varepsilon_{nl}\rvert < \Delta U_{if}(R_0^i)$, $R_{nl} > R_0^i$, $U_i(R_{nl}) < 0$	$0 \leqslant E \leqslant \lvert\varepsilon_{nl}\rvert$	—	—	b_E^i	R_{nl}
		$R_0^f < b_E^i < R_0^i < R_{nl} < R_E$			
	$\lvert\varepsilon_{nl}\rvert \leqslant E \leqslant \Delta U_{if}(R_0^f)$	R_E	R_{nl}	b_E^i	R_E
		$R_0^f < b_E^i < R_E < R_{nl}$, $R_0^i < R_{nl}$			
	$\Delta U_{if}(R_0^f) \leqslant E < \infty$	b_E^i	R_{nl}	—	—
		$0 < b_E^i < R_0^f < R_0^i$			
$\Delta U_{if}(R_0^i) < \lvert\varepsilon_{nl}\rvert <$ $< \Delta U_{if}(R_0^f)$, $R_0^f < R_{nl} < R_0^i$, $0 < V_i(R_{nl}) < \lvert\varepsilon_{nl}\rvert$	$0 \leqslant E \leqslant U_i(R_{nl})$	—	—	—	—
		$R_{nl} < R_E$, $R_0^f < R_{nl} < R_0^i$			
	$U_i(R_{nl}) \leqslant E \leqslant \lvert\varepsilon_{nl}\rvert$	—	—	b_E^i	R_{nl}
		$R_0^f < b_E^i < R_{nl} < R_0^i$, $R_{nl} < R_E$			
	$\lvert\varepsilon_{nl}\rvert \leqslant E < \Delta U_{if}(R_0^f)$	R_E	R_{nl}	b_E^i	R_E
		$R_0^f < b_E^i < R_E < R_{nl} < R_0^i$			
	$\Delta U_{if}(R_0^f) \leqslant E$	b_E^i	R_{nl}	—	—
		$0 < b_E^i < R_0^f < E_{nl} < R_0^i$			
$\Delta U_{if}(R_0^f) < \lvert\varepsilon_{nl}\rvert <$ $< U_i(R_{nl})$, $R_{nl} < R_0^f$	$0 \leqslant E \leqslant U_i(R_{nl})$	—	—	—	—
		$R_{nl} < b_E^i$			
	$0 \leqslant U_i(R_{nl}) \leqslant E$	b_E^i	R_{nl}	—	—
		$0 < b_E^i < R_{nl} < R_0^f$			

3.3. Ionization in the one electron term of quasi-molecular ion

When ionization proceeds without changing of the electronic state $\varphi_f = \varphi_i$ of quasi-molecular ion BA^+, the main contribution into cross sections of processes (1a) and (1b) gives the interaction

$$\langle\varphi_i \mid V \mid \varphi_i\rangle = -\frac{\mathbf{D}^i(\mathbf{R})\,\mathbf{r}_e}{r_e^3} = -\frac{\mu}{M_{A^+}}\frac{\mathbf{R}\mathbf{r}_e}{r_e^3} - \frac{\mathbf{D}_2^i(\mathbf{R})\,\mathbf{r}_e}{r_e^3} \tag{70}$$

of the outer electron e^- with the total dipole moment $\mathbf{D}^i(\mathbf{R}) = \mathbf{D}_1^i(\mathbf{R}) + \mathbf{D}_2^i(\mathbf{R})$ of this ion (relative its center of mass) in the considered electronic state φ_i. The first term in (70) determinedby the contribution $\mathbf{D}_1^i(\mathbf{R}) = \mathbf{R}_{A^+} = \mu\mathbf{R}/M_{A^+}$ into the dipole moment $\mathbf{D}^i(\mathbf{R})$ of the positive Coulomb center A^+ may be presented in the form $V_1^i(\mathbf{r}_e, \mathbf{R}) = -(\nabla_{\mathbf{R}} U_i(R) \times \mathbf{r}_e)/M_{A^+}$. Thus, it corresponds to the inertia force $\mathbf{F}_e = m_e \ddot{\mathbf{R}}_{A^+}$, which acts on the Rydberg electron (mass m_e = 1 a.u.) due to the acceleration of the ion core A^+ at its collision with atom B. The second term in (70) corresponding to the dipole moment $\mathbf{D}_2^i(\mathbf{R})$, contribution results from the shift (polarization) of inner electrons of quasi-molecula $BA^+ + e^-$ relative nuclei as particles A^+ and B approach each other (for more details, see [2, 10, 11]).

The nuclear transition matrix elements in (70) using the Erenfest theorem $d^2\mathbf{r}_e/dt^2 = -\nabla_{\mathbf{r}_e}(-1/r_e) = \mathbf{r}_e/r_e^3$ for the operator of the outerelectron e^- radius vector in the Coulomb field of BA^+ ion and the dipole selection rules may be presented in the following form:

$$\frac{1}{2l+1}\sum_{m,m'l'} |\langle \chi_{EJ}^i | V_{1,nlm}^{i,\varepsilon l'm'} | \chi_{E'(v),J}^i \rangle|^2 = \left[\frac{l+1}{2l+1}|r_{nl}^{\varepsilon,l+1}|^2 + \frac{l}{2l+1}|r_{nl}^{\varepsilon,l-1}|^2\right]\times$$

$$\times \frac{\omega^4}{3}|D_{E,E'(v)}^i(J)|^2 = \frac{\omega^3 c}{4\pi^2}\sigma_{nl}^{ph}(\omega)|D_{E,E'(v)}^i(J)|^2. \qquad (71)$$

Here c = 137 a.u. is the velocity of light; $\sigma_{nl}^{ph}(\omega)$ is the photoionization cross section of highly excited atom A(nl) (see, for example, [33]) by the quantum of frequency $\omega = |\varepsilon_{nl}| + \varepsilon$; $r_{nl}^{\varepsilon l'} = \langle R_{nl}(r_e)|r_e|R_{\varepsilon l'}(r_e)\rangle$ is the radial matrix element of r_e coordinate over Coulomb wave functions of the outer electron e^-; and $D_{E,E'(v)}^i(J) = \langle \chi_{EJ}^i(R)|D^i(R)|\chi_{E'(v),J}^i(R)\rangle$ is the radial matrix element of the dipole moment of BA^+ ion for free-free ($E \to E'$) or free-bound ($E \to v$) transitions of A^+ and B nuclei in the one electron therm $U_i(R)$.

We shall use the results of quasi-classical calculations [10] for the quantity $D_{E,E'(v)}^i(J)$ for transitions between the states with energies close to the dissociation boundary E, E', $|E_{vJ}| \ll E_0$ in the term with a sufficiently deep well $E_0 = |U_i(R_e)|$. We shall also take into account that in this case the main contribution to the cross sections of processes (1a) and (1b) is made by the orbital moments $0 \le J \le J_{cap}$ for which the capture of atom B by the atomic residue A^+ of the Rydberg atom A(nl) into the region of small internuclear distances $R \le \rho_{cap} = J_{cap}/q$ is possible. Then, substituting in (56) and (57) the integration over the emitted electron energy $d\varepsilon$ or over ion BA^+ vibrational levels $d\upsilon$ by the integration over all possible transition frequencies $d\omega$ and using (41) and the law of conservation of energy we obtain for the total ionization cross section $\sigma_{nl} = \sigma_{nl}^{d.i.} + \sigma_{nl}^{a.i.}$ of atom A(nl) the expression

$$\sigma_{nl}(E) = c\sigma_{cap}^{A+B}(E) \int_{|\varepsilon_{nl}|}^{E_0+E} \sigma_{nl}^{ph}(\omega)\, d^2(\omega) \exp(-2\omega\tau_0)\, \omega d\omega. \tag{72}$$

Here τ_0 is the collision time of B and A^+ particles in the left turning point $R = b_E^i \approx R_0^i$ on the repulsive branch of term $U_i(R) \ge 0$ which coincides with the reciprocal value of the lower vibrational quantum $\tau_0 = \omega_e^{-1}$ of BA^+ ion; $d(\omega)$ is the function of frequency (for more details, see [10]) which corresponds to the increment of the dipole moment $(dD^i/dR)_{R_\omega} \Delta R_\omega$ on the characteristic size $\Delta R_\omega = U_i'(R_\omega)/U''_i(R_\omega)$ of the term change at the point R_ω, which vicinity makes the main contribution to the transition with a given frequency ω; $\sigma_{cap}^{A+B} = \pi\rho^2_{cap}$ is the cross section of B atom capture by A^+ ion which at the degree approximation of the right attraction branch of the term $U_i(R) = -c_\nu/R^\nu$ $(R \gg R_e)$ is determined by the expression

$$\sigma_{cap}^{A+B}(E) = \pi \left(\frac{\nu}{\nu-2}\right)^{(\nu-2)/\nu} \left(\frac{\nu c_\nu}{2E}\right)^{2/\nu}. \tag{73}$$

Let us use the results of quasi-classical calculations [37] for values $r_{nl}^{\varepsilon, l\pm1}$ at the ionization from the selectively excited Rydberg nl level at $n \gg 1$, $K \ll 1$, $l \le n$, $1/K$

$$r_{nl}^{\varepsilon, l\pm1} = -\frac{l^2}{3^{1/2}\pi n_{eff}^{3/2}\omega}\left[K_{2/3}\left(\frac{\omega l^3}{3}\right) \mp K_{1/3}\left(\frac{\omega l^3}{3}\right)\right], \tag{74}$$

where $K_\lambda(Z)$ are Mcdonald functions (modified Bessel's functions the third kind, see [36]). One can obtain a simple analytic formula for the atom ionization cross section σ_{nl}^{ph}(71)) in the considered frequency region $\omega l^3/3 \ll 1$ using Eq.(74)

$$\sigma_{nl}^{ph}(\omega) = \frac{2^{4/3}}{3^{2/3}} \Gamma^2(2/3) \frac{1}{cn_{eff}^3 \omega^{7/3}}. \tag{75}$$

As a result, using 72) - (75) we obtain the following formula for the total rate constant @of direct (1a) and associative (1b) ionization processes of A(nl) atom with l ≪ n at the Maxwell velocity distribution

$$K_{nl}(T) = \frac{2^{8/3}\Gamma^2\ (^2/_3)\ \Gamma\,(2-2/\nu)}{3^{2/3}\pi^{1/2}n_{eff}^3}\, v_T\sigma_{cap}^{A+B}(T)\, d^2(|\varepsilon_{nl}|)\,\tau_0^{1/3}\Gamma\,(-1/3,\ 2|\varepsilon_{nl}|\tau_0), \tag{76}$$

where $v_T = (2T/\mu)^{1/2}$, and @ is the incomplete gamma-function (see [36]).

To determine the total cross section @ and the rate constant @ of (1a) and (1b) ionization processes in the case of equal population of nlm-sublevels within a given Rydberg n level of the atom one should use in (72) the known expression for the corresponding photoionization cross section (see, for example, [33])

$$\sigma_n^{ph}(\omega) = \frac{8\pi}{3^{3/2}c}\ \frac{G_n(\omega)}{n^5\omega^3}. \tag{77}$$

Here $G_n(\omega)$ is the Gaunt factor equal to unity in the Kramers approximation. As a result, we arrive at the following expression:

$$K_n(T) = \frac{32\pi^{1/2}\Gamma\,(2-2/\nu)}{3^{3/2}n^3}\, v_T\sigma_{cap}^{A+B}(T)\, d^2(|\varepsilon_n|)\, E_2(2|\varepsilon_n|\tau_0), \tag{78}$$

where @ is the integral exponential function of 2-nd order (see [36]).

At the deduction of formulas (76) and (78), the upper limit of integration over frequencies in (72) was substituted by ∞ which is justified for Rydberg levels $n \gg (2E_0)^{-1/2}$ with the bond energy $|\varepsilon_{nl}|$ (or $|\varepsilon_n|$) small compared to the potential well depth $E_0 = |U_i(R_e)|$. The second restriction on the nvalue arises from the side of high levels $n \ll (\rho_{cap}/v)^{1/2}$.

Fig. 2, a, b shows the results of specific calculations of $K_{nl}(T)$ and $K_n(T)$ total rates constants of direct and associative ionization processes (1a), (1b) performed by formulas (76) and (78) for thermal collisions of hydrogen Rydberg atoms H(nl) (at l << n) or H(n) with He($^1S_0$) atoms.

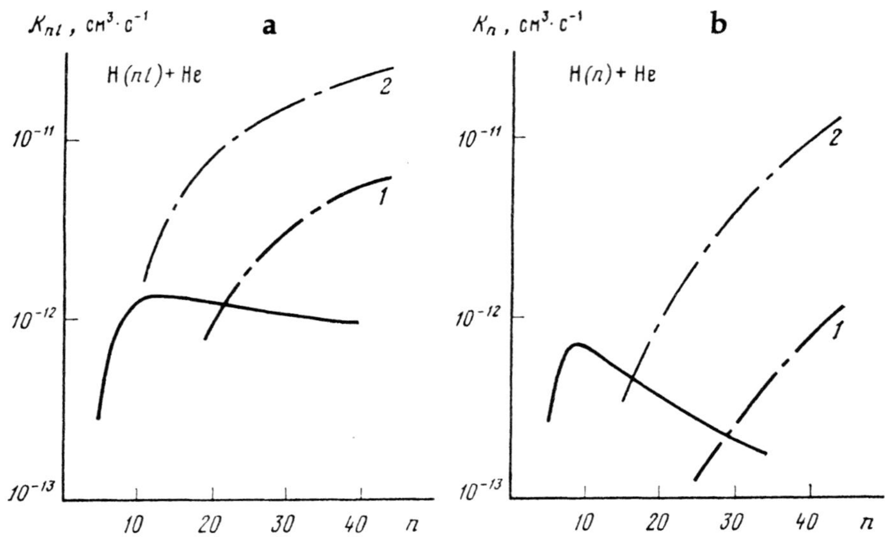

Fig. 2. Ionization rate constants $K_{nl}(T)$ and $K_n(T)$ ($cm^3 sec^{-1}$) of Rydberg hydrogen atoms H(nl) at $l \ll n$ (a) and H(n) (b) at the collisions with He atoms at T = 300 K (curves 1) and T = 1000 K (curves 2).
Solid curves correspond to the mechanism of scattering of the perturbing atom He by the atom residue (proton) H^+ (the calculation by formulas (76) and (78)), dashed-dotted curves correspond to the mechanism of scattering by the quasi-free electron e^- (the calculation by formulas (39) and (41) at L_{He} = 1,19 a.u.)

In the case of high levels $n \gg (\rho_{cap}/\upsilon)^{1/2}$ the direct ionization process (1a) of atom A(nl) occures at the great internuclear distances $R_\omega \geq \rho_{cap}$ on the right attraction branch of $U_i(R)$ term. In this case the associative ionization (1b) is inefficient. The process (1a) cross section for such levels can be calculated due to condition $\omega\rho_{cap}/\upsilon \ll 1$ (where $\omega_{nl,\varepsilon} \sim |\varepsilon_{nl}|$) in the framework of the sudden perturbation theory (see [32, 38]) using the model of the Rydberg electron shaking [2, 10, 11]

$$\sigma_{nl}(E) = \int_0^{E-|\varepsilon_{nl}|} [d\sigma_{nl,\varepsilon}(E)/d\varepsilon]\, d\varepsilon, \quad E \geqslant |\varepsilon_{nl}|,$$

$$\frac{d\sigma_{nl,\varepsilon}}{d\varepsilon} = \frac{1}{2l+1} \sum_{m,m'l'} \int |\langle nlm | \exp(i\mathbf{P}\mathbf{r}_e) | \varepsilon l'm' \rangle|^2 |f_{A^+B}(q, \theta_{qq'})|^2 \, dO_{qq'}. \tag{79}$$

Here $\mathbf{P} = m_e \Delta \mathbf{V}_{A^+}$ is the electron momentum ($m_e = 1$ a.u.) corresponding to the change of the atomic residue $\Delta \mathbf{V}_{A^+} = (\mathbf{q}' - \mathbf{q})/M_{A^+}$ velocityat its collision with the perturbing particle B; f_{A^+B} is the amplitude of the elastic scattering of A^+ and B particles. It follows from (79) that in the limit of very high values of $n \gg M_{A^+}/\mu v$ (when $Pn^2 \gg 1$) the ionization cross section σ_{nl} of atom A(nl) in the mechanism involved asymptotically tends to the value $\sigma_{el}^{A^+B} = \int |f_{A^+B}|^2 dO_{qq'}$ of the total cross section of the elastic scattering of the perturbing atom B by the atomic residue A^+.

4. IONIZATION IN THE BUFFER INERT GAS

Let us calculate ionization rate constants (1) for the case of collisions of X(nl) Rydberg atoms of inert gas with $Y(^1S_0)$ atoms of the lighter inert gas in the ground state. Fig.3 and Table 2 present the typical form and basic parameters of lower electronic terms of heteronuclear ions of inert gases(for which the Hund bond type C is realized, see [32]) obtained in [39, 40]. The ground $X|j = 3/2, \Omega_f = 1/2\rangle$ and the first excited $A_1|j = 3/2, \Omega_i = 3/2\rangle$ electronic terms $U_f(R) = U_{j\Omega_f}(R)$ and $U_i(R) = U_{j\Omega_i}(R)$, between which transitions occure in the processes of direct (1a) and associative (1b) ionization of the Rydberg electron are splitted due to the electrostatic interaction between X^+ and Y particles. They correlate with the ground state $X^+[n_0p^5(^2P_{3/2})] + Y(^1S_0)$ of isolated ion and atom of inert gases and have different values of projections $\Omega_i = 3/2$ and $\Omega_f = 1/2$ of the total electronic momentum $\mathbf{j} = \mathbf{L} + \mathbf{S}$ ($j_i = j_f = j = 3/2$) of quasi-molecular ion YX^+ on the internuclear axis R (n_0 is the main quantum number of the main electronic shell of X^+ ion, i.e. $n_0 = 3$, 4 and 5 for Ar^+, Kr^+ and Xe^+, respectively). The second excited term $A_2|j' = 1/2, \Omega' = 1/2\rangle$, correlating with the state $X^+[n_0p^5(^2P_{1/2})] + Y(^1S_0)$, does not take part in the ionization mechanism at thermal energies of collisions $T \sim 0.3 - 0.08$ eV $\ll \Delta_{jj'}$.

because of the great value of spin-orbital splitting of $n_0p^5(^2P_{3/2})$ and $n_0p^5(^2P_{1/2})$ states of X^+ ion (see Table 1).

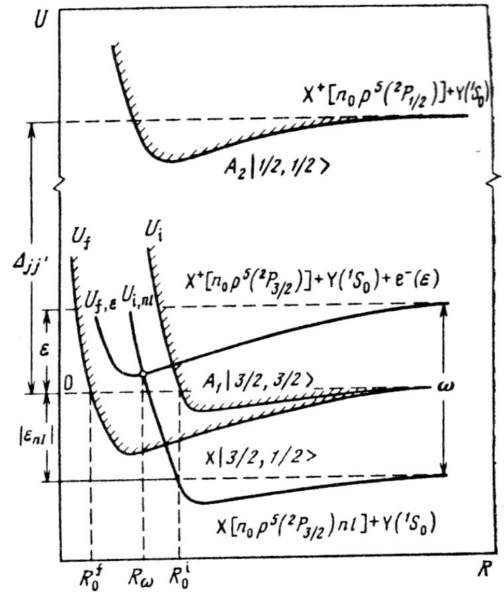

Fig.3. Typical form of lower electronic terms $U_i(R)$ and $U_f(R)$ of the heteronuclear ions YX^+ of inert gases and of corresponding electronic terms $U_{i,nl}(R) = U_i(R) + \varepsilon_{nl}$ and $U_{f\varepsilon}(R) = U_f(R) + \varepsilon$ of quasimolecule $YX^+ + e$.

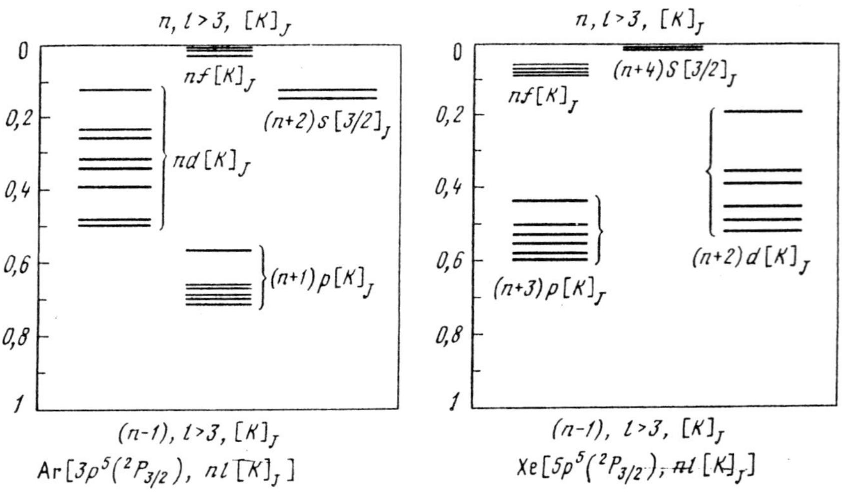

Fig.4. The stucture of Rydberg states nl $[K]_J$ of Ar and Xe atoms converging to the ionization boundary $n_0p^5(^2P_{3/2})$.

ion	A, eV	B, a.u.	term	E_0, эВ	R_e, a. e.	R_0, a. e.
$HeXe^+$	1090	1,84	U_i	0,019	5,95	5,25
			U_f	0,1	4,5	3,85
$KrXe^+$	4900	1,67	U_i	0,13	7,35	6,45
			U_f	0,361	6,2	5,4

Table 2. Parameters of approximations (see formula (82)) of the splitting energy $U_{if}(R)$ (in electron volts) of the ground $U_f(R) = U_i(R) - U_{if}(R)$ and the first excited $U_i(R)$ electronic terms $|f\rangle = X|3/2, 1/2\rangle$ and $|i\rangle = A_1|3/2, 3/2\rangle$ of heteronuclear ions $HeXe^+$ and $KrXe^+$ from [39, 40]. The energy of spin-orbit splitting of ${}^2P_{j=3/2}$- and ${}^2P_{j'=1/2}$ states of the main $5p^5$ shell of Xe^+ ion is $\Delta_{jj'} = 1.3$ eV

The calculation of electronic matrix elements (54) of the Coulomb interaction [7] (46) of the outer (r_e) electron with all inner ($r_{æ}$) electrons of the heteronuclear quasi-molecule $YX^+ + e^-$ can be performed for the bound-free transition $|i, nlm\rangle \to |f, \varepsilon l'm'\rangle$ similarly to the case of the bound-bound transition $|i, nlm\rangle \to |f, n'l'm'\rangle$ considered in [13] (see Appendix). As a result, we get for the sum $\sum_{m,m'} |\langle n_0 j\Omega_i | \langle nlm | V | \varepsilon l'm'\rangle n_0 j\Omega_f\rangle|^2$

$$\sum_{mm'} |V_{i,nlm}^{f,\varepsilon l'm'}|^2 = \frac{2\gamma_{ll'}}{25 n_{eff}^3}, \quad \gamma_{ll'} = (2l+1)(2l'+1)\begin{pmatrix} l & 2 & l' \\ 0 & 0 & 0 \end{pmatrix}^2 \times \tag{80}$$

$$\times \sum_m \begin{pmatrix} l & 2 & l' \\ -m & -1 & m+1 \end{pmatrix}^2 \left[\frac{2^{l+l'+2}\langle r_v^{l+l'+2}\rangle_{n_0 L}}{(2l+1)!\,(2l'+1)!\,(l+l'+5)} + f_0 \frac{\langle r_v^2\rangle_{n_0 L}}{(l+1/2)^3}\delta_{ll'}\right]^2.$$

Here $\langle r_v^s\rangle_{n_0 L} = \int_0^\infty r_v^s R_{n_0 L}^2(r_v)\, r_v^2 dr_v$ is the multipole moment of vacancy of s order in the $X^+(n_0 p^5)$ ion main shell averaged over the Hartry-Fok radial wave function $R_{n_0 L}(r_v)$ with the main quantum number n_0 and the orbital moment $L = 1$, and f_0 is a constant on the order of unity ($f_0 < 1$) determined by formulas (15) and (16) in [13].

(Footnote 7: For transitions between different terms $U_i(R)$ and $U_f(R)$ the Coulomb part of interaction (46) of the outer electron e^- with Z_X and Z_Y nuclei of YX^+ ion does not make a contribution to the matrix element $V_{i, nlm}^{f, \varepsilon l'm'}$ because of the orthogonality of wave functions $\varphi_i(r_{æ}, R)$ and $\varphi_f(r_{æ}, R)$).

Substituting expression (80) to the general formula (69) for the total rate constant of ionization of highly excited nl levels at collisions between various atoms of inert gases $X[n_0p^5(^2P_{3/2}), nl]$ and $Y(^1S_0)$ we obtain

$$K_{nl}(T) = \frac{16\pi^2\gamma_l}{25\,(2l+1)\,n_{\rm eff}^3}\,\frac{g_f}{g_{\rm tot}}\int^{R_{nl}} \exp\left[-\frac{U_i(R)}{T}\right]\Phi(R)\,R^2\,dR, \qquad (81a)$$

where @. Note that the normalization of the rate constant $K_{nl}(T)$ is changed in (81a) as compared with (69), i.e. factor g_f/g_{tot} was introduced instead of the statistical weights ratio g_f/g_i (and $g_i = g_{j\Omega_i} = 2$; $g_f = G_{j\Omega_f} = 2$ and $g_{tot} = g_i + g_f = 4$). In this case to determine by (81a) the total number of ionization acts (1a) and (1b) in a unit volume per a unit time one should multiply the quantoty K_{nl} by concentration N_Y of buffer gas atoms $Y(^1S_0)$ and the total concentration $N_{X^{**}}(j = 3/2, nl)$ of Rydberg atoms X^{**} in the state $n_0p^5(^2P_{j=3/2})nl>$ with two possible values $\Omega_i = 3/2$ and $\Omega_f = 1/2$ of theprojection of moment $j = 3/2$ of the atomic residue X^+ electrons on the internuclear axis R.

It should be remembered that highly excited levels $X\{n_0p^5(^2P_j)nl[K]_J\}$ of atoms of inert gases are also characterised by the quantum number $K = j + 1, j + 1 - 1,$, $|j - 1|$ (corresponing to the momentum $K = J + 1$, where $j = 3/2$ and $j' = 1/2$ are the values of the total momentum of the atomic residue X^+) and by the total atom momentum $J = K \pm 1/2$ in addition to the main quantum number n and orbital quantum number l of the outer electron e^- (see [33]). For heavy elements Ar, Kr and Xe, sublevels with different values of K and J are considerably splitted in energy within nl level at l = 0,1,2 and have large quantum defects $\delta_{l[K]_J}$ (nl states with $l > 3, 4, \ldots$ are practically hydrogren-like, i.e. $\delta_l = 0$, see Fig. 4). To determine the ionization rate constant $K_{nl[K]J}(T)$ at the selective excitation of such $n_0p^5(^2P_{j=3/2})nl\ [K]_J$ sublevels of X^{**} atom one should only change corresponding statistical weights of the states in (81a).

Parameter γ_l determining the value of the electronic matrix element (see (80)) turns out to be maximal for np states ($\gamma_{>}$ = 0) and decreases abruptly with increasing l (for more details, see [13]. Hence, if all Rydberg sublevels for a given n value are occupied equally, the main contribution to ionization is due to transitions np $\to$ εp with l' = l = 1. Thus, the corresponding ionization rate constant $K_n(T) = \sum_{l=0}^{n-1} [(2l+1)/n^2] K_{nl}(T)$ has the form

$$K_n(T) = \frac{16\pi^2\gamma}{25n^6} \frac{g_f}{g_{tot}} \int_0^{R_n} \exp\left[-\frac{U_i(R)}{T}\right] \Phi(R) R^2 dR, \qquad (81b)$$

where $\gamma = \sum_{ll'} \gamma_{ll'} \approx \gamma_{11}$, a $\Delta U_{if}(R_n) = |\varepsilon_n| = 1/2n^2$.

Specific calculations of the rate constants $K_{nl}(T)$ and $K_n(T)$ of ionizations processes (1a) and (1b) were performed in this paper using formulas (81a) and (81b) for Rydberg atoms $Xe[5p^5(^2P_{3/2})nl]$ at thermal collisions with He($^1S_0$) and Kr($^1S_0$) atoms of buffer gas. In this case the following approximations were used

$$U_i(R) = E_0^i [\exp(-2y) - 2\exp(-y)], \quad y = \ln 2 \left(\frac{R - R_0^i}{R_e^i - R_0^i}\right),$$
$$\Delta U_{if}(R) = U_i(R) - U_f(R) = A\exp(-BR) \qquad (82)$$

for the ground $U_f(R) = U_{j\Omega_f}(R)$ and first excited $U_i(R) = U_{j\Omega_i}(R)$ terms of heteronuclear ions $HeXe^+$ and $KrXe^+$ (see Fig. 3) of inert gases with parameters (see Table 2) obtained from numerical calculations [39, 40]. For quantities

$\gamma = \sum_{ll'} \gamma_{ll'}$ and $\gamma_l = \sum_{l'} \gamma_{ll'}$ (see (80)), the values γ_s = 0, γ_p = 2.04 a.u., γ_d = 0.102 a.u. were used calculated for Xe atom $[5p^5(^2P)nl]$ in [13]. Corresponding calculation results for the cases of selectively excited Rydberg np-levels and nlm-sublevels equally populated within a given n level are presented in Figs. 5 and 6. One can see that maximum values of ionization rate constants are of order of $K_n \sim 10^{-13} - 10^{-12}$ cm^3/sec and $K_{np} \sim 10^{-11} - 10^{-10}$ cm^3/sec. Because of the abrupt γ_l decrease with increasing of the orbital momentum l, ionization rate constants K_{nd}, K_{nf} etc. turn out to be considerably lower than K_{np} values. The ionization of ns-levels does not occure because of the selection rules, i.e. K_{ns} = 0.

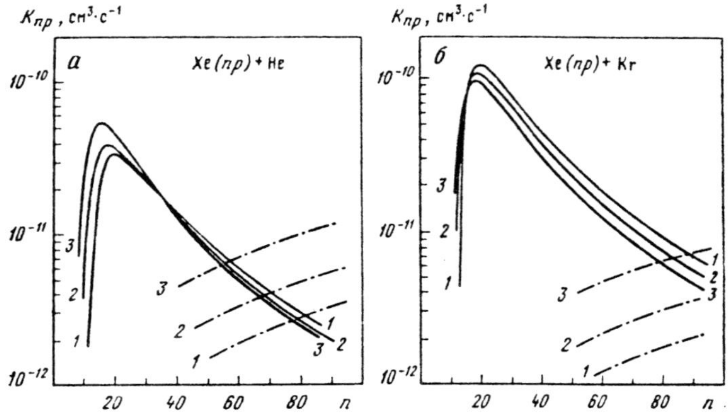

Fig. 5. Ionization rate constants $K_{np}(T)$ ($cm^3 sec^{-1}$) of selectively excited Rydberg np-levels of Xe atoms $[5p^5(^2P_{3/2})np]$ at thermal collisions (T = 300 K (1), 500 K (2) and 1000 K (3) with He (a) and Kr (b) buffer gas atoms
Solid curves correspond to the mechanism of scattering of He or Kr perturbing atoms by the atomic residue Xe^+ $[5p^5(^2P_{3/2})$ (calculation by formulas (81a) and (81b)). Dashed-dotted curves are results of calculations by formulas (39) and (41) in the quasi-free electron model (L_{He} = 1.19 a.u., L_{Kr} = -3.6 a.u.).

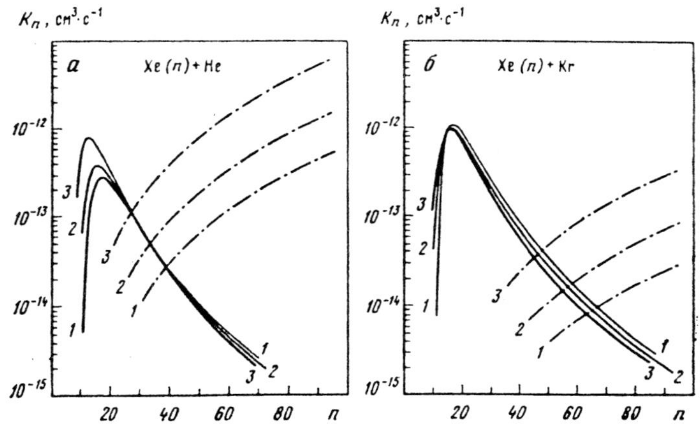

Fig. 6. Ionization rate constants $K_n(T)$ ($cm^3 sec^{-1}$) of Rydberg n levels (under conditions of equally populated lm-sublevels) of Xe atoms $[5p^5(^2P_{3/2})$ at thermal collisions with He and Kr atoms
Designations correspond to that in Fig. 5

Figs. 5 and 6 also present results of calculations of ionization rate constants performed by formulas (39) and (41) in Section 2 in the framework of competitive mechanism caused by scattering of perturbing He and Kr atoms by quasi-free electron e^- of Xe(n) atom. One can see that the Fermi mechanism begins dominate only in the region of sufficiently high values of n.

5. IONIZATION IN THE OWN INERT GAS

The general formulas (61), (65), and (69) for cross sections and ionization rate constants obtained in Section 3 are also valid for the case of Rydberg atoms X(nl) collisions with X atoms of the own gas in the ground state. It is known (see [5 - 8]) that in this case ionization processes (1a) and (1b) are caused by dipole transitions between symmetrical $U_g(R)$ and antisymmetrical $U_u(R)$ terms of the quasi-molecular ion X_2^+, so that the electronic matrix element $V = r_e D(r_{\varkappa})/r_e^3$ of interaction of the outer and inner electrons of quasi-molecule $X_2^+ + e^-$ has the form (83)

$$V_{i,nlm}^{f,\varepsilon l'm'}(R) = \left\langle \psi_{nlm}(r_e) \left| \left\langle \varphi^i_{X_2^+}(\mathbf{r}_\varkappa, \mathbf{R}) \left| \frac{\mathbf{r}_e \cdot \mathbf{D}(\mathbf{r}_\varkappa)}{r_e^3} \right| \varphi^f_{X_2^+}(\mathbf{r}_\varkappa, \mathbf{R}) \right\rangle \right| \psi_{\varepsilon l'm'}(\mathbf{r}_e) \right\rangle =$$
$$= -\omega_{if}^2 \langle nlm | \mathbf{r}_e \mathbf{D}_{if}(R) | \varepsilon l'm' \rangle .$$

Here @ is the matrix element of the dipole moment @ of inner electrons over electronic wave functions @ and @ of symmetrical and antisymmetrical terms of X_2^+ ion (electron coordinates r_e and $r_{\varkappa}$ are measured from the center of mass of this ion) and ω_{if} is the transition $|i\rangle \rightarrow |f\rangle$ frequency (see (59)).

Substituting (83) to (69a) and summing over m, m', l (similar to the case of transitions within a single term, see (71)), we obtain for the rate constant of dipole ionization of highly excited X(nl) atom by X atoms of the own gas the expression

$$K_{nl}(T) = \frac{2cg_f}{g_{tot}} \int_0^{R_{nl}} R_\omega^2 D_{if}^2(R_\omega)\, \omega_{if}^3 \sigma_{nl}^{ph}(\omega_{if}) \exp\left[-\frac{U_i(R)}{T}\right] \Phi(R_\omega)\, dR_\omega, \quad (84)$$

where $\omega_{if} = \Delta U_{if}(R\omega)$ and a point R_{nl} is determined by the relation (62).

The result (84) describes the case when the upper term $U_i(R)$ of X_2^+ ion along with repulsive[8] region $U_i(R) > 0$ (at $R < R_0^i$) has also the attraction region $U_i(R) \leq 0$ (at $R \geq R_0^i$) caused, for example, by the polarization interaction ($U_i(R) = -\alpha_X/2R^4$ at $R \to \infty$). Substituting in (84) quasi-classical expressions (75) and (77) for the Rydberg atom ionization cross sections and using the asymptotic approximation $D_{if}(R) = R/2$ for the matrix element of the transition $|i\rangle \to |f\rangle$ dipole moment, we obtain the following formulas for the rate constants K_{nl} and K_n of ionization processes (1a) and (1b) from the selectively excited nl level

$$K_{nl}(T) = \frac{2^{1/3}\Gamma^2(2/3)\, g_f}{3^{2/3} n_{\mathrm{eff}}^3 g_{\mathrm{tot}}} \int_0^{R_{nl}} \Delta U_{\mathrm{if}}^{2/3}(R) \exp\left[-\frac{U_i(R)}{T}\right] \Phi(R) R^4 dR \qquad (85a)$$

and for the ionization in conditions of equally populated nlm-sublevels within a given level n

$$K_n(T) = \frac{4\pi g_f}{3^{3/2} n^5 g_{\mathrm{tot}}} \int_0^{R_n} \exp\left[-\frac{U_i(R)}{T}\right] \Phi(R) R^4 dR. \qquad (85b)$$

At thermal energies $T \sim 0.03 - 0.08$ ev $\ll \Delta_{ij}$, of collisions of heavy atoms of inert gases $X[n_0p^5(^2P_{3/2})nl] + X(^1S_0)$, the mechanism considered may have two channels $|i_1\rangle \to |f_1\rangle$ and $|i_2\rangle \to |f_2\rangle$ of both direct (1a) and associative (1b) ionization of Rydberg nl-levels. They are caused by dipole transitions in the quasi-molecular X_2^+ ion between its gerade (g) and ungerade (u) four lower terms $|i_1\rangle = |i(3/2, 1/2)_g\rangle$,

(Footnote 8: In the region of quantum numbers @ the value of $\Phi(R)$ in formula (84) is equal to 1 for all acceptable values of the internuclear distance $0 < R < R_{nl}$ (because $R_{nl} < R_0^i$ and, respectively, $U_i(R) > 0$, see (69b)). In this case it follows directly from (84) (using the asymptotic approximation $D_{if}(R) \approx R/2$ and after the transition to the integration over the emitted electron energy $d\varepsilon = d\omega = \Delta F_{if}(R)dR$) the known expression (see formula (24) in [8]) for the rate constant of the ionization process).

$|f_1\rangle = |I(3/2, 1/2)_u\rangle$, $|i_2\rangle = |I(3/2, 3/2)_u\rangle$, $|f_2\rangle = |I(3/2, 3/2)_g\rangle$ with different values $\Omega_{i_1} = \Omega_{f_1} = 1/2$ and $\Omega_{i_2} = \Omega_{f_2} = 3/2$ of the total momentum j = 3/2 projection on the internuclear axis R (Fig. 7). Due to large values of the interchange splitting of gerade and ungerade terms $\Delta U_{i_1 f_1}(R)$ $\Delta U_{i_2 f_2}(R)$ ionization processes (1a) and (1b) occure in the asymptotic region of distances R (where the bond of C type by Hund is realized). In this case dipole matrix elements of the transitions over the wave jΩ functions of jΩ representation (as in the case of LS-coupling) are reasonably well described in the approximation R/2 ([42])

$$\mathbf{D}_{i_1 f_1} = \left\langle I\left(^3/_2, {}^1/_2\right)_g \left| \sum_{\varkappa=1}^{N} \mathbf{r}_\varkappa \right| I\left(^3/_2, {}^1/_2\right)_u \right\rangle = \mathbf{R}/2,$$

$$\mathbf{D}_{i_2 f_2} = \left\langle I\left(^3/_2, {}^3/_2\right)_u \left| \sum_{\varkappa=1}^{N} \mathbf{r}_\varkappa \right| I\left(^3/_2, {}^3/_2\right)_g \right\rangle = \mathbf{R}/2. \qquad (86)$$

Therefore, at ionization of Rydberg levels of atoms $X[n_0p^5 \times (^2P_{3/2})nl$ by atoms $X(^1S_0)$ of the own inert gas, the total rate constant of processes (1a) and (1b) has to be calculated by formula (85) considering two possible types (see (86)) of dipole transitions, i.e.

$$K_{nl}(T) = K_{nl}^{i_1, f_1}(T) + K_{nl}^{i_2, f_2}(T), \qquad (87)$$

Statistical weights of terms $U_{i_k}(R)$ and $U_{f_k}(R)$ (where k = 1, 2) of homonuclear X_2^+ ions of inert gases correlating with the $X^+[n_0p^5(^2P_{3/2})] + X(^1S_0)$, state are equal to $g_{i_k} = g\left(j\Omega_{i_k}\right) = 2$ $g_{f_k} = g\left(i\Omega_{f_k}\right) = 2$, $g_{tot} = 2(2j+1) = 8$. respectively. In specific calculations of ionization rate constants by formulas (85) - (87) in the asymptotic region of distances, we shall use the following approximations of terms and their splittings:

$$U_{i_k}(R) = \eta_k R^{\varkappa_k} \exp(-\xi_k R) - \alpha_X/2R^4, \qquad (88a)$$

$$\Delta U_{i_k f_k}(R) = \zeta_k \exp(-\lambda_k R), \qquad (88b)$$

with parameters obtained from the comparision with the results of available numerical calculations [41, 42]. Here α_X is the atom X polarizability in the ground $^1S_0$ state. For Xe_2^+ ion the corresponding values are: $\xi_1 = \xi_2 = 0.945$ a.u.; $\eta_1 = 0.606$ a.u., $\eta_2 = 5.07$ a.u.; $\varkappa_1 = 1.59$, $\varkappa_2 = 0.22$; $\alpha_{Xe} = 27.06$

a.u.; ζ_1 = 21.7 a.u., ζ_2 = 27.5 a.u.; λ_1 = 0.879 a.u., λ_2 = 1.1 a.u. ($\Delta U_{i_k f_k}$, U_{i_k}, and R are measured in atomic units).

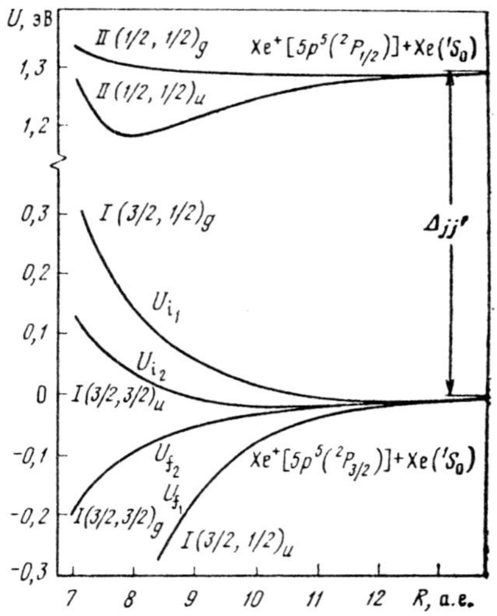

Fig. 7. Form of lower electronic terms of homonuclear Xe_2^+ ion in the asymptotic region of internuclear distances R

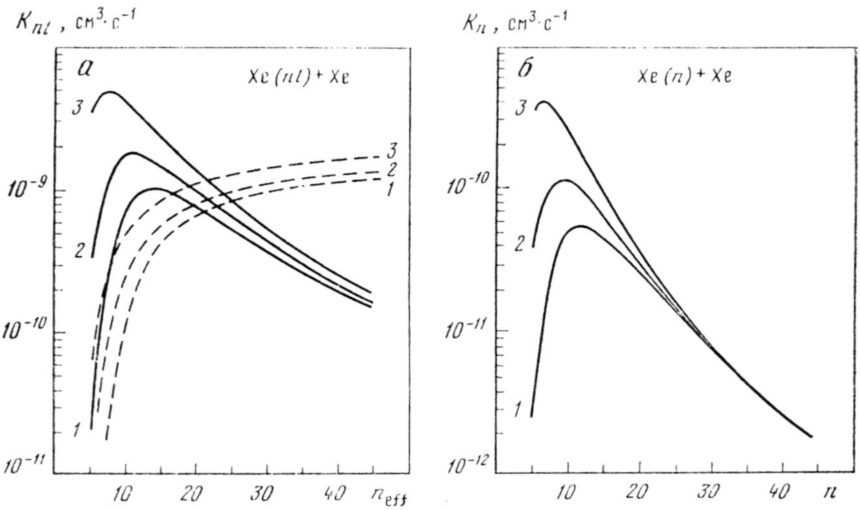

Fig. 8. Ionization rate constants $K_{nl}(T)$ and $K_n(T)$ cm^3sec^{-1}) of Rydberg $Xe[5p^5(^2P_{3/2})_n^{nl}$ atoms at thermal collisions (T = 300 K (1), 500 K (2), and 1000 K (3)) with Xe atoms of the own gas.

Solid curves: calculation by formulas (85a), (85b) in the framework of perturbation theory (see also (87), (88)). Dashed curves: calculation by formula (89)

Calculation results for ionization rate constants of Rydberg nl and n levels of $Xe[5p^5(^2P_{3/2})nl]$ atoms at thermal collisions with $Xe(^1S_0)$ atoms of the own gas obtained in the framework of perturbation theory by formulas (85), (87) and (88) are presented in Fig. 8, a, b (solid curves). For comparision, this figure also shows maximum possible values of ionization rate constants[9]

$$K_{nl}^{\max}(T) = \pi R_{nl}^2 \langle v \rangle_T W_{nl},$$
$$W_{nl} = \begin{cases} \exp[-U_i(R_{nl})/T], & R_{nl} < R_0^i, \\ 1 + |U_i(R_{nl})|/T, & R_{nl} \geqslant R_0^i, \end{cases} \tag{89}$$

for mechanism caused by the intersection of terms (see dashed curves in Fig. 8. a). One can see that at ionization from selectively excited nl-levels ($l \ll n$) in the region $n \lesssim 20$ the application of perturbation theory results in overstated values of $K_{nl}(T)$. The formula (85) becomes applicable in the region of sufficiently high values of $n \gtrsim 20$. Therefore, values of $K_{nl}(T)$ at temperatures $T \sim 300 - 1000$ K turn to be of order of $10^{-10} - 10^{-9}$ $cm^3 sec^{-1}$. Similarly, the formula (85b) for ionization, when nlm-sublevels within a given n-level are equally populated, considerably overstates real values of $K_n(T)$ for the Xe(n) + Xe collision process at $n \lesssim 20$ and becomes only applicable at high values of n.

6. CONCLUSION

It follows from the above examination that Rydberg atoms ionization at collisions with neutral atomic particles can be effected by various physical mechanisms. The relative contribution of a specific mechanism to the total cross section

(Footnote 9: The result (89) is obtained assuming that the ionization probability $W_{nl}^{max}(\rho, E) = 1$ for all aiming parameters $\rho \leq \rho_{max} \approx R_{nl}[1 - U_i(R_{nl})/E]^{1/2}$, so that $\sigma_{nl}^{max} \approx \pi\rho^2_{max}$. The integration of the quantity $v\sigma_{nl}^{max}(E)$ over the Maxwell distribution of kinetic energy of colliding atoms ($E_{min} \leq E < \infty$, where $E = \mu v^2/2$ and $E_{min} = \max(0, U_i(R_{nl}))$ leads to the formula (89)).

or ionization rate constant depends on the type of colliding atoms A(nl) and B, their masses M_A and M_B< the relative velocity v (gas temperature T), the region of principal quantum numbers n and orbital momentum l. However, one can notice as a whole that at thermal energies in the experimentally investigated region of n, the role of scattering of perturbing B atoms by the atomic residue A^+ of highly excited A(nl) atom, as a rule, turns out to be more essential in the ionization processes than the scattering by the quasi-free electron e^-. Therein, in particular, lies the essential difference of the ionization processes (1), (2) from broadening, shift and quasi-elastic quenching of Rydberg levels by neutral atomic[10] particles which, as a rule, are determined by the Fermi mechanism (see, for example, [1] and [18, 28, 29]).

The mechanism of scattering of perturbing Y atom by the quasi-free e^- electron becomes essential only for the ionization in the region of sufficiently high values of n ≥ 40 - 80 (see Figs. 5, 6). For B and A(n) atoms with small reduced mass this occures at thermal energies already at n ≥ 20 - 30 (see Figs. 1, 2). In this case the role of the mechanism increases with increasing value of the elastic scattering cross section σ^{el}_{eV} of ultraslow e^- electron by the perturbing B atom.

In the region of relatively low values 5 ≤ n ≤ 20 - 30 the ionization processes (1), (2) occure especially efficiently in the cases when the electronic terms of quasi-molecule $BA^+ + e^-$ forming at the scattering of the disturbing atom by the atomic residue A^+ may intersect. This is illustrated by the well-known ionization mechanism in the own gas (see [5 - 8] for alkali metals and Section 5 for atoms of inert gases) and also by the ionization mechanism in the buffer inert gas considered in Section 4 (see Figs. 5, 6, 8)

(Footnote 10: Large values of the ionization cross sections at collisions with B molecules in the mechanism of scattering by the quasi-free e^- electron are explained by the compensation of the electron energies difference $\Delta\varepsilon_{if} = |\varepsilon_{nl}| + \varepsilon$ from the inner molecular degrees of freedom).

The physical peculiarity of direct (1a) and associative (1b) ionization processes in the buffer inert gas (besides noted ones in Section 4 and Appendix) lies in the fact that the perturbing Y atom does not influence practically on the character of interaction of Rydberg e^- electron and the atomic residue X^+ determining the main contribution to the cross sections $\sigma_{nl}^{d.i}$ and $\sigma_{nl}^{a.i}$. Thus, the role of the perturbing Y particle is primarily to ensure a resonance between the splitting energy $\Delta U_{if}(R_\omega)$ of ion YX^+ terms and the energy $\Delta\varepsilon_{if} = |\varepsilon_{nl}| + \varepsilon$ of the bound-free transition of e^- electron. At collisions of identical X(nl) and X atoms, contrastingly, the dipole interaction between e^- electron and quasi-molecular X_2^+ ion that results in the ionization id determined directly by the presence of X atom.

The comparision of calculated rate constants of ionization of Rydberg nl and n levels of xenon atoms by atoms of the buffer He and Kr) and own (Xe) gases shows that their ratio turns out to be of order of $K_n^Y/K_n^X \sim 10^{-2} - 10^{-1}$. This means that in the mixture of inert gases Xe/Y (where Y = He, Ar, Kr) with concentrations $[Xe]/[Y] \leq 10^{-2}$ the ionization of Rydberg levels of Xe(n) atoms arise from the collisions with atoms Y of lighter buffer gas. Collisions with buffer gas atoms (He) are found also to be rather essential at the energy relaxation among highly excited atoms Xe(n) levels in the process of recombination of electrons with atomic Xe^+ ions of the mixture Xe/He plasma ([Xe] ≪ [He]) excited by the impulsing discharge (see [12, 13]). In conclusion, notice that investigations of various collisional and radiation processes in theplasma of mixtures of inert gases Xe/He, Xe/Ar, and Xe/Kr excited by the electron beam or in the discharge is of interest for the kinetics of powerful xenon laser in IR region (see, for example, [43]).

APPENDIX

Let us formulate here the main idea of the calculation of electronic matrix elements $V_{i,nlm}^{f,\varepsilon l'm'}$ (see (54)) for the free-bound transition $nl \rightarrow \varepsilon l'$ of the outer electron e^- followed by the change of the electronic state $A_1|j = 3/2, \Omega_i = 3/2\rangle \rightarrow X|j = 3/2, \Omega_f = 1/2\rangle$ of heteronuclear ion YX^+ of inert gas.

A. It is convenient to reduce the Coulomb part of the electron-electron interaction in (46) to the interaction $V_{ev} = |r_e - r_v|^{-1}$ of the outer electron e^- with the one-electron vacancy (with a coordinate r_v) in $X^+[n_0p^5(^2P_{3/2})]$ ion, while the interaction with electrons from the closed shell of $Y(^1S_0)$ atom can be neglected. As the wave functions of the initial $|i\rangle$ and final $|f\rangle$ states of molecular YX^+ ion in a zero approximation, we shall use the corresponding wave functions @≠ and @ of the vacancy in the atomic $X^+[n_0p^5(^2P_{3/2})]$ ion which by means of Klebsh-Gordon relations (see [32, 33]) are reduced to the coordinate @ and spin $\eta_{s\mu}$ wave functions of the n_0p-shell (in the $LS\Lambda\mu$ representation) with different values of projection $\Lambda_i = 1$, $\Lambda_f = 0$ and $\mu = \pm 1/2$ of the vacancy orbital momentum ($L = 1$) and spin ($S = 1/2$) on the internuclear axis R. As a result, after expansion of an operator $V_{ev} = -|r_e - r_v|^{-1}$ in a series in multipoles and calculation of angular integrals of spherical functions, we get for the electronic matrix element @ the expression

$$V_{i,nlm}^{f,\varepsilon l'm'} = \left(\frac{2}{3}\right)^{1/2} \langle n_0 L\Lambda_i|\,\langle nlm\,|\,r_{ev}^{-1}\,|\,\varepsilon l'm'\rangle\,|\,n_0 L\Lambda_f\rangle = \left(\frac{2}{3}\right)^{1/2}(2L+1)\,(2l+1)^{1/2},$$

$$(2l'+1)^{1/2}\, i^{l'-l} \sum_{k=0}^{\infty}\sum_{q=-k}^{k}(-1)^{k-q+m}\begin{pmatrix} L & k & L\\ 0 & 0 & 0\end{pmatrix}\begin{pmatrix} L & k & L\\ -\Lambda_i & q & \Lambda_f\end{pmatrix}\times$$

$$\times\begin{pmatrix} l & k & l'\\ 0 & 0 & 0\end{pmatrix}\begin{pmatrix} l & k & l'\\ -m & -q & m'\end{pmatrix} I_{nl,\varepsilon l'}^{(k)}. \qquad \text{(II.1)}$$

Here @ is the radial transition matrix element of order of k and $r_> = \max(r_e, r_v)$ and $r_< = \min(r_e, r_v)$. For the transitions considered with the change of the projection $\Omega_i - \Omega_f = 1$ of the vacancy momentum $j = 3/2$ (and, respectively, $\Lambda_i - \Lambda_f = 1$, $L = 1$) a contribution to (II.1) makes only one member of the sum with $k = 2$ and $q = 1$. It follows herefrom the selection rules: $l' = l$ ($l \neq 0$), $l' = l \pm 2$ and $m' = m + 1$.

B. Therefore, the calculation of quantity $V^{f,\varepsilon l'm'}_{i,nlm}$ is reduced to the calculation of the radial matrix element

$$I^{(k=2)}_{nl,\varepsilon l'} = \int_0^\infty R^2_{n_0 L}(r_v)\, r_v^2\, dr_v \times$$

$$\times \left\{ \int_0^{r_v} \frac{r_e^2}{r_v^3} R_{nl}(r_e) R_{\varepsilon l'}(r_e)\, r_e^2\, dr_e + \int_{r_v}^\infty \frac{r_v^2}{r_v^3} R_{nl}(r_e) R_{\varepsilon l'}(r_e)\, r_e^2\, dr_e \right\} \qquad \text{(II.2)}$$

for the short-range $(0 \le r_e \le r_v)$ and long-range $(r_v < r_e < \infty)$ quadrupole parts of the Coulomb interaction of the outer electron with vacancy of the type $r_<^2/r_>^3$. The calculation of the first ("short-range") term in (II.2) is performed using known expansion (see [44]) of Coulomb wave functions of the discrete $R_{nl}(r_e)$ and continuous $R_{\varepsilon l'}(r_e)$ spectra in a series in the vicinity of the coordinate origin $r_e = 0$. The second ("long-range") term in (II.2) is calculated using the quasi-classical approximation for the $R_{nl}(r_e)$ and $R_{\varepsilon l'}(r_e)$ wave functions. In this case the expression obtained for $I^{(2)}_{nl,\varepsilon l'}$ differs from the case of bound-bound transition (see [13] only by the normalization constant for $R_{\varepsilon l'}$, i.e. @. Thus, using formulas (II.1) and (II.2) and results from [13], we obtain the expression (80) for the sum of electronic matrix elements @ of interest to us.

ABSTRACT

The theory of direct and associative ionization of Rydberg atoms at thermal collisions with neutral particles of the buffer or own gas has been developed. Some physical mechanisms of ionization due to the scattering of the perturbing particle both by the highly excited electron and by the core of the Rydberg atom have been investigated. In particular, the simple formulas relating the ionization cross section with the differential cross section for the electron-perturber scattering and with the momentum distribution function of the Rydberg electron in the initial nl-state have been derived from the impulse approximation. The main attention is paid to the consideration of the nonadiabatic effects of energy exchange between the Rydberg electron and inner electrons or nuclei of quasi-molecular ion. Numerical calculations of the ionization rate constants have been carried out for the Rydberg hydrogen and xenon atoms at thermal collisions with He, Kr, and Xe atoms in the ground state.

REFERENCES

[1] Rydberg States of Atoms and Molecules (Eds. R.F. Stebbings and F.B. Dunning), Univ. Press, Cambridge, (1983); Mir, Moscow, (1985) (in Russian).

[2] V.S. Lebedev and V.S. Marchenko, Collisional and radiative processes involving highly excited states of atoms and molecules, In Kinetics of the Low-Temperature Plasma and Gaseous Lasers, Nauka, Moscow, (1984) p. 79 (Proc. Lebedev Phys. Inst., vol. 145), Engl. Transl.: J. Sov. Laser Res. 7, (1986) 489.

[3] A.N. Klyucharev and M.L. Yanson, Elementary Processes in the Plasma of Alkali Metals, Energoatomizdat, Moscow, (1988) 224 p.

[4] V.A. Smirnov and A.A. Mikhailov, Inelastic collisions of of highly excited atoms, Opt. Spectrosk., 30, (1971) 964.

[5] A.Z. Devdariani et al., Collisional ionizationn processes of Rydberg states of alkali atoms, Pis'ma Zh. Tekhn. Fiz., 4, (1978) 1013.

[6] R.K. Janev and A.A. Mikhailov, Resonant ionization in slow-atom Rydberg atom collisions, Phys. Rev. A - Gen. Phys., 21, (1980) 819.

[7] A.A. Mikhailov and R.K. Janev, Ionization in atom - Rydberg atom collisions: ejected electron energy spectra and reaction rate coefficients, J. Phys. B: Atom. Mol. Phys., 14, (1981) 1639.

[8] E.L. Duman and I.P. Shmatov, Ionization of highly excited atoms in the own gas, Zh. Eksp. Teor. Fiz., 78, (1980) 2116.

[9] V.S. Lebedev, V.S. Marchenko, and S.I. Yakovlenko, Excitation and ionization of Rydberg atom states at the collision of a neutral particle with a core, Izv. Akad. Nauk SSSR, ser. fiz., 45, (1981) 2995.

[10] V.S. Lebedev and V.S. Marchenko. Transitions between highly excited atom states at the flying of a neutral particle nearby its atomic residue, Zh. Eksp. Teor. Fiz., 84, (1983) 1623.

[11] V.S. Lebedev and V.S. Marchenko, Direct and associative ionization of the highly excited atom at the flying of a neutral particle nearby its atomic residue, Khim. Fiz., 3, (1984) 210.

[12] V.A. Ivanov, V.S. Lebedev, and V.S. Marchenko, Recombination of electrons with atomic ions in the buffer inert gas, Pis'ma Zh. Tekh. Fiz., 14, (1988) 1575.

[13] V.A. Ivanov, V.S. Lebedev, and V.S. Marchenko, Collisional quenching of Rydberg atomic levels and electron-ion recombination in the buffer inert gas, Zh. Eksp. Teor. Fiz., 94, (1988) 86.

[14] M.R. Flannery, Semiquantal theory of heavy-particle excitation, deexcitation and ionization by neutral atoms, Ann. Phys, 61, (1970) 465.

[15] L.P. Pitaevskii, Recombination of electrons in monoatomic gas, Zh. Eksp. Teor. Fiz., 42, (1962) 1326.

[16] D.R. Bates and S.P. Khare, Recombination of positive ions and electrons in dense neutral gas, Proc. Phys. Soc., 85, (1965) 231.

[17] M.R. Flannery, Theoretical treatment of collisions of Rydberg atoms with neutral atoms and molecules: Semiquantal, impulse, and multistate-orbital theories, Phys. Rev. A - Gen. Phys., 22, (1980) 2408.

[18] V.S. Lebedev and V.S. Marchenko, Nonresonance collisional mixing of Rydberg atom levels in the model of Fermi pseudopotential, Zh. Eksp. Teor. Fiz., 88, (1985) 754.

[19] L. Petitjean and F. Gounand, Simple analytical formulas for collisional l-mixing, n-changing, and ionization of Rydberg atoms with neutral particles at thermal energies, Phys. Rev. A - Gen. Phys., 30, (1984) 2946.

[20] B. Kaulakys, Analytical expressions for cross sections of Rydberg-neutral inelastic collisions, J. Phys. B: Atom. Mol. Phys., 18, (1985) L167.

[21] G.F. Chew and M.L. Goldberger, The scattering of elementary particles by complex nuclei: A generalization of the impulse approximation, Phys. Rev. 87, (1952) 778.

[22] M. Goldberger and K. Watson, Theory of collisions, Mir, Moscow, (1967).

[23] R. Newton, Theory of scattering of waves and particles, Mir, Moscow, (1969) 608 p.

[24] M. Matsuzawa, Ionization of long-lived highly excited atoms by collision with molecules, J. Chem. Phys. 55, (1971) 2685.

[25] A.K. Kazanskii and I.I. Fabrikant, Scattering of slow electrons by molecules, Usp. Fiz. Nauk, 143, (1984) 601.

[26] R.K. Nesbet, Variational methods in electron-atom scattering theory, Plenum Press, N.Y., (1980).

[27] M. Matsuzawa, Validity of the impulse approximation in Rydberg-neutral collisions, J. Phys. B: Atom. Mol. Phys., 17, (1984) 795.

[28] V.S. Lebedev and V.S. Marchenko, Collisional quenching of Rydberg atoms in the presence of resonance on the quasi-discrete level of the perturbing particle, Zh. Eksp. Teor. Fiz. 91, (1986) 428.

[29] V.S. Lebedev and V.S. Marchenko, Impact broadening of Rydberg atomic levels in inelastic collisions with alkali-metal atoms, J. Phys. B: Atom. Mol. Phys., 20, (1987) 6041.

[30] S.T. Butler and R.M. May, Production of highly excited neutral atoms for injection into plasms devices, Phys. Rev. A - Gen. Phys. 137, (1965) 10.

[31] B.M. Smirnov, Ions and excited atoms in plasma, Atomizdat, Moscow, (1974).

[32] L.D Landau and E.M. Livshiz, Quantum Mechanics, Nauka, Moscow, (1974) 750 p.

[33] I.I. Sobelman, Introduction to the Theory of Atomic Spectra, Nauka, Moscow, (1977) 320 p.

[34] E.E. Nikitin and S.Ya. Umanskii, Nonadiabatic Transitions at Slow Atomic Collisions, Atomizdat, Moscow, (1979) 272 p.

[35] L.I. Gudzenko and V.S. Lebedev, Multipole corrections to electronic terms of highly excited molecule, In Kinetic

Models in Laser Physics and Theory of Vibrations, Nauka, Moscow, (1980) p. 107 (Proc. Lebedev Phys. Inst. Vol. 120)

[36] Handbook of Mathematical Functions (Eds. M. Abramowitz and I.A. Stegan), Nauka, Moscow, (1979) 832 p.

[37] S.P. Goreslavskii, N.B. Delone, and V.P. Krainov, Probabilities of radiative transitions between highly excited atomic states, Zh. Eksp. Teor. Fiz. 82, (1982) 1789.

[38] A.M. Dykhne and G.L. Yudin, Shaking of the quantum system and the character of transitions stimulated by it, Usp. Fiz. Nauk 125, (1978) 377.

[39] D. Hausamann and H. Morgner, The heteronuclear rare gas ions: A simple model for the determination of the potential curves, Mol. Phys. 54, (1985) 1085.

[40] B. Brunetti et al., Potential energy curves for the $HeAr^+$ and $NeAr^+$ ions, Chem. Phys. Lett, 126, (1986) 245.

[41] W.R. Wadt, The electronic states of Ar_2^+, Kr_2^+, Xe_2^+. 1. Potential curves with and without spin-orbit coupling, J. Chem. Phys. 68, (1978) 402.

[42] W.R. Wadt, The electronic states of Ne_2^+, Ar_2^+, Kr_2^+, and Xe_2^+. 2. Absorption cross sections for the $1(1/2)_u \rightarrow 1(3/2)_g$, $1(1/2)_g$, $2(1/2)_g$ transitions, Ibid 73, (1980), 3915.

[43] N.G. Basov and V.A. Danilychev, Lasers on condensed and compressed gases, Usp. Fiz. Nauk, 148, (1986) 55.

[44] G. Bete and E. Solpiter, Quantum Mechanics of Atoms with One or Two Electrons, Fizmatgiz, Moscow, (1960) 557 p.

Low-Frequency Bremsstrahlung in the Resonance Scattering

An.V. Vinogradov

INTRODUCTION

According to the Low theorem [1], the amplitude of photon bremsstrahlung in the limit of low frequencies is expressed via the elastic scattering amplitude with a precision of the term of expansion in frequency next to the main infrared term. In the case of short-range interaction such a relation results from the general principles of relative quantum theory [1,2], also it can be obtained by the Lippman-Schwinger equation [1, 3] or using wave functions asymtotics at infinity when evaluating the matrix element of bremsstrahlung [4]. The relation between two amplitudes with a precision beyond the Low theorem is known in the opposite special cases of Born approximation and S-scattering of slow electrons [5 - 7].

For the dipole matrix element of the induced bremsstrahlung or photon absorption at the elastic electron scattering the Low result may be presented in two forms equivalent in the limit $\omega \to 0$:

$$M_{\mathbf{p}',\mathbf{p}} = \frac{ie\pi\hbar}{(m\omega)^2}\left[\mathbf{E}(\mathbf{p}-\mathbf{p}')f(\bar{\varepsilon};\, q^2) \pm \mathbf{E}(\mathbf{p}'+\mathbf{p})\frac{\hbar\omega}{2}\frac{\partial f(\bar{\varepsilon},\, q^2)}{\partial\bar{\varepsilon}} + O(\omega^2)\right], \quad (1)$$

$$M_{\mathbf{p},'\mathbf{p}} = \frac{ie\pi\hbar}{(m\omega)^2}[\mathbf{E}\mathbf{p}f(\varepsilon_p \pm \hbar\omega;\, q^2) - \mathbf{E}\mathbf{p}'f(\varepsilon_p;\, q^2) + O(\omega^2)]. \quad (2)$$

Here E is the electric field vector, f is the elastic scattering amplitude, $\varepsilon_p = p^2/2m$, $p' = (p^2 \pm 2m\hbar\omega)^{1/2}$, $q^2 = (p' - p)^2$, $\bar{\varepsilon} = (\varepsilon_{p'} + \varepsilon_p)/2$.

The role of terms of ω^{-1} order in formulas (1) and (2) was discussed in detail by Brown and Goble [2]. The presence of these terms extends possibilities of formulas (1), (2) application and is very essential for some problems. For example, at the calculation of plasma absorption coefficient, the main infrared terms in the cross sections of the induced bremsstrahlung and absorption cancel each other and only terms of the next order make the contribution.

The disadvantage of the original deduction of formulas (1) and (2) (see [1, 2]) based only on the general principles of the theory is the fact that it does not allow to indicate the specific physical value of frequency ω at which formulas (1), (2) are valid. One can only argue that if the corresponding elastic scattering problem has some characteristic parameters of the energy dimension, then energy of the quantum $\hbar\omega$ has to be far less than the least of them. In the case of bremsstrahlung at the electron scattering by an atom this requirement means that energy of the quantum has to be low compared to the excitation energy of ε_a atom, kinetic energy of the electron and also compared to the characteristic energy scale of the elastic scattering amplitude change. The physical meaning of the first mentioned requirement consists in neglecting the contribution of the atom electrons into bremsstrahlung process[1] having at $\hbar\omega \approx \varepsilon_a$ a resonance character [8]. It is quite apparent that the bremsstrahlung amplitude can not be expressed via the elastic scattering amplitude at $\hbar\omega = \varepsilon_a$. The fullfilment of other two conditions is necessary to make formulas (1) and (2) equivalent.

In this paper we present the deduction of formula (2) which allows not only to obtain the terms of ω^{-2} and ω^{-1} order but alsoto estimate rather generally the terms of ω^0 order [9, 10]. The only assumption used is the short-range

(Footnote 1: We shall use below the term "bremsstrahlung process", if the statement concerning it is valid both for the bremsstrahlung absorption and spontaneous and induced bremsstrahlung emission).

character of interaction between electron and the scattering center, and the only parameter characterizing the interaction and used in the estimation of terms of ω^0 order, except scattering amplitudes, is its radius. In this case we obtain the upper estimation for the part of the bremsstrahlung process amplitude associated with the region of space where the motion of the scattering electron is not free. The results of the paper allow to state that formula (2) has far greater region of application than (1) and in some cases can be used at the finite value of ω. In particular, if there is a resonance with a width Γ in the elastic scattering, the formula (2) is valid in the vicinity of the resonance at $\hbar\omega \gg \Gamma$, while the formula (1) is valid only at $h\omega \ll \Gamma$.

Studies to elucidate conditions of application and the precision of various formulas of the low-frequencsy approximation for bremsstrahlung processes have been performed before. For this purpose formulas of the low-frequency approximation in [11, 12] have been compared with the results of corresponding calculations for exactly solvable models. A certain interest for the formulas (1), (2) is also provoked by the experiments on the direct observation of elementary induced bremsstrahlung processes at the electron scattering by atoms [13 - 16] which allow to check these formulas directly.

Of special note is the problem of application of formulas of low-frequency approximation in the case of resonance scattering. It has been discussed in the context of the nuclear physics problems [17 - 19], physics of the electron-atom collisions[11, 12] and in connection with the calculation of spectral emission characteristics of weakly ionized plasma [20]. In papers [17, 18] it was correctly suggested that formula (2) is valid at $\hbar\omega > \Gamma$. Hence, it follows from (2) that one resonance in the amplitude of the elastic scattering results in the appearance of two resonances in the matrix element of the single-quantum bremsstrahlung process shifted relative to each other by the value of $\hbar\omega$. The experimental splitting of the isolated resonance in the cross section of the electron elastic scattering by argon atom into two resonances at the induced bremsstrahlung in qualitative agreement with (2) was observed by Langhans [21].

Formulas presented in [11, 12, 20] differ from (2) at $\omega \to 0$ by terms of ω^{-1} order, which is indicative of the presence of errors in them. The reasons for the appearance of errors in these papers will be considered below. Contrary to the previous results, the results of the present paper allow not only to justify rigorously the applicability of (2) in the case of resonance scattering at $\hbar\omega > U$, but also to find physical parameters consraining the ω value. Notice that the appearence of two resonance singularities in cross sections of bremsstrahlung processes instead of one singularity in the elastic scattering cross section is of considerable interest since it means that radiative transitions in the continuous spectrum acquire characteristic features of transitions between discrete energy levels. Specifically, there is a principal possibilty to use bremsstrahlung at the resonance scattering to obtain the negative absorption coefficient of considerable absolute value due to transitions in the continuous spectrum [22].

References to the other experimental and theoretical works and the discussion of the applicabilty of formulas (1), (2) are presented in a review [23]. A review of earlier works is given in [4].

1. MATRIX ELEMENT AND PROBABILITY OF THE SINGLE-QUANTUM INDUCED BREMSSTRAHLUNG PROCESS

To calculate the probability of the single-quantum bremsstrahlung process in dipole approximation one has to calculate the matrix element

$$M_{\mathbf{p}',\mathbf{p}} = {}^1\!/_2 \langle \Psi^-_{\mathbf{p}'} \,|\, e\,\mathbf{Er} \,|\, \Psi^+_{\mathbf{p}} \rangle, \tag{3}$$

where $\Psi^{\pm}_{\mathrm{p}}$ are known in the scattering theory wave functions of the continuous spectrum [24]. Presenting the wave functions as an expansion in partial waves

$$\Psi_{\mathbf{k}}^{\pm}(r) = \frac{1}{4\pi}\sum_{l=0}^{\infty}(2l+1)\,\psi_l^{\pm}(k,r)\,P_l\left(\cos\left(\widehat{\mathbf{k},r}\right)\right),\ \mathbf{k} = \mathbf{p}/\hbar,$$

and integrating over angular variables in (3), we get

$$M_{\mathbf{p}',\mathbf{p}} = \frac{e}{8\pi}\sum_l\{E_z P_l(\cos\theta)\,[lR_{l,l-1} + (l+1)\,R_{l,l+1}] + \\ + E_y \mathrm{R}_l^1(\cos\theta)\,[R_{l,l+1} - R_{l,l-1}]\}, \tag{4}$$

where θ is the scattering angle and z axis alignes with direction of p,

$$R_{l,l'} = (-1)^l\int_0^{\infty} r^3\psi_l(k',r)\,\psi_{l'}(k,r)\,dr,\ \ \psi_l(k,r) \equiv \psi_l^{+}(k,r). \tag{5}$$

The probability of the induced bremsstrahlung W at the scattering of the electron with momentum p into the totalsolid angle is:

$$W = \frac{e^2E^2mp'}{3\cdot 2^6\pi^3\hbar^4}\left[A + BP_2\left(\cos\left(\widehat{\mathbf{E},\mathbf{p}}\right)\right)\right], \\ A = \sum_l\{l\,|R_{l,l-1}|^2 + (l+1)\,|R_{l,l+1}|^2\}, \\ B = \sum_l(2l+1)^{-1}\{l\,(l+1)\,|R_{l,l-1}|^2 + \\ + (l+1)(l+2)\,|R_{l,l+1}|^2 + 3l\,(l+1)\left[R_{l,l-1}R_{l,l+1}^{*} + \text{к. с.}\right]\}. \tag{6}$$

2. THE LOW-FREQUENCY LOW APPROXIMATION FOR MATRIX ELEMENTS $R_{l,l\pm 1}$

Low has shown in [1] that the analog of formula (1) for single-quantum spontaneous bremsstrahlung in the relative quantum theory results from the gauge invariance and does not require using dynamical equations for its justification. In the same paper Low notes that this formula can be obtained using the Lippman-Schwinger equation. The deduction of formula (1) by the Lippman-Schwinger equation for the case of the potential scattering of two particles was subsequently performed by Geller [3].

In papers [11, 12, 20] the Lippman-Schwinger equation was used again to obtain the low-frequency approximation. However, as it was already noted in the Introduction, the formulas obtained in these works agree with (1), (2) only in the main infrared part and disagree in the part proportional to ω^{-1}. To make clear the reason of the disagreement, we present the deduction of (2) assuming that the interaction between the electon and atom is described by the local short-range potential V. Using the Lippman-Schwinger equation for $\Psi_{\mathbf{p}}^{\pm}$ functions

$$\Psi_{\mathbf{p}}^{\pm} = \varphi_{\mathbf{p}} + (\varepsilon_p - H_0 \pm i\delta)^{-1} V \Psi_{\mathbf{p}}^{\pm}, \tag{7}$$

it is easy to transform the matrix element (3) to the form

$$M_{\mathbf{p}',\mathbf{p}} = -\frac{ie}{2m\hbar\omega^2}[\Lambda_1 + \Lambda_2], \tag{8}$$

$$\Lambda_1 = \mathbf{Ep}\langle\Psi_{\mathbf{p}'}^{-}|V|\varphi_{\mathbf{p}}\rangle - \mathbf{Ep}'\langle\varphi_{\mathbf{p}'}|V|\Psi_{\mathbf{p}}^{+}\rangle, \tag{9}$$

$$\Lambda_2 = (\varepsilon_{p'} - \varepsilon_p)\sum_{\mathbf{p}_1}\frac{\langle\Psi_{p'}^{-}|V|\varphi_{p_1}\rangle}{\varepsilon_{p'} - \varepsilon_{p_1}}\mathbf{Ep}_1\frac{\langle\varphi_{\mathbf{p}_1}|V|\Psi_{p}^{+}\rangle}{\varepsilon_p - \varepsilon_{p_1}}. \tag{10}$$

Potential matrix elements in (8) - (10) are scattering matrices

$$\langle\Psi_{\mathbf{p}'}^{-}|V|\varphi_{\mathbf{p}}\rangle = T(\varepsilon_{p'};\ \mathbf{p}',\mathbf{p}),\ \langle\varphi_{\mathbf{p}'}|V|\Psi_{\mathbf{p}}^{+}\rangle = T(\varepsilon_p;\ \mathbf{p}',\mathbf{p}) \tag{11}$$

outside the energy surface. Considering further passing to the limit $\omega \to 0$, we shall assume that $p' \to p$. Since Λ_2 is already of the first order of smallness compared to Λ_1 then accurate to the terms of the second order of smallness we have

$$\Lambda_2 = (\varepsilon_{p'} - \varepsilon_p)\sum_{\mathbf{p}_1}\frac{T(\varepsilon_p;\ \mathbf{p}',\mathbf{p}_1)\,\mathbf{Ep}_1 T(\varepsilon_p;\ \mathbf{p}_1,\mathbf{p})}{(\varepsilon_p - \varepsilon_{p_1})^2}. \tag{12}$$

Geller deduction of (1) is based on the possiblity to transform Eq. (12) to the form

$$\Lambda_2 = (\varepsilon_{p'} - \varepsilon_p)\, m\mathbf{E}\nabla_{\mathbf{p}} T(\mathscr{E};\ \mathbf{p}+\mathbf{q},\mathbf{p})|_{\mathscr{E}=\varepsilon_p},\ \mathbf{q} = \mathbf{p}' - \mathbf{p}, \tag{13}$$

which is derived from (12) as a result of Eq.(7) and is valid at rather low value of E. Following further to Geller, we shall consider scattering matrices in (11), (13) as the functions of scalar variables:

$$\mathscr{E} = \varepsilon_{p'} \text{ или } \varepsilon_p, \; \Delta' = \mathscr{E} - \varepsilon_{p'}, \; \Delta = \mathscr{E} - \varepsilon_p,$$
$$T(\mathscr{E};\; \mathbf{p}', \mathbf{p}) \equiv T(\mathscr{E}, q^2, \Delta', \Delta).$$

At $\Delta = 0$, $\Delta' = 0$ we obtain the scattering matrix on the energy surface proportional to the elastic scattering amplitude:

$$T(\mathscr{E}, q^2, 0, 0) = -2\pi\hbar^2 m^{-1} f(\mathscr{E};\, q^2). \tag{14}$$

Let us expand Λ_1 in small values Δ, Δ':

$$\Lambda_1 = \mathbf{Ep}T(\varepsilon_{p'}, q^2, 0, 0) - \mathbf{Ep}'T(\varepsilon_p, q^2, 0, 0) + \\ + (\varepsilon_{p'} - \varepsilon_p)\,[\mathbf{Ep}\partial T(\varepsilon_{p'}, q^2, 0, \Delta)/\partial\Delta\,|_{\Delta=0} + \mathbf{Ep}'\partial T(\varepsilon_p, q^2, \Delta', 0)/\partial\Delta'\,|_{\Delta'=0}]. \tag{15}$$

In the terms of new variables Eq. (13) takes the form

$$\Lambda_2 = -(\varepsilon_{p'} - \varepsilon_p)\left[\mathbf{Ep}\frac{\partial T(\mathscr{E}, q^2, \Lambda, \Lambda')}{\partial\Lambda} + \mathbf{Ep}\frac{\partial T(\mathscr{E}, q^2, \Lambda, \Lambda')}{\partial\Delta'}\right]. \tag{16}$$

Summing up (15) and (16) one can readily see that the terms involving derivatives of the scattering matrix cancel each other in the first order of smallness in ω. As a result, substituting (14) - (16) to (8) we obtain the formula (2).

Notice that it is rather difficult from the presented deduction of formula (2) to indicate concrete physical parameters of smallness defining its applicability. To do this, one has, obviously, to study the behaviour of scattering matrix beyond the energy surface. It is also rather difficult to indicate the concrete physical value of frequency for which $|\Lambda_1| \gg |\Lambda_2|$. Hence, if Λ_2 is omitted in (8) then the expression obtained in this way will be less exact than (2). Its use for the expression of matrix element $M_{p',p}$ in terms of the elastic scattering amplitudes, as has been done in [20], will wittingly result in error in the term proportional to ω^{-1}.

The essential moment in deriving (2) is the choice of the quantity q^2 as an independent variable. More often, however, the elastic scattering amplitude is considered as a function of energy and scattering angle cosine:

$$f(\mathscr{E};\, q^2) = f(\mathscr{E};\, \cos\theta), \; \cos\theta = 1 - q^2/4m\mathscr{E}. \tag{17}$$

Since the amplitudes in (2) are defined at the same values of q^2 and different values of @, the scattering angles corresponding to them will be different as well. Substituting to (17)

$$q^2 = p^2 + (p')^2 - 2pp' \cos\theta$$

and considering an accuracy required, one can readily bring (2) to the form

$$M_{\mathbf{p}',\mathbf{p}} = \frac{ie\pi\hbar}{(m\omega)^2}\Big[\mathbf{Ep}f(\varepsilon_{p'};\ \cos\theta) - \mathbf{Ep}'f(\varepsilon_p;\ \cos\theta) - \mathbf{E}(\mathbf{p}+\mathbf{p}')\,\zeta(\cos\theta - 1)\times \\ \times\frac{\partial f(\varepsilon_p;\ \cos\theta)}{\partial\cos\theta} + O(\omega^2)\Big],\ \zeta = \pm\frac{m\hbar\omega}{p^2}. \qquad (18)$$

In papers [11, 12] an approximation is called low-frequency if only two first terms are considered in the right-hand side of (18). As it follows from (18) such approximation is true only with a precision of main infrared terms. Formulas of the low-frequency approximation for the radial matrix elements $R_{1,1\pm1}$ obtained in [11] has the similar drawback.

Using Eq. (18) one can find the form of matrix elements $R_{1,1\pm1}$ in the limit of low frequencies. For this purpose we expand scattering amplitudes in the series in Legendre polynomials:

$$f(\mathscr{E},\cos\theta) = \sum_l (2l+1)\,f_l(\mathscr{E})\,P_l(\cos\theta). \qquad (19)$$

Substituting (19) in (18) and using recurrent relations for Legendre polynomials we obtain after a number of transformations

$$M_{\mathbf{p}',\mathbf{p}} = \frac{ie\pi\hbar p}{(m\omega)^2}\sum_l\big[E_zP_l(\cos\theta)\,a_l - E_yP_l^1(\cos\theta)\,b_l\big], \qquad (20)$$

$$a_l = l\,[f_l(\varepsilon') - f_{l-1}(\varepsilon) - l\zeta f_{l-1}(\varepsilon)] + (l+1)\,[f_l(\varepsilon') - f_{l+1}(\varepsilon) + (l+1)\,\zeta f_{l+1}(\varepsilon)],$$

$$b_l = f_{l+1}(\varepsilon) - f_{l-1}(\varepsilon) + (2l+1)\,\zeta f_l(\varepsilon) - l\zeta f_{l-1}(\varepsilon) - (l+1)\,\zeta f_{l+1}(\varepsilon).$$

When deducing (20) we assumed that $f_l(\varepsilon') = f_l(\varepsilon) \pm \hbar\omega f_l'(\varepsilon)$ and $p' = p(1+\zeta)$ and neglected all the terms of ω^0 order. Comparing (20) and (4) it is easy to find the formulas defining in the low-frequency limit matrix elements $R_{1,1\pm1}$ in terms of partial scattering amplitudes:

$$R_{l,l+1} = \frac{i8\pi^2\hbar p}{(m\omega)^2}[f_l(\varepsilon') - f_{l+1}(\varepsilon) + (l+1)\zeta f_{l+1}(\varepsilon) - l\zeta f_l(\varepsilon) + O(\omega^2)], \quad (21)$$

$$R_{l,l-1} = \frac{i8\pi^2\hbar p}{(m\omega)^2}[f_l(\varepsilon') - f_{l-1}(\varepsilon) - l\zeta f_{l-1}(\varepsilon) + (l+1)\zeta f_l(\varepsilon) + O(\omega^2)]. \quad (22)$$

Note that in the case typical for the electron-atom collisions when a certain finite number of partial scattering amplitudes is known, it is more convenient to use formulas (4), (6) together with (21), (22) than general formulas (1) and (2).

3. GENERALIZED FORMULAS OF THE LOW-FREQUENCY APPROXIMATION FOR MATRIX ELEMENTS $R_{l,l\pm1}$ AND CONDITIONS OF THEIR APPLICABILITY

We shall assume that the interaction between the electron and scattring center is short-range and may be neglected at r > a. Then at r > a wave functions are expressed in terms of scattering phases:

$$\begin{aligned} \psi_l(k,r) &= (i)^l[A_l(k)j_l(kr) + B_l(k)n_l(kr)], \\ A_l(k) &= 4\pi e^{i\delta_l}\cos\delta_l = 4\pi(1+ikf_l(k)), \\ B_l(k) &= -4\pi l^{i\delta_l}\sin\delta_l = -4\pi k f_l(k). \end{aligned} \quad (23)$$

In (23) $j_l(kr)$, $n_l(kr)$ are Bessel´s spherical functions. Let us divide the integration region in (5) into the inner $0 \le r \le a$ and external[2] $a < r < \infty$ parts:

$$R_{l,l\pm1} = R^{<}_{l,l\pm1} + R^{>}_{l,l\pm1}, \quad (24)$$

$$R^{>}_{l,l\pm1} = (-1)^l \int_a^\infty r^3\psi_l(k',r)\psi_{l+1}(k,r)\,dr, \quad (25)$$

$$R^{<}_{l,l\pm1} = (-1)^l \int_0^a r^3\psi_l(k',r)\psi_{l\pm1}(k,r)\,dr. \quad (26)$$

(Footnote 2: A quite general approach to the calculation of radiative matrix elements in thr external region is presented in [25])

Substitute (23) into (26):

$$R^{>}_{l,l\pm 1}=(\pm i)\Bigg[A_l(k')A_{l\pm 1}(k)\int_a^{\infty} j_l(k'r)\,j_{l\pm 1}(kr)\,r^3\,dr+ \qquad (27)$$

$$+B_l(k')B_{l\pm 1}(k)\int_a^{\infty} n_l(k'r)\,n_{l\pm 1}(kr)\,r^3\,dr+$$

$$+A_l(k')B_{l+1}(k)\int_a^{\infty} j_l(k'r)\,n_{l\pm 1}(kr)\,r^3\,dr+$$

$$+B_l(k')A_{l\pm 1}(k)\int_a^{\infty} n_l(k'r)\,j_{l\pm 1}(kr)\,r^3\,dr\Bigg].$$

Integrals in (27) can be considered, for example, as limits of the type

$$\lim_{\lambda\to 0}\int_a^{\infty}\varphi(r)\,e^{-\lambda r}\,dr,$$

where $\varphi(r)$ is any of the integrands in (27). Using asymptotic formulas for functions $j_l(x)$, $n_l(x)$ at $x \gg 1$

$$j_l(x)=\frac{\sin(x-l\pi/2)}{x}+\frac{l(l+1)\cos(x-l\pi/2)}{2x^2}+\ldots, \qquad (28)$$

$$n_l(x)=(-1)^{l+1}\left[\frac{\cos(x+l\pi/2)}{x}-\frac{l(l+1)\sin(x+l\pi/2)}{2x^2}+\ldots\right],$$

one can easily see that for the two first integrals in (29) the limit of the type

$$\lim_{k'\to k}\lim_{\lambda\to 0}\int_a^{\infty}\varphi(r)\,e^{-\lambda r}\,dr$$

exists and has a finite value, while for the next two integrals it becomes infinite. Since the wave functions are limited at the point r = 0 and are continuous the integral in (26) is also finite at k` = k. Because of this, according to the Low theorem. in the whole expression (24) only two last terms in (27) in the limit of low frequencies contain the terms proportional to $\omega^{-2}$ and $\omega^{-1}$. Since the appearance of singularities in two last integrals in (29) at k` → k is associated with their divergence in the infinity, to obtain the main members of the $R_{1,1\pm 1}$ expansion, when $\omega \to 0$, it will suffice to calculate these integrals substititing in them two first terms of the asymptotic expansion of the wave functions product at $r \to \infty$

$$(\pm i)\int_a^\infty j_l(k'r)\,n_{l\pm 1}(kr)\,r^3\,dr = -(\pm i)\int_a^\infty n_l(k'r)\,j_{l\pm 1}(kr)\,r^3\,dr =$$
$$= \frac{i}{2k'k}\left[\frac{1}{(k'-k)^2} - \frac{(l\pm 1+1)(l\pm 1) - l(l+1)}{2(k'-k)k}\right]. \qquad (29)$$

Substituting in (29) $(k'-k)^{-2} = (k\zeta)^{-2} + (k\zeta)^{-1}$, $k' = k(1+\zeta)$, we find

$$R^{>}_{l,l+1} = \frac{i8\pi^2}{k^3\zeta^2}\left[f_l(k') - f_{l+1}(k) - l\zeta f_l(k') + (l+1)\zeta f_{l+1}(k)\right], \qquad (30)$$

$$R^{>}_{l,l-1} = \frac{i8\pi^2}{k^3\zeta^2}\left[f_l(k') - f_{l-1}(k) + (l+1)\zeta f_l(k') - l\zeta f_{l-1}(k)\right]. \qquad (31)$$

Formulas (30) and (31) coincide with formulas (21) and (22), the only difference being that it is not nessesary to consider amplitudes in (30), (31) to be expanded in $f_l(k') \equiv f_l(\varepsilon_p \pm \hbar\omega)$ frequency. It follows from the deduction presented that formulas (30) and (31) are valid at the conditions

$$\frac{\hbar\omega}{\varepsilon_p} \ll 1, \qquad \frac{m\omega a^3}{\hbar} \ll 1,$$

since integrals in (27) depend on frequency only via these parameters. The advantage of the deduction of formulas (30) and (31) presented in this section is, in our opinion, that it allows to estimate the value of frequency at which these formulas can be considered to related to the natrix element (24) as a whole. To do this, one, obviously, has to know the value of $R^{<}_{l,l\pm 1}$. Let us show that $|R^{<}_{l,l\pm 1}|$ can be evaluated knowing only scattering amplitudes and the interaction radius a. We wright the Cauchy-Bjunikowskii inequality for $|R^{<}_{l,l\pm 1}|$:

$$|R^{<}_{l,l\pm 1}| \leqslant a\left[\int_0^a r^2|\psi_l(k',r)|^2\,dr \int_0^a r^2|\psi_{l\pm 1}(kr)|^2\,dr\right]^{1/2}. \qquad (32)$$

Using the Schrodinger equation, one can express integrals in the right-hand side of (32) through the wave functions at r = a:

$$I_l = \int_0^a r^2|\psi_l(k,r)|^2\,dr = \frac{a^2}{2k}\left[\frac{\partial\psi_l}{\partial r}\frac{\partial\psi_l^*}{\partial k} - \psi_l\frac{\partial^2\psi_l^*}{\partial k\,\partial r}\right]. \qquad (33)$$

Note that although we have used so far designations corresponding rigorously only to the case of the potential scattering, the estimation of $R^{<}_{l,l\pm1}$ based on (32) and (33) is also valid in more general case, when at r > a the single-particle description is already impossible.

Substituting (23) into (33) we obtain

$$I_l(k) = 8\pi^2\hbar(km)^{-1}\tau_l(k), \tag{34}$$
$$\tau_l(k) = v^{-1}[d\delta_l(k)/dk + a^2kF_l(k)], \quad v = p/m,$$

$$F_l(k) = \cos^2\delta_l\left[x\left(j_l'(x)\right)^2 - j_l(x)\left(xj_l'(x)\right)'\right] + \sin^2\delta_l\left[x\left(n_l'(x)\right)^2 - \right. \tag{35}$$
$$\left. - n_l(x)\left(xn_l'(x)\right)'\right] - \sin\delta_l\cos\delta_l\left[2xj_l'(x)\,n_l'(x) - j_l(x)\left(xn_l'(x)\right)' - \right.$$
$$\left. - n_l(x)\left(xj'_l(x)\right)'\right], \quad x = ka, \tag{36}$$

where $\tau_l(k)$ is a half of the time the scattered electrons spend in the inner region r < a [26]. Now the inequality(32) takes the form

$$|R^{<}_{l,l\pm1}| \leqslant \frac{8\pi^2\hbar^2}{m^2}\left[\frac{a^2}{vv'}\,\tau_l(k')\,\tau_{l\pm1}(k)\right]^{1/2}. \tag{37}$$

Formulas (36) - (37) give the upper estimate of $R^{<}_{l,\pm1}$ and can be used to determine the range of applicability of (30),(31). In this case it is important to remember the role of terms of ω^{-1} order in (30), (31). These terms are essential for the calculation of difference between the cross sections of photon absorption and induced emission, where by virtue of compensation of the main orders they make the contributon of the same order as the terms of ω^{-2} order in (30), (31). Therefore, it is enough to require only the main infrared terms in (30), (31) to be great in comparison with $|R^{<}_{l,l\pm1}|$, whileterms of ω^{-1} order have to be considered great compared to $|(R^{<}_{,l\pm1})^{+} - (R^{<}_{l,l+-1})^{-}|$ (+ corresponds to the photon absorption, - corresponds to its emission). One can easily estimate the latter expression similar to (32) - (37):

$$|(R^{<}_{l,l\pm1})^{+} - (R^{<}_{l,l\pm1})^{-}| \leqslant a\left\{k\,|\zeta|\,I_{l\pm1}(k)\left[\frac{dI_l(k^{+})}{dk^{+}} - \frac{dI_l(k^{-})}{dk^{-}}\right]\right\}^{1/2} + O(\omega^2),$$
$$k^{\pm} = [k^2 \pm 2m\omega/\hbar]^{1/2}.$$

As an illustration let us consider the problem of applicability of (30), (31) to the resonance scattering. Suppose one of the energies $\varepsilon_{p'}$, $\varepsilon_{p} \pm \hbar\omega$ coincides with the resonance energy in a certain partial amplitude f_l. The time of scattering corresponding to the resonance is

$$\tau_l = \hbar/\Gamma. \tag{38}$$

The second time τ_l' in (37) depends strongly on the electron velocity. One can easily obtain $F_l(ka) \approx (ka)^{-1}$ using formulas (28) at $ka \gg 1$. Considering that at $ka \gg 1$ $d\delta_l/dk \sim k^{-2}$ far away from the resonance (see [24, par.131]), we see that for fast electrons τ_l coincides with the transit time:

$$\tau_{l'} = a/v, \quad ka \gg 1. \tag{39}$$

At $ka \ll 1$

$$F_l(k) = \cos^2\delta_l \frac{2\,(ka)^{2l+1}}{(2l+1)!!\,(2l+3)!!} + \sin^2\delta_l \frac{2\,[(2l-1)!!]^2}{(1-2l)\,(ka)^{2l+1}} - \sin\delta_l\cos\delta_l \times \tag{40}$$

$$\times \left[\frac{2l+1}{(ka)^2} - \frac{2}{2l+1}\right],$$

$$\delta_l = \alpha(ka)^{2l+1} \ll 1. \tag{41}$$

Subsituting (40), (41) into (35) (at $l = 0$ in (41) one should also take into consideration the nextexpansion term of $(ka)^3$ order), we obtain

$$\tau_{l'} = av^{-1}(ak)^{2l'+2}G_{l'}, \quad ka \ll 1,$$

where G_l is a constant determined from (40), (41).

Using formulas (42), (38), (39), (40) it is easy to find sufficient conditions at which formulas (30) and (31) will describe the splitting of resonance singularities in the cross section of single-quantum induced bremsstrahlung observed in [21] at $\hbar\omega > \Gamma$. In this case resonance terms of ω^{-2} order in (30), (31) have to be great compared to $|R^{<}_{l,l\pm1}|$ and the condition $\hbar\omega \gg \Gamma$ has to be met simultaneously:

at $ka \ll 1$

$$\Gamma < \hbar\omega \ll \hbar\left[G_{l'}\frac{\Gamma}{\hbar}\left(\frac{v}{a}\right)^3\right]^{1/4}(ka)^{-\frac{l'+1}{2}} = \left[G_{l'}ka\Gamma\varepsilon_a^3\right]^{1/4}(ka)^{-l/2}, \tag{42}$$

at ka >> 1

$$\Gamma < \hbar\omega \ll \hbar\left[\frac{\Gamma}{\hbar}\left(\frac{v}{a}\right)^3\right]^{1/4} = \left[(ka)^3\Gamma\varepsilon_a^3\right]^{1/4}, \tag{43}$$

where $\varepsilon_a = \hbar^2/ma^2$. In the typical case of scattering of the electron by the atom $\Gamma/\varepsilon_a \sim 10^{-3}$. Hence, there is rather wide range of frequencies obeying inequalities (42) and (43).

In conclusion consider the case when at ka ≪ 1, k´a ≪ 1 the contribution of one of partial scattering amplitudes proves to be dominant. In this case it will suffice to consider two latter terms in (27) (similar to the limit $\omega \to 0$) assuming the lower limits of integration in the corresponding integrals to be zero. Below we present formulas obtained in this way considering separately S [5-7] and P scattering amplitudes

$$R_{0,1} = \frac{i8\pi^2}{k^3\zeta^2} f_0(k'), \qquad R_{1,0} = -\frac{i8\pi^2}{k^3\zeta^2}(1+2\zeta)^{1/2} f_0(k); \tag{44}$$

$$R_{1,0} = \frac{i8\pi^2}{k^3\zeta^2}\left[(1+2\zeta)^{1/2} f_1(k') + \zeta(1+2\zeta)^{-1/2} f_1(k')\right], \tag{45}$$

$$R_{0,1} = -\frac{i8\pi^2}{k^3\zeta^2} f_1(k)(1-\zeta),$$

$$R_{1,2} = \frac{i8\pi^2}{k^3\zeta^2} f_1(k')(1+2\zeta)^{-1/2}, \qquad R_{2,1} = -\frac{i8\pi^2}{k^3\zeta^2} f_1(k)(1+2\zeta).$$

Note that formulas (44), (45) do not depend on parameters determining the region of their applicability. In view of this they are valid in anarbitrary order inω. The substitution of (44) into (4) is equivalent to the substitution of total scattering amplitudes in (2) by S-amplitudes.

4. NEGATIVE ABSORPTION AT THE FREE-FREE TRANSITIONS NEAR THE SCATTERING RESONANCE

At the electron scattering in the presence of the electromagnetic field, transitions both with absorption and emission of photon may occure. For this reason the condition of existence of the inverse distribution in a certain region of the continuous spectrum energies proves to be insufficient to obtain the negative absorpton [27]. For the negative absorption to occure at the transitions in the continous spectrum, the emission probability in this energy region have to exceed the absorption probability. However, for the most scattering mechanisms the situation is reversed. The possibility of the negative absorption is known only forthe electron scattering by the Coulomb potential at the additional condition of strong anisotropy of the distribution function in the momentum space [28, 29].

We show that at the presence of scattering resonances the negative absorption becomes possible even at the isotropic distribution of electrons.

In the case of isotropic distribution the absorption coefficient is determined by the formula

$$\beta = \frac{4\pi}{I} N_a \int_0^\infty d\varepsilon \varepsilon^{1/2} g(\varepsilon) \left[\overline{W}_p^+ - \overline{W}_p^- \right] \hbar\omega,$$

where I is the incident radiation intensity, N_a is the concentration of the scattering centers, $\overline{W}_p^{\pm}$ are the probabilities of absorption and induced emission averaged over the angle between p and E, $g(\varepsilon)$ is the distribution function of electrons over energies normalized by the condition

$$4\pi \int_0^\infty d\varepsilon \varepsilon^{1/2} g(\varepsilon) = N_e.$$

We shall restrict our consideration only to S-type scattering assuming the electrons to be slow. Using formulas (6) and (44) one obtains the following expression for the absorption coefficient

$$\beta = \frac{4\pi e^2}{m\omega^2 c} N_e \nu_{eff}(\omega), \tag{46}$$

$$\nu_{eff} = \frac{(4\pi)^2 N_a}{3} \frac{1}{N_e} \int_0^\infty d\varepsilon \varepsilon^{1/2} \frac{p}{m} \xi^{-1} \{ [(1+\xi)^{3/2} - (1-\xi)^{3/2}] |f_0(k)|^2 + \tag{47}$$

$$+ (1+\xi)^{1/2} |f_0(k^+)|^2 - (1-\xi)^{1/2} |f_0(k^-)|^2 \} g(\varepsilon),$$

where $\xi = \hbar\omega/\varepsilon$. Let us substitute expression for S-amplitude of scattering near the resonance [24] into integral (47)

$$f_0(k) = -\alpha - \frac{\Gamma/2}{k\,[\varepsilon - \varepsilon_2 + i\Gamma/2]}, \quad \varepsilon = \frac{(\hbar k)^2}{2m}. \tag{48}$$

According to the preceding sections, the validity of formula (47) does not restricted by the condition $\hbar\omega < U$. At the calculation of the integral (47) we shall neglect the change of $g(\varepsilon)$ function at the resonance width scale. Then, inserting (48) into (47) we get

$$\nu_{eff}(\omega) = \frac{16\pi}{3} N_a \alpha^2 \bar{v}(\omega) + \frac{4\pi^2 N_a v_r}{3k_r^2}\left\{\frac{\varepsilon_r}{\hbar\omega}\left[\left(1 + \frac{\hbar\omega}{\varepsilon_r}\right)^{3/2} - \left(1 - \frac{\hbar\omega}{\varepsilon_r}\right)^{3/2}\right] n_e(\varepsilon_r) + \right.$$
$$\left. + \left(\frac{\varepsilon_r}{\hbar\omega} - 1\right) n_e(\varepsilon_r - \hbar\omega) - \left(\frac{\varepsilon_r}{\hbar\omega} + 1\right) n_e(\varepsilon_r + \hbar\omega)\right\}, \tag{49}$$

where

$$\bar{v}(\omega) = \frac{\pi}{N_e}\int_0^\infty d\varepsilon\,\varepsilon^{1/2} g(\varepsilon)\, v\xi^{-1}\left[(1 + \xi)^{1/2}(2 + \xi) - (1 - \xi)^{1/2}(2 - \xi)\right],$$

$n_e(\varepsilon)$ are the relative numbers of electrons involved in the resonance bremsstrahlung processes

$$n_e(\varepsilon) = 2\pi\Gamma\varepsilon^{1/2} g(\varepsilon)/N_e.$$

The first term in the right-hand side of (49) corresponds to the contribution of the common potential scattering, the second one is caused by the resonance. Let us designate the first term in (49) as ν_{eff}^{pot}. The second term can be simplified in inverse cases $\varepsilon_r > \hbar\omega$ and $\varepsilon_r < \hbar\omega$. At $\varepsilon_r > \hbar\omega$ we get:

$$\nu_{eff}(\omega) = \nu_{eff}^{pot} + \frac{4\pi^2 N_a v_r}{3k_r^2}\left\{[3n_e(\varepsilon_r) - n_e(\varepsilon_r - \hbar\omega) - n_e(\varepsilon_r + \hbar\omega)] + \right.$$
$$\left. + \frac{\varepsilon_r}{\hbar\omega}[n_e(\varepsilon_r - \hbar\omega) - n_e(\varepsilon_r + \hbar\omega)] + O\left(\frac{\hbar\omega}{\varepsilon_r}\right)\right\}. \tag{50}$$

At $\varepsilon_r < \hbar\omega$ only processes of the resonance bremsstrahlung absorption by electrons with energy ε_r and emission by electrons with energy $\varepsilon_r + \hbar\omega$ may occure. Hence, the formula (49) takes the form

$$\nu_{eff}(\omega) = \nu_{eff}^{pot} + \frac{4\pi^2 N_a v_r}{3k_r^2}\left\{\left[\left(\frac{\hbar\omega}{\varepsilon_r}\right)^{1/2} + \frac{3}{2}\left(\frac{\varepsilon_r}{\hbar\omega}\right)^{1/2}\right] n_e(\varepsilon_r) - \left(1 + \frac{\varepsilon_r}{\hbar\omega}\right) n(\varepsilon_r + \hbar\omega) + + O\left(\left(\frac{\varepsilon_r}{\hbar\omega}\right)^{3/2}\right)\right\}. \tag{51}$$

Formulas (46), (49) - (51) demonstrate clearly the principal possibility of the negative absorption due to free-free transitions near the scattering resonance at the presence of the inverse distribution of electrons.

ABSTRACT

New derivation is given for the formula which express the soft photon bremsstrahlung matrix element through the elastic scattering amplitudes depending on the initial and final energy of the radiating particle. For the part of the bremsstrahlung matrix element connected with the area where the motion of the radiating particle is not free and which can not be expressed via the scattering amplitudes the upper estimation based on the Cauchy-Bjunikowsky inequality is obtained. The only parameter involved in this estimation, besides scattering amplitudes is the interaction radius of colliding particles. The basic conclusion is that the solution of the elastic scattering problem allows one not only to compute the leading soft-photon terms of the bremsstrahlung matrix element, but also to estimate the limiting value of the photon frequency for which such connecton exists. For example, the problem of the applicability of the obtained formulas for resonance scattering is considred.

The possibility in principle of the negative absorption due to free-free transitions near the scattering resonance is also discussed.

REFERENCES

[1] F.E. Low, Bremsstrahlung of very low-energy quanta in elementary particle collisions, Phys. Rev. 110, (1958) 974.

[2] L.S. Brown and R.L. Goble, Soft photons and the classical limit, Ibid. 173, (1968) 1505.

[3] L. Heller, Soft-photon theorem for bremsstahlung in a potential model, Ibid. 174, (1968) 1580.

[4] R.B. Johnston, Free-free radiative transitions - a survey of theoretical results, J. Quant. Spectrosc. Radiat. Transfer. 7, (1967) 815.

[5] I.M. Shmushkevich, Electromagnetic radiation at collisions of electrons with protons, Zh. Eksp. Teor. Fiz. 19, (1949) 353.

[6] O.B. Firsov and M.I. Chibisov, Bremsstrahlung of slow electrons on neutral atoms, Zh. Eksp. Teor. Fiz. 39, (1960) 1170.

[7] B.A. Zon, Absorption of optical emission by weakly ionized gases, Zh. Eksp. Teor. Fiz. 77, (1979) 44.

[8] V.M. Buimistrov and L.I. Trakhtenberg, Bremsstrahlung cross section at the electron scattering by the hydrogen atom, Zh. Eksp. Teor. Fiz. 69, (1975) 108.

[9] An.V. Vinogradov, Conditions for applicability of the low-frequency approximation in the problem of one photon bremsstrahlung, Preprint FIAN, 197 (1980) 19 p.

[10] An.V. Vinogradov, Conditions for applicability of the low-frequency approximation in the problem of one photon bremsstrahlung, Proc. of the XY Intern. Conf. on Phenomena in Ionized Gases, Minsk, 1981, p. 1617.

[11] H. Kruger and N. Schultz, Laser-induced free-free electronic transitions near a resonance, J. Phys. B.: Atom. Mol. Phys. 9, (1976) 1899.

[12] C. Jung and H. Kruger, On the accuracy of soft-photon approximation for resonance free-free transitions, Ztschr. Phys. A. 287, (1978) 7.

[13] D. Andric and L. Langhans, Measurements of free-free transitions in e-Ar scattering, J. Phys. B: Atom. Mol. Phys. 9, (1976) L1495.

[14] A. Weingartshofer et al., Direct observation of multiphoton processes in laser-induced free-free transitions, Phys. Rev. Lett. 39, (1977) 269.

[15] D. Andric and L. Langhans, Measurement of the free-free cross section of e-Ar scattering, J. Phys. B: Atom. Mol. Phys. 11, (1978) 2355.

[16] A. Weingartshofer et al., Experiments on multiphoton free-free transitions, Phys. Rev. A - Gen. Phys. 19, (1979) 2371.

[17] H. Feshbach and D.R. Yennie, Radiation of low energy quanta in nuclear reactions, Nucl. Phys. 37, (1962) 150.

[18] S. Barshay and T. Yao, Emission of low-energy photons in scattering and decay processes and a possible method for determining electromagnetic properties of short-lived states, Phys. Rev. 171, (1968) 1708.

[19] L.A. Ponomarev, Fast change of the adron amplitude and γ-quantum bremsstrahlung, Yad. Fiz. 12, (1970) 570.

[20] V.A. Kas´yanov and A.N. Starostin, Bremsstrahlung at the resonance scattering in molecular gas, Fiz. Plasmy 4, (1978) 120.

[21] L. Langhans, Resonance structure in the free-free cross section of e-Ar scattering, J. Phys. B: Atom. Mol. Phys. 11, (1978) 2361.

[22] An.V. Vinogradov, Negative absorption due to free-free transitions near the scattering resonance, Abstr. of Contrib. Pap. 2nd Intern. Conf. on Multiphoton Processes, Budapest, 1980, F9.

[23] M. Cavrila and M. Van der Wiel, Free-free radiative transitions of electron-atom system, Comments Atom. Mol. Phys. 8, (1978) 1.

[24] L.D. Landau and E.M. Lifshiz, Quantum Mechanics, Nauka, Moscow, (1974) 752 p.

[25] M.J. Seaton, Outer-region contributions to radiative transition probabilities, J. Phys. B. Atom. Mol. Phys. 19, (1986) 2601.

[26] A.I. Baz´, Ya.B. Zel´dovich, and A.M. Perelomov, Scattering, Reactions and Decays in Nonrelativistic Quantum Mechanics, Nauka, Moscow, (1971) 544 p.

[27] F.V. Bunkin, A.E. Kozakov, and M.V. Fedorov, Interaction of the intensive optical radiation with free electrons: (Nonrelativistic case), Usp. Fiz. Nauk. 107, (1972) 559.

[28] D. Marcuse, Stimulated emission of bremsstrahlung, Bell Syst. Techn. J. 41, (1962) 1557.

[29] An.V. Vinogradov, On the Drude formula in semiconductors, Zh. Eksp. Teor. Fiz. 70, (1976) 999.

ON THE ELEMENTARY DECAY PROCESS OF THREE-LEVEL SYSTEMS

D.Yu. Kuznetsov

Introduction

The quantum-mechanical model of the interaction of atoms with the field has been studied in papers [1, 2]. These papers present some arguments in favour of the applicability of such models. Not analysing the applicability of the rotary wave approximation and the problem of making the effective Hamiltonian to be dimensionless, we shall assume that the velocity of light and Plank constant are equal to 1 and modes interacting efficiently with the atomic system have been selected in the field, and corresponding coupling constants have been calculated. For brevity, this atomic system will be called simply an "atom", although, of course, it can be an ion, molecule and more complex cluster.

We shall assume that atom and field are described by the Hamiltonian

$$H_0 = \int a_k^+ a_k k\, dk - E u^+ u + F v^+ v, \tag{1}$$

and their interaction is described by the Hamiltonian

$$H_1 = \int \eta \left(a_k^+ w^+ u + a_k u^+ w\right) dk + \int \zeta \left(a_k^+ w^+ v + a_k v^+ w\right) dk + \\ + \int \xi \left(a_k^+ v^+ u + a_k u^+ v\right) dk, \tag{2}$$

where

$$a_k^+ a_p - a_p a_k^+ = \delta(k-p), \qquad w^+ w + w w^+ = 1, \\ u^+ u + u u^+ = 1, \ \ v^+ v + v v^+ = 1, \tag{3}$$

and all other commutators a, a^+, u, u^+, v, v^+, ω, ω^+ are equal to zero.

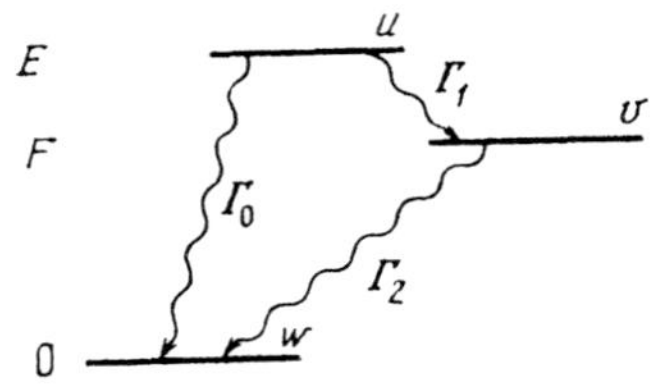

Fig 1. Diagram of atom levels.

Hamiltonian $H = H_0 + H_1$ describes a three-level atom interacting with continuum of field modes: each of these modes can be treated as a harmonic oscillator. Fig.1 presents the diagram of the atom levels. Parameters η, ξ, ζ in (2) are the coupling constants. In the approximation ofnarrow spectral lines the y are considered to be independent on k. It will be clear below that $\Gamma = \Gamma_0 + \Gamma_1 = 2\pi\eta^2 + 2\pi\xi^2$, $\Gamma_2 = 2\pi\zeta^2$ are the decay rates of u and v levels.

In this paper we shall restrict our consideration to three specific cases when one of the constants η, ξ, ζ is zero. This means that the direct transition between one pair of levels is forbidden. In this case one can obtain the Schrodinger equation solution by the Weisscopf method [3]

$$id\Psi/dt = H\Psi \qquad (4)$$

in the visible analytic form.

1. Λ-TYPE TRANSITIONS (THE CASE OF $\zeta = 0$)°

At $\zeta = 0$ the atom relaxes from u level to one of the stable levels w or v emitting one photon. It is convenient to search the equation (4) solution in the form

$$\Psi = h(t)\, e^{-iEt} u^{+} |0\rangle + \int g_0(t)\, e^{-i(E-q)t} a^{+}_{q+E-F} v^{+} |0\rangle\, dq + \int f_q(t)\, e^{-i(E+q)t} \times$$
$$\times\, a^{+}_{q+E} w^{+} |0\rangle\, dq, \qquad (5)$$

where $|0\rangle$ is a vacuum, i.e. $a_k|0\rangle = u|0\rangle = v|0\rangle = \omega|0\rangle = 0$.

Substituting (5) into (4) we obtain the system

$$idh/dt = \int (\eta f_q + \xi g_q) e^{-iqt} dq,$$
$$idg_q/dt = \xi h e^{iqt}, \quad idf_q/dt = \eta h e^{iqt}. \tag{6}$$

For the initial condition $h(0) = 1$, $g_q(0) = 0$, $f_q(0) = 0$ (which corresponds to the decay from the upper excited state) the system has a solution in the Weisscopf approximaton

$$h = \exp\{-(\Gamma/2)t\},$$
$$g_q = i\xi \frac{1 - \exp\{(iq - \Gamma/2)t\}}{iq - \Gamma/2}, \quad f_q = i\eta \frac{1 - \exp\{(iq - \Gamma/2)t\}}{iq - \Gamma/2}, \tag{7}$$

where $\Gamma = \Gamma_0 + \Gamma_1$, $\Gamma_0 = 2\pi\eta^2$, $\Gamma_1 = 2\pi\xi^2$.

In the limit $t \to \infty$ the photon emitted at the transition from u level to one of the lower levels has a spectrum in the form of the Lorentzian bell-like profile localized near energy E in the case of transiton to the lower level and near energy E - F in the case of transition to the intermediate level. These bells have the same width which is determined by the total decay rate Γ. The relation between Γ_0 and Γ determines only relative areas of these bells, i.e. probabilities of atom transitions to levels ω and v.

Note that $a_k^+\omega^+|0\rangle$ and $a_k^+v^+|0\rangle$ states are not identical and, hence, the interference between the two Lorentzian profiles does not occure even in the case when they overlap.

2. V-TYPE TRANSITION (THE CASE OF $\xi = 0$)

This case is interesting when $E \approx F$ and the upper levels overlap, otherwise, the second level will in no way manifests itself at the transition fron the upper level and the spectrum of the photon emitted will be determined by a single Lorentzian profile. Therefore, at $\xi = 0$ it would appear reasonable to seek the soluton of (4) in the form

$$\Psi = h(t) e^{-iEt} u^+|0\rangle + g(t) e^{-iFt} v^+|0\rangle + \int f_q(t) e^{-i(G+q)t} w^+|0\rangle, \tag{8}$$

where $G = (E + F)/2$. Substituting (8) in (4), we get the system

$$idh/dl = \int \eta f_q \exp\{-i(q+\varepsilon)l\}\,dq, \qquad (9)$$
$$idg/dt = \int \zeta f_q \exp\{-i(q-\varepsilon)t\}\,dq,$$
$$idf_q/dt = \eta h \exp\{i(q-\varepsilon)t\} + \zeta g \exp\{i(q+\varepsilon)t\},$$

where $\varepsilon = (E - F)/2$. Let $f_q(0) = 0$. Then in the Weisscopf approximation the solution of system (9) has the form

$$h = \rho\cdot\left((R_- + \gamma + i\varepsilon)h(0) + \sqrt{\gamma\beta}\,g(0)\right)e^{(R_+ + i\varepsilon)t} -$$
$$-\rho\cdot\left((R_+ + \gamma + i\varepsilon)h(0) + \sqrt{\gamma\beta}\,g(0)\right)e^{(R_- + i\varepsilon)t}, \qquad (10a)$$
$$g = \rho\cdot\left(\sqrt{\gamma\beta}\,h(0) - (R_+ + \gamma + i\varepsilon)g(0)\right)e^{(R_+ - i\varepsilon)t} -$$
$$-\rho\cdot\left(\sqrt{\gamma\beta}\,h(0) - (R_- + \gamma + i\varepsilon)g(0)\right)e^{(R_- - i\varepsilon)t},$$
$$f_q = \frac{-\rho}{\sqrt{\pi}}\,\frac{(R_+ - i\varepsilon)h(0) - (R_+ + i\varepsilon)g(0)}{R_+ + iq}\left(1 - e^{(R_+ + iq)t}\right) +$$
$$+\frac{\rho}{\sqrt{\pi}}\,\frac{(R_- - i\varepsilon)h(0) - (R_- + i\varepsilon)g(0)}{R_- + iq}\left(1 - e^{(R_- + iq)t}\right),$$

where

$$\rho = (R_- - R_+)^{-1},\ R_\pm = -(\gamma+\beta)/2 \pm [(\gamma+\beta)^2/4 - \varepsilon^2 + i\varepsilon(\gamma-\beta)]^{1/2},$$
$$\gamma = \pi\eta^2 = \Gamma_0/2,\ \beta = \pi\zeta^2 = \Gamma_2/2. \qquad (10b)$$

In the limit of interest to us $t \to \infty$

$$f_q = -\frac{\eta(q+\varepsilon)h(0) + \zeta(q-\varepsilon)g(0)}{(R_+ + iq)(R_- + iq)}. \qquad (11)$$

At the transition from the upper excited state, the spectral curve has a hole (theoretically to zero) localized at the place of the second excited state. Fig.2 shows the characteristic form of such a curve. The curve is based on formulas (10b), (11) at $h(0) = 1$, $g(0) = 0$, $\gamma = 2\varepsilon$, $\beta = 0{,}3\varepsilon$. If the initial condition is given in the form of superposition of two excited states, one can obtain the superstable state when

$$h(0) = -g(0)\sqrt{\beta/\gamma}, \qquad f_q = \sqrt{\frac{\gamma\beta}{\gamma+\beta}}\,\frac{2\varepsilon}{(R_+ + iq)(R_- + iq)}, \qquad (12a)$$

and the superradiating state when

$$h(0) = g(0)\sqrt{\beta/\gamma}, \qquad f_q = \sqrt{\frac{\gamma\beta}{\gamma+\beta}}\,\frac{2}{(R_+ + iq)(R_- + iq)}. \qquad (12b)$$

In the case of $\beta \approx \gamma \gg \varepsilon$ the emission spectrum width differs essentially from β, γ. Conversly, at $\varepsilon \gg \beta$, γ the spectrum takes the form of superposition of two nonoverlaped bells with half-widths β and γ.

Let us discuss the possibility of the experimental realization of such systems. For the Hamiltonian (2) to be applicable, it is important that both excited states would interact with the same modes of the field. The system of two similar ions in a Paul trap (see [4]) can be an appropriate simulation of such an atom in the case when the distance between ions exceeds substantially the atomic scale (on order of 10^{-8} cm^{-1}) but is considerably less than the emission wavelength (for example, on order of 10^{-4} cm^{-1}). Note that superradiating pairs of atoms similar to the (12b) case were described in [5]. Some problem consists in exciting only one of two close atoms (ions): the usual π pulse will excite efficiently both ions. The problem may be avoided using the difference between the coupling constants $\eta = \sqrt{\gamma/\pi}$ and $\zeta = \sqrt{\beta/\pi}$.

The same quasi-classical wave package may be $(2m + 1)\pi$ pulse for one level and $2n\pi$ pulse for another one.

If the splitting of the upper system levels is caused by the external (classical) field, one can prepare a system in the excited state in a strong field (when levels do not overlap) and then attenuate adiabatically the external field and observe the relaxation.

In this case the emission spectrum of such excited system will resemble Fig. 2.

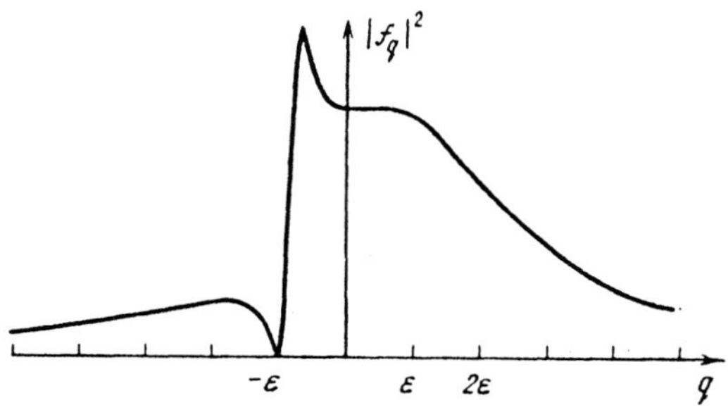

Fig.2. Form of the observed emission spectrum at the decay of the upper excited state by the v-type scheme constructed the formula (11) at $\gamma = 2\varepsilon$, $\delta = 0.3\varepsilon$, $g(0) = 0$, $h(0) = 1$

3. CASCADE TRANSITION AND FREQUENCY-TEMPORAL OSCILLATIONS (THE CASE OF $\eta = 0$)

At $\eta = 0$ only two-photon decay may occure. In this case it is convenient to seek the solution of Eq.(4) in the form

$$\Psi = h(t)\, e^{-iEt}u^+ |0\rangle + \int g_k(t)\, e^{-i(F+k)t} v^+ a_k^+ |0\rangle\, dk + \\ + \int\int f_{k,p}(t)\, e^{-i(k+p)t} a_k^+ a_p^+ w^+ |0\rangle\, dk\, dp. \quad (13)$$

The substitution of (13) to (4) gives

$$i dh/dt = \int \xi g_k \exp\{i(E - F - k)t\}\, dk, \quad (14)$$

$$i dg_k/dt = \int \zeta (f_{k,p}) + f_{p,k} \exp\{i(F - k)t\}\, dp + \xi h \exp\{-i(E - F - k)t\}, \quad (15)$$

$$i df_{k,p}/dt = \mu \xi g_k \exp\{-i(F - k)t\} + (1 - \mu)\zeta g_p \exp\{-i(F - p)t\}, \quad (16)$$

where μ is an arbitrary parameter. It is convenient to assume the equality $f_{k,p} = f_{p,k}$; to fit it to (16) one should assume $\mu = 1/2$.

Decay from the upper level corresponds to the initial condition

$$h(0) = 1, \quad g_k(0) = f_{k,p}(0) = 0. \quad (17)$$

The Weisscopf solution of the system (14) - (17) has the form

$$h(t) = \exp\{-\alpha t\}, \quad (18)$$

$$g_k(t) = -iR(\exp\{i(k - E + F)t - \alpha t\} - \exp\{-\beta t\}), \quad (19)$$

$$f_{k,p}(t) = {}^1\!/_2 \tilde{f}_{k-E+F,p-F}(t) + {}^1\!/_2 \tilde{f}_{p-E+F,k-F}(t),$$

where $R = \xi/(ik + \alpha - \beta)$, $\alpha = \pi\xi^2 = \Gamma_1/2$, $\beta = \pi\zeta^2 = \Gamma_2/2$, (20)

$$\tilde{f}_{g,r}(t) = R\,\frac{1 - \exp\{(iq + ir - \alpha)t\}}{iq + ir - \alpha} - R\,\frac{1 - \exp\{(ir - \beta)t\}}{ir - \beta}. \quad (21)$$

The distribution of emitted photons is determined by the function

$$\tilde{f}_{q,r} = \lim_{t\to\infty} \tilde{f}_{q,r}(t) = \frac{\sqrt{\alpha\beta}}{(iq + ir - \alpha)(ir - \beta)}. \quad (22)$$

One can measure $|\tilde{f}_{q,r}|^2$ using the spectral instrument. The Fourier transform module squared

$$f(x, y) = \frac{1}{2\pi}\int \exp\{-iqx - iry\}\, \tilde{f}_{q,r}\, dq\, dr = \\ = \sqrt{\alpha\beta}\, \theta(x)\, \theta(y - x) \exp\{-\alpha x - \beta(y - x)\} \quad (23)$$

can be measured by recording the time when emitted photons reach detectors. The meaning of formula (23) is clear: at first the upper level u decays with the rate of $\Gamma_1 = 2\alpha$ and then the intermediate level v decays with the rate of $\Gamma_2 = 2\beta$. One can check that the normalization has not been lost:

$$\int\int |\tilde{f}_{q,r}|^2 \, dq \, dr = \int\int |f(x, y)|^2 \, dx \, dy.$$

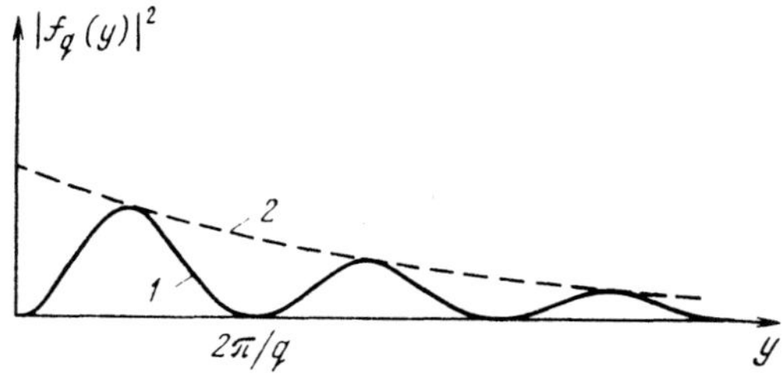

Fig. 3. The probability density of registration of the second photon at the instant of time y provided that the first photon has a frequency shift q (a curve 1) constructed for the case $\Gamma_1 = \Gamma_2 = \Gamma$ by formula (25); a curve 2 corresponds to the envelope $2(\Gamma/q)^2 e^{-\Gamma y}$

It is of more interest is the case when the first photon of energy of E - F order arrives at the spectral instrument, while the second photon (with the energy of F order) falls directly on the detector. After a sequence of tests such a scheme measures the function modulus squared

$$f_q(y) = \frac{1}{\sqrt{2\pi}} \int e^{-iqy} \tilde{f}_{q,r} \, dr.$$

Performing an integration one obtains

$$|f_q(y)|^2 = \frac{(2\pi)^{-1} \Gamma_1 \Gamma_2}{q^2 + (\alpha - \beta)^2} (e^{-2\alpha y} + e^{-2\beta y} - 2e^{-(\alpha+\beta)y} \cos(qy)). \qquad (24)$$

At $\alpha = \beta$ ($\Gamma_1 = \Gamma_2 = \Gamma$) the formula (24) is simplified:

$$|f_q(y)|^2 = (2\pi)^{-1} (\Gamma/q)^2 e^{-\Gamma y} (1 - \cos(qy)). \qquad (25)$$

This function is presented in Fig. 3. The oscillation period is determined by the first photon frequency shift q = k

- (E - F). One may say in some sense that there are beats occuring between the atom state and that of the field with slightly different energy.

The oscillating behaviour of the two-photon wave function may be explained on the basis of nearly classical arguments. Suppose the reduction of the two-photon wave package occures at the second photon registration. In this case the first photon is emitted in a short time interval. The appropriate classical oscillator emits the wave package during $0 \leq t < y$ time interval; then the Fourier component amplitude with the frequency shift q is proportional to

$$c_q(y) = \int_0^y e^{iqt}\, dt = (1 - e^{iqy})/(iq). \qquad (26)$$

Classical spectrum $|c_q(y)|^2$ determined by formula (26) involves the same oscillating factor $(1 - \cos(qy))$ as in the quantum theory (the formula (25)).

Correlatons between photons emitted at the cascade transition may be considered as a realization of the Einstein-Podolskii-Rosen paradox (for noncascade transitions this paradox is described, for example, in [6]).

Note that distinction between F and E - F energies is not necessary if the photons involved can be distinguished by other characteristics, for example, by polarization. If the polarization is the same then photons have to be of "different colors", otherwise one can not say that one of the photons is "the first" and the other is "the second".

Few words can be said on the preparation of an atom in the excited state. This may be done either by two short π pulses at F and E - F frequencies (or by the equivalent pulse on the frequencies mixture) or by relaxation from the forth (higher than u) atom level with an obligatory registration of the photon emitted at the relaxation (the photon must turn the clock on to registrate the moment the last photon is emitted). Hence, the latter case deals with three-photon correlations.

CONCLUSION

The spectral distribution of photons emitted at the elementary decay of simple quantum-mechanical system was studied by an example of three-level atom. Even the rotary wave approximation allows to trace the influence of two close (neighbouring) levels on the emission spectrum. If there are, at least, two emitted quanta then a certain correlation between them occures. Similar consideration may be performed in the case of many levels: if all cascades formed by the allowed transitions involve the same number of transitions then the solution may be obtained in the approximation of a certain number of quanta. The more general case when multi-quantum states are admixed (for example, at $\eta \neq 0$, $\xi \neq 0, \zeta \neq 0$) requires a special consideration. From the physical results of the paper one may set aside the following:

1. At the decay of one of the nearby levels connected with the ground state by the identical transitions, the hole appears in the spectral curve corresponding in frequency to the other level (see Fig. 2). The hole can shift at the decay of the superposition state (Eq.(11)). To study such dependences experimentally the special preparation of superpositional states is necessary.

2. At the cascade decay of three-level system two emitted photons are correlated in a certain way, the moment of the second photon registration is correlated specifically with the energy of the first photon. This relation is described by Eq. (25) and illustrated in Fig. 3. It can be detected with the coincidence curcuit. It would be interesting to observe the oscillations, as at Fig.3, in the experiment with ions in a trap.

The author thanks A. V. Masalov, S.M. Kharchev, and V.A. Shcheglov for useful discussions and interest to this work.

ABSTRACT

The quantum-mechanical system (atom + field) is considered under assumption that one level is connected with two others by radiative transitions. The solutions of the Schrodinger equation are analyzed. The large-time limit of the solutions determines the distribution of the emitted photons. In some conditions this distribution has a nontrivial structure. Some ways to observe this structure in an experiment are discussed.

REFERENCES

[1] S.V. Lavande and B.N. Jagtap, Cooperative effects in a system of strongly driven three-level atoms, Phys. Rev. A - Gen. Phys. 39, (1989) 694.

[2] M.H. Mahran, A.-R. Samman, and A.-S.F. Obada, Quantum effects in 3-level atoms, J. Mod. Opt. 36, (1989) 53.

[3] V. Weisscopf and E. Wigner, Berechnung der naturlichen Lichtteorie, Ztschr. Phys. 63, (1930) 54.

[4] R. Blumel et al., Chaos and order of laser-cooled ions a Paul trap, Phys. Rev. A - Gen. Phys. 40, (1989) 808.

[5] F.T. Arecchi and E. Courtens, Cooperative phenomena in resonant electronagnetic propagation, Ibid. 2, (1970) 1370.

[6] D.N. Klyshko, The Einstein-Podolskii-Rosen paradox for the energy-time variables, Usp. Fiz. Nauk 158, (1989) 327.

SUBJECT INDEX